AF323162

RADIO FREQUENCY IDENTIFICATION
FUNDAMENTALS AND APPLICATIONS,
DESIGN METHODS AND SOLUTIONS

Radio Frequency Identification Fundamentals and Applications, Design Methods and Solutions

Edited by
Cristina Turcu

Radio Frequency Identification Fundamentals and Applications, Design Methods and Solutions

Edited by Cristina Turcu

CBS, Edition 2016

Published by InTech

Janeza Trdine 9, 51000 Rijeka, Croatia

Notice

Statements and opinions expressed in the chapters are these of the individual contributors and not necessarily those of the editors or publisher. No responsibility is accepted for the accuracy of information contained in the published chapters. The publisher assumes no responsibility for any damage or injury to persons or property arising out of the use of any materials, instructions, methods or ideas contained in the book.

Additional hard copies can be obtained from orders@intechopen.com

Radio Frequency Identification Fundamentals and Applications, Design Methods and Solutions,
Edited by Cristina Turcu
p. cm.
ISBN 978-953-7619-72-5

Preface

In January 2009, IN-TECH publisher printed a book entitled "Development and Implementation of RFID Technology". Approaching a variety of aspects concerning RFID (Radio Frequency IDentification) systems, the book focused on several key issues such as new design solutions for RFID antennas, the typology of readers and tags, ways to maintain security and privacy in RFID applications, the selection of appropriate encryption algorithms, etc.

The number of applications for RFID systems has increased each year and various research directions have been developed to improve the performance of these systems. Therefore IN-TECH publisher has decided to continue the series of books dedicated to the latest results of research in the RFID field and launch a new book, entitled "Radio Frequency Identification Fundamentals and Applications, Design Methods and Solutions", which could support the further development of RFID.

Chapter 1 comprises reviews of recent works in current passive UHF RFID systems to provide guidance regarding the RFID system design and deployment. The chapter proposes a variety of issues, problems and solutions such as: UHF RFID radio links using the link budget concept to calculate forward-link and reverse-link interrogation ranges; reader hardware design considerations; phase diversity and quadrature signal combining, phase noise with range correlation effect, and transmitter leakage reduction methods; deployment issues including reader-to-reader interference.

Chapter 2 is dedicated to design considerations for the digital core of an EPC Class 1 Gen 2 (C1G2) RFID tag.

Chapter 3 proposes a brief introduction to RFID systems, and then focuses on the design of efficient space-filling antennas for passive UHF RFID tags.

The fourth chapter introduces the concept of RFID systems and the relevant parameters for proper antenna design. It also approaches the expressions for the phase constants, propagation constants and the characteristic (or Bloch) impedance of a wave propagating down an infinite transmission line to introduce the concept of LH-propagation. Subsequently, the design of several meta-material-based antennas for passive UHF RFID tags is summarized.

Chapter 5 proposes an in-depth investigation of the requirements for the antenna part of UHF RFID tags, with focus on antenna design, characterization and optimization from the perspectives of both costs involved and technical constraints. A special attention is given to antennas that could be manufactured if one follows more or less standard manufacturing techniques available in the packaging industry. The chapter also presents some new ideas on how to utilize the antenna structure itself as a sensor for measuring different physical properties within the logistic chain.

Chapter 6 focuses on the operation theory of the RFID system. The antenna in RFID system is discussed, and the designing considerations of the antennas for RFID applications are presented. Also the design, simulation and implementation of some commonly used antennas in the RFID system are investigated.

Chapter 7 deals with the design strategy and process integration for a small on-chip-antenna with a small RFID tag on a chip-area 0.64 x 0.64 mm at 2.45 GHz for communication in near field.

Chapter 8 presents some considerations over the design of an RFID tag.

Chapter 9 discusses active RFID tags system energy analysis as excitable linear bifurcation system.

In Chapter 10, several types of tag antennas which are mountable on metallic platforms are introduced and analyzed. It is generally known that metallic objects strongly affect the antenna performance by lowering the efficiency of tags. Therefore tag antennas have to be designed to enable tags to be read near and on metallic objects without severe performance degradation.

Chapter 11 also deals with problems raised by the use of RFID technologies in metal environments and proposes various solutions. Thus, the authors explain the basics of the inductive coupling method, the detuning and the shielding effects due to metals. Additionally, a new system that is able to work at ultra-low frequencies (ULF) and through a metallic shielding is proposed. Finally, the properties of the low frequencies and the new ULF systems are compared.

Chapter 12 refers to the development of metallic coil identification system based on RFID technologies. This type of system was developed for the supply chain management in the iron and steel industry.

Chapter 13 presents a TransCal software-based system design approach for inductively coupled transponder systems. The authors discuss three design examples to show the advantages and limits of their approach.

The broad objective of Chapter 14 is to show an integrated process flow for the integration of gas sensors onto flexible substrates together with an RFID transponder to get a Flexible Tag Microlab innovative system for food logistic applications.

Chapter 15 gives additional insight into the inks to be used in printing RFID antennas, their properties, their performance, benefits and drawbacks, and future concerns. In addition, some attention was given to adhesives, which are necessary to bond the die or die strap to the antenna.

Chapter 16 describes how inkjet printing techniques can be used for the fabrication of conductive tracks on a polymer substrate; these techniques can be applied to manufacture RFID tags.

Chapter 17 introduces a Wi-Fi RFID active tag called Tag4M with the functionality of a multifunctional input/output measurement device. This tag offers a combination of Wi-Fi radio and measurement capabilities for sensors and actuators that generate output as voltage, current, or digital signal. Tag4M is suitable for prototyping wireless sensor measurements, as well as for educational purposes such as teaching wireless measurement using the existing Wi-Fi infrastructure.

The final chapter of this book presents the technology, design and implementation of an inductively-coupled passive 64-bit organic RFID tag, which is fully functional at 13.56 MHz.

One of the best ways of documenting in the domain of RFID technology is to analyze and learn from those who have trodden the RFID path. And this book is a very rich collection of articles written by researchers, teachers, engineers, and technical people with strong background in the RFID area.

I wish to sincerely acknowledge the efforts of all scientists that contributed to this book. In addition, I would like to express my appreciation to the team at InTech that has fulfilled its mission with the highest degree of dedication again.

Editor

Cristina TURCU
Stefan cel Mare University of Suceava
Romania

Contents

Hardware Design and Deployment Issues in UHF RFID Systems

Byung-Jun Jang
Kookmin University, Seoul
Korea

1. Introduction

Recently, radio frequency identification (RFID) have created emerging applications for tracking, sensing, and identifying various targets in wide-ranging areas such as supply chain, transportation, airline baggage handling, medical and biological industry, and homeland security. RFID systems with a variety of radio frequencies and techniques have been introduced. Among them, ultra-high frequency (UHF) band passive RFID systems that operate in the 860 – 960MHz band have drawn a great deal of attention because of its numerous benefits, such as cost, size, and increased interrogation range. In particular, the interrogation range of the UHF RFID system is comparatively large, due to the use of a travelling electromagnetic (EM) wave to transfer power and data. The increased interrogation range makes it possible for UHF RFID systems to revolutionize commercial processes, such as supply chain management. Several major supply chain companies such as Wal-Mart and Tesco plan to mandate the use of an UHF RFID system in their supply chains (Finkenzeller, 2003).

UHF band passive RFID system based on modulated backscatter has a unique characteristic, quite distinct from those encountered in most other radio systems which involve active transceivers on both sides of the link (wireless LAN, Bluetooth, etc). Because tag has no internal power supply, RFID reader must always supply the power in order to communicate with tags. This puts a different emphasis on the radio link, hardware design, and deployment aspects (Nikitin & Rao, 2008).

In this chapter, we review recent works in current passive UHF RFID systems to provide guidance regarding RFID system design and deployment. We cover the following topics.

- UHF RFID radio links using the link budget concept to calculate forward-link and reverse-link interrogation ranges.
- Hardware design considerations at the reader: phase diversity and quadrature signal combining, phase noise with range correlation effect, and transmitter leakage reduction methods.
- Deployment issues including reader-to-reader interference

The organization of this chapter is as follows. Section 2 analyzes the RFID link characteristics and shows the necessity of link budget concepts to calculate the RFID interrogation range. The hardware issues in an RFID reader are discussed in Section 3 along with recently published research results. Section 4 shows the RFID deployment issues with emphasis on reader-to-reader interference in dense reader environments. Finally, the conclusions are presented in Section 5.

2. RFID link budget

A communication link, as is well known, encompasses the entire communication path from the transmitter (TX), through the propagation channel, and up to the receiver (RX). In a typical wireless communication system, illustrated in Fig. 1(a), there are forward and reverse links. The forward link is the communication link from a base station (BS) to a mobile station (MS), whereas the reverse link is the opposite communication link, from MS to BS. Because BS and MS can simultaneously transmit data to each other through the forward and reverse links, a typical communication link is called full duplex. In addition, the power levels of the two links have few differences. Therefore, the forward link coverage is almost the same as that of the reverse link, although the transmit power and sensitivity of both links are a little different (Dubkin, 2008).

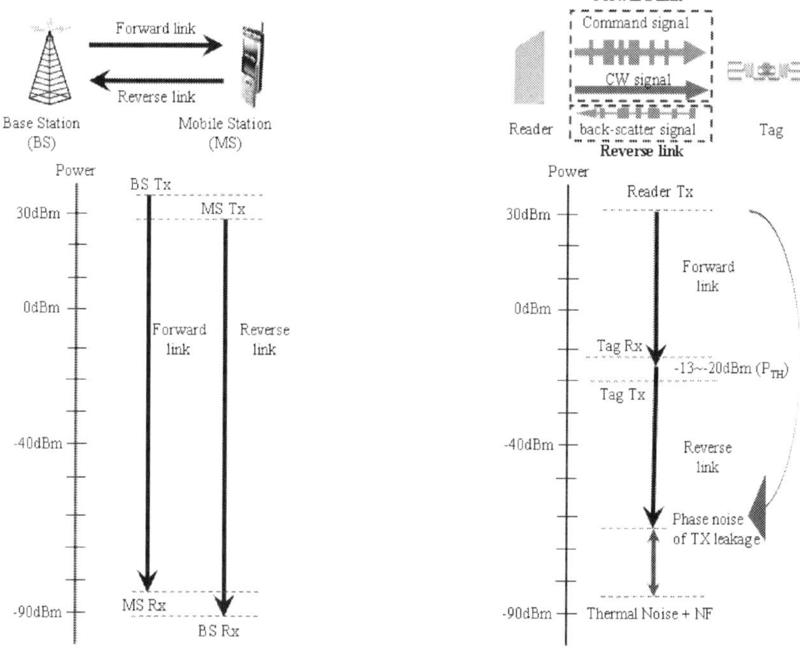

(a) Typical wireless communication system   (b) RFID system

Fig. 1. Comparison of link characteristics between a typical wireless system and an UHF RFID system

On the other hand, UHF RFID links, as illustrated in Fig. 1(b) are different from typical wireless links. An RFID system is generally comprises two components: reader and tag. The reader, sometimes called the interrogator, is made up of a TX/RX module with one or more antennas. The tag consists of a microchip for storing data and an antenna to transmit stored data. Tags are normally categorized into active and passive types by the presence or absence of an internal power supply. Because the passive tag has no power supply of its own, it obtains energy from the continuous wave (CW) signal transmitted by a reader. In addition, the passive tag transmits its data by backscattering the CW signal. In other words, the data transmission from tags to the reader is done by reflecting the wave energy back to the reader. Therefore, an RFID link is half duplex: reader to tag and then tag to reader. This means that RFID links are intrinsically unbalanced. Moreover, the reverse link is highly correlated with the forward link, because the tag's transmit power is determined by the reader's transmit power (Yoon & Jang, 2008).

These link characteristics of the UHF RFID system can be easily calculated using the link budget concept, which is the wireless communication system designer's primary tool for estimating the cell coverage.

2.1 Forward link budget calculation

In the forward link, the power received by the RFID tag, P_{RX}, can be found by applying the Friis EM wave propagation equation in free space:

$$P_{RX}(r) = \left(\frac{\lambda}{4\pi r}\right)^2 P_{TX}G_T G_R \tag{1}$$

where

λ : the wavelength in free space

r : the operational distance between an RFID tag and the reader

P_{TX} : the signal power feeding into the reader antenna by the transmitter

G_R : the gain of the reader antenna

G_T : the gain of the tag antenna

One portion of the power P_{RX} is absorbed by the tag for direct current (DC) power generation, and the other portion of P_{RX} is backscattered for the reverse link. In order to deliver enough power to turn the tag's microchip on, the absorption power for DC power generation must be larger than the minimum operating power required for tag operation, P_{TH}. For example, the forward link budget which has amplitude shift keying (ASK) backscatter modulation is given by:

$$P_{RX}(r) = \frac{1 - m^4}{(m+1)^2}\left(\frac{\lambda}{4\pi r}\right)^2 P_{TX}G_T G_R \geq P_{TH} \tag{2}$$

where m means the modulation depth.

The forward-link interrogation range (FIR) using the forward link budget calculation is depicted in Fig. 2. The FIR is proportional to the square root of the transmitted effective isotropic radiated power (EIRP), $P_{TX}G_T$, and the tag antenna's gain, G_R, and is inversely proportional to the square root of the tag's power threshold level, P_{TH}. From experience, it is known that the threshold power level required to turn on a tag ranges from 10uW (-20dBm) to 50uW (-13dBm) (Karthasu & Fischer, 2003). The modulation depth, m, is chosen to be an average value between 0.1 and 0.9.

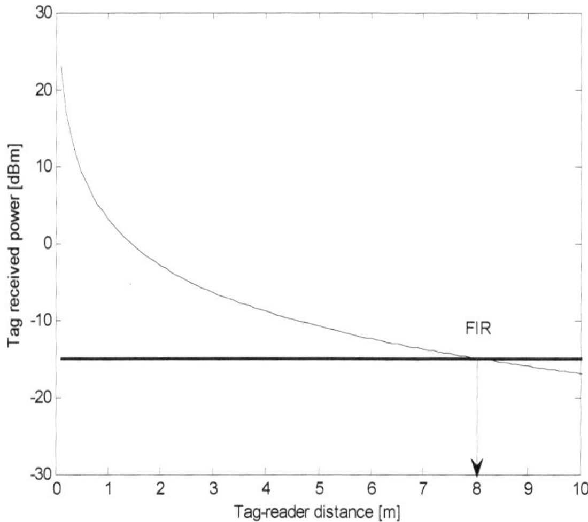

Fig. 2. Forward link budget of an UHF RFID system with center frequency of 915MHz, receive antenna gain of 2.15dBi, of -15dBm, and transmit EIRP of 4W

2.2 Reverse link budget calculation

In the reverse link, the backscattered signal from a tag should be strong enough so that the reader's demodulation output signal will meet the system's minimum signal-to-noise-ratio (SNR_{min}) requirement. This is very similar to typical wireless communication system links. However, because the CW signal always exists in a reverse-link to turn the tag on, the TX leakage level plays an important role in determining the reverse-link budget. Fortunately, the DC offset due to TX leakage is removed from a baseband bandpass filter. Nonetheless, the phase noise of the TX leakage, N_{PN}, on the receiving bandwidth is unfortunately not removed by the filter. Therefore, it may be much stronger than the thermal noise, to a degree that the reverse link budget mainly depends on the phase noise of the TX leakage. On the other hand, in a typical wireless communication system, the phase noise of the TX leakage within the receiving bandwidth is normally not a major problem, because duplexing techniques, such as frequency division duplexing (FDD) and time division duplexing (TDD), are applied.

Figure 3 shows a link budget example in the stationary reader case according to tag-reader distance. The reverse-link interrogation range (RIR) is defined as the maximum distance at which the tag's backscattered signal meets the minimum reader sensitivity condition. As showin in Fig. 3, the forward link is determined by a tag threshold voltage, the reverse link is mainly determined by the phase noise of TX leakage.

2.3 Interrogation range

The performance of an UHF RFID system is usually characterized by its interrogation range, which is defined as the maximum distance at which an RFID reader can recognize a tag. This can be divided into two categories: the FIR and the RIR. Since the actual interrogation

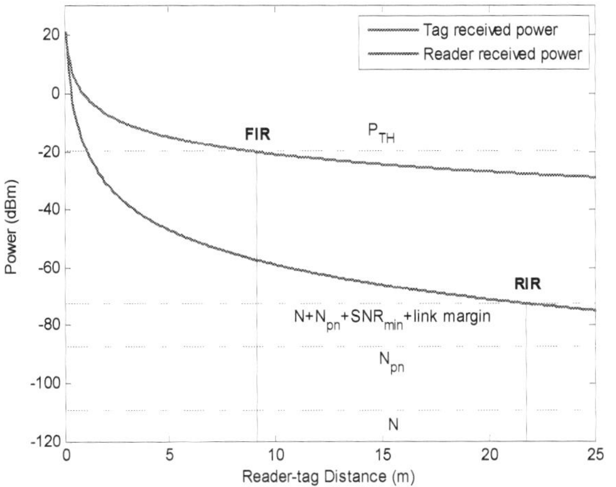

Fig. 3. Reverse link budget of an UHF RFID system (N: thermal noise) (Yoon & Jang, 2008)

range is determined by the smaller value of FIR and RIR, both values should be considered simultaneously when deploying UHF RFID systems. As shown in Fig. 3, FIR has a smaller value than RIR in the case of a well-designed reader. However, RIR may be much more significant than the FIR in environments such as warehouses because of interference from other readers. Also, the interrogation range of a battery-assisted tag is determined by the RIR only.

3. Hardware design issues in the UHF RFID reader

In order to discuss hardware design issues in the UHF RFID reader, let us consider an UHF RFID system model using a direct-conversion I/Q demodulator, as shown in Fig. 4. The reader is composed of local oscillator (LO), a transmitter, a receiver and an antenna. The power amplifier (PA) amplifiers the LO signal to achieve a high power level. The amplified

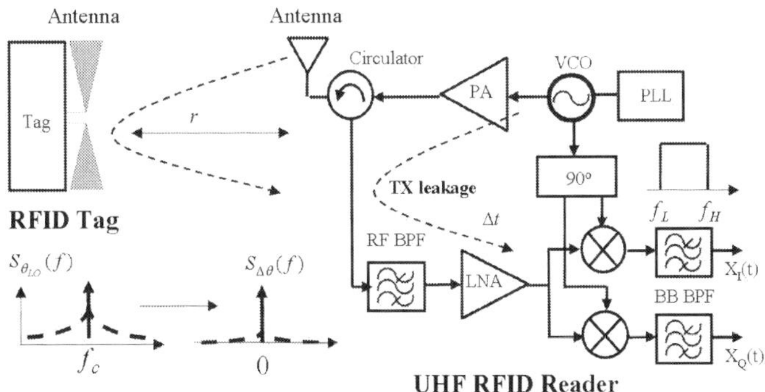

Fig. 4. Architecture of an UHF RFID system and block diagram of a reader and a tag

signal feeds into the reader antenna via the circulator and then radiates into the air. The reader antenna simultaneously receives the backscattered signals from the tag. The antenna can be configured in two ways: two antennas or one antenna with a circulator. The circulator is a non-reciprocal three-port device, where the signals travel from the transmitter port to the antenna port or from the antenna port to the receiver port. In practice, the circulator cannot entirely isolate the transmitter from the receiver, due to the inherent leakage between its ports. Generally, TX leakage is between -20 to -50dB (Jang & Yoon, 2008a).

3.1 Phase diversity and optimal I/Q signal combining

As shown in Fig. 4, the same LO provides two identical frequency signals, one for the transmitter and the other for the receiver. The LO signal for the receiver is further divided using a power splitter to provide two orthonormal baseband outputs, I and Q signals. Because the received signal and the LO signal have the same frequency, the absolute phase of the received signal influences the amplitude of the down-converted signal. Therefore, some sort of phase diversity using I and Q signals should be provided to demodulate the tag signal (Jang, 2008).

Figure 5 shows the simulation results of normalized I and Q signal power at the quadrature receiver for the case of tag moving. For this simulation, the tag located 1 meter below the reader antenna is assumed to move up to 5m away from the reader. The complex plot forms a spiral-like shape due to the periodic received signal power variation.

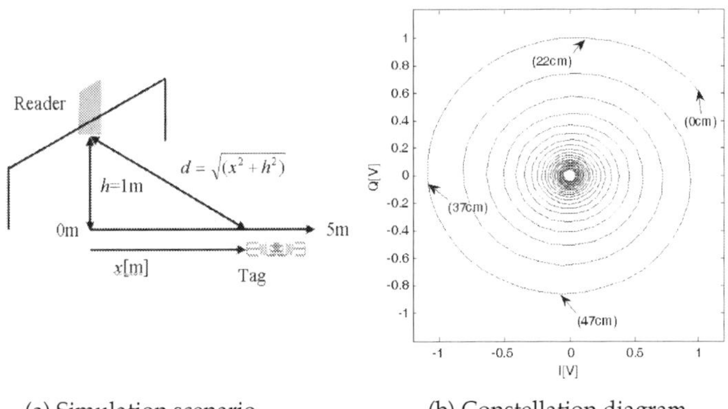

(a) Simulation scenario   (b) Constellation diagram

Fig. 5. Received signal variation characteristics of an UHF RFID receiver with respect to the reader-tag distance

Using the quadrature receiver, the demodulator can choose the higher of the tag signals to retrieve the tag's data. This is called selection diversity. Now, the reader can select the better of the quadrature (I and Q) channel outputs and overcome the limitation of a single channel receiver. Figure 6 shows the performance of selection diversity compared with the I and Q channel signals with respect to phase value from zero to π. In selection diversity, two extreme instances, i.e., 'minimum' and 'optimum' occur every $\lambda / 8$ meters, as the tag moves away from the reader antenna. At 900 MHz, these minimum points occur every 4.2cm. For the optimum instance, the tag signal can be demodulated without loss. However, for the minimum instance, the tag signal can be reduced with a 3dB loss in power. In order to overcome this 3dB loss of selection diversity, various I/Q combining techniques can be used. For example, the power combining technique can be used in the ASK case. On the other hand, signal combining with phase shift keying (PSK) is not as easy as ASK. Recently, arctangent combining and principal component combining (PCC) have been suggested (Jang, 2008).

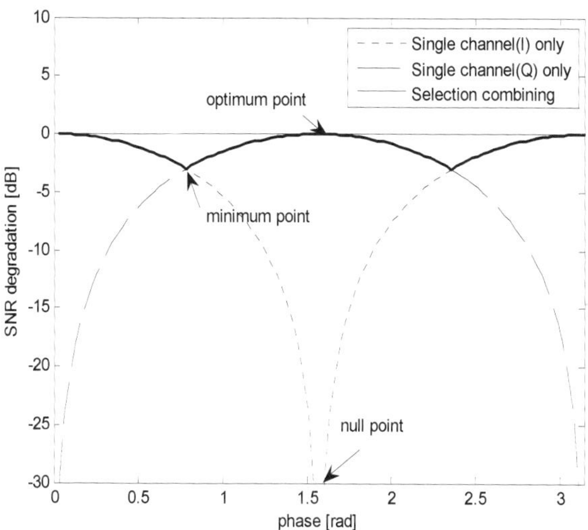

Fig. 6. SNR degradation for various receiver combining techniques

3.2 Phase noise and range correlation effects

Phase noise is an important parameter in designing RFID systems since it can have a significant influence on system performance. Because a LO is generally used for both CW signal generation and the down-converting operation, the phase noise of the received signal is correlated with that of the LO signal. The correlation level is inversely proportional to the time difference between the two signals. In an UHF RFID system, this time difference is very small (several nsec) due to the short tag-reader distance, and so phase noise is reduced by the correlation effect. In an RFID application, this phase noise reduction phenomenon is called the range correlation effect (Jang & Yoon, 2008b).

The baseband power spectral density (PSD), $S_{\Delta\theta(t)}(f)$, for LO phase noise with the offset frequency Δf_c and a round-trip delay of Δt is given by (Droitcour et al., 2004):

$$S_{\Delta\theta(t)}(f) = S_{\theta_{LO}(t)}(f)\, 4\sin^2\left(4\pi\frac{r\Delta f_c}{c}\right)^2 \tag{3}$$

where $\Delta\theta(t) = \theta_{LO}(t) - \theta_{LO}(t - \Delta t)$ and $\theta_{LO}(t)$ is the phase noise of the LO signal.

The term in parenthesis embodies the range correlation effect on the baseband spectrum. Assuming that the typical values for r and f_o are 8m and 160kHz, respectively, the value of $r\Delta f_c / c$ will be on the order of 10^{-3}. So the range correlation effect will dramatically reduce the PSD of the LO phase noise.

Figure 7 shows an example of a typical PSD of the LO itself and the phase noise reduction effects due to the range correlation with a round-trip delay of 1m. The typical PSD of the LO is selected considering state-of-the-art UHF RFID LO performance. The effect of the range correlation on the phase noise for different offset frequencies was estimated by (3). For example, at an offset frequency of 10Hz, the phase noise is reduced by 130dB.

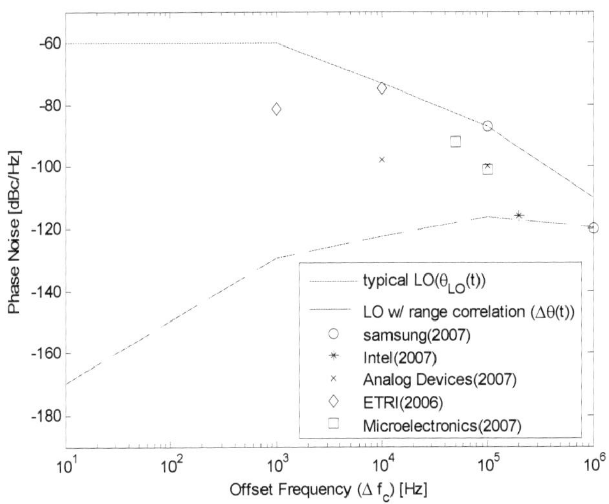

Fig. 7. LO phase noise as a function of offset frequency

In addition, the phase noise may affect the symbol-error-rate (SER) performance in an RFID system. Figure 8 shows the SER performance of the PSK modulation and FM0 coding with phase noise as a function of range correlation. Without the range correlation, the SER performance is worse for the case of typical LO phase noise, as shown in Fig. 7. This degradation is worse for a small modulation phase noise. However, the phase noise of the LO with range correlation effects is almost identical to the SER performance in AWGN environments because of the phase noise reduction by range correlation. For a real LO using a phase-locked loop (PLL), the power spectral density of the phase noise is filtered by the transfer function of the PLL, and the phase noise effects on the error performance are even small. Unlike PSK modulation, phase noise has no effects on ASK modulation, because there is no information in the carrier's phase (Jang & Yoon, 2008b).

3.3 TX leakage reduction methods

Finally, some difficult technical problems arise from TX-to-RX leakage because the RFID reader transmits CW and simultaneously receives back-scattered data from tags. The strong TX leakage into the receiver side degrades the reader performance in relation to the sensitivity of the receiver and its interrogation range. In detail, the low noise amplifier

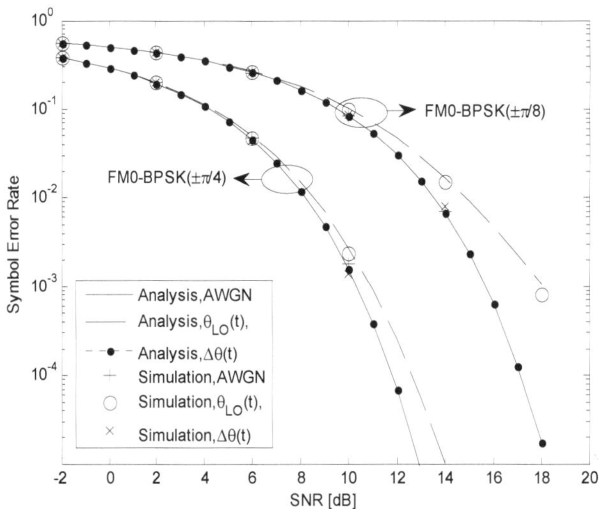

Fig. 8. SER performance of FM0-BPSK signal as a function of phase noise and range correlation effect (Jang & Yoon, 2008b)

(LNA) of the receiver can be saturated by this strong TX leakage, decreasing the dynamic range of LNA. A DC offset problem is also caused by self mixing at the mixer in the reader receiver.

To alleviate the TX leakage problem, the strong TX signal should be separated from the RX signal as much as is possible to achieve higher performance from the RFID reader. The simplest solution is to separate the TX and RX antennas. However, the size and cost of the reader hardware will increase. A circulator of ferrite material or an active CMOS circulator may lighten this burden, but the cost is still high, and isolation of these circulators is insufficient to meet some required criteria. A directional coupler may, therefore, be a better choice given its simplicity and low cost (Kim et al., 2006).

4. Deployment Issues

In supply-chain applications, tens or hundreds of RFID readers will be in operation within close range of each other, which may cause serious interference problems.

There are three types of UHF RFID interference: multiple-tag-to-reader interference (tag collision), multiple-reader-to-tag interference (tag interference), and reader-to-reader interference (reader interference or frequency interference) as shown in Fig. 9.

Multiple-tag-to-reader interference arises when multiple tags are simultaneously energized by a reader and reflect their respective signals back to the reader. Due to a mixture of scattered waves, the reader cannot differentiate individual IDs from the tags: therefore, anti-collision mechanisms such as those known as binary-tree and ALOHA are needed to resolve multiple-tag-to-reader interference (Dubkin, 2008), (EPCglobal, 2004). Multiple reader-to-tag interference happens when a tag is located at the intersection of two or more reader interrogation ranges and the readers attempt to communicate with the tag simultaneously. This can cause a tag to behave and communicate in undesirable ways. Multiple reader-to-tag interference can be solved simply by separating reader intterrogation ranges.

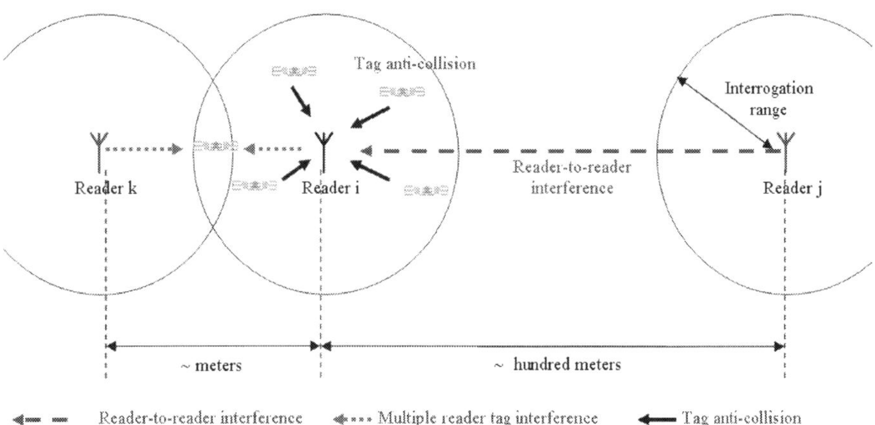

Fig. 9. Three types of Interference in UHF RFID systems

The last type of interference, reader-to-reader interference, is induced when a signal from one reader reaches other readers (Birari & Iyer, 2005). This can happen even if there is no intersection among reader interrogation ranges. As the signal transmitted from distant readers may be strong enough to impede accurate decoding of the signals that are back-scattered from adjacent tags, reader-to-reader interference can cause serious problems in UHF RFID system deployment (Kim et al., 2008), (Kim et al., 2009). Moreover, the interference is potentially magnified in a dense reader environment, which can involve hundreds of readers in one warehouse or manufacturing facility. Many attempts to mitigate reader-to-reader interference have been made. They are normally based on standard multiple access mechanisms such as frequency-division multiple access (FDMA), time-division multiple access (TDMA), or carrier-sense multiple access (CSMA). For example, the electronic product code for global class 1 generation 2 (EPCglobal C1G2) includes spectrum management of an UHF RFID operation in a dense reader environment. According to EPCglobal C1G2, reader transmit signals and tag back-scattered signals are separated in a spectral domain (EPCglobal, 2004).

Additionally, careful consideration of the positioning and type of RFID reader antenna selected are important for reader-to-reader interference (Leong et al., 2006). The situation can also be improved by using reader synchronization and frequency channelling. Actual field testing will be carried out in the future, especially in warehouses, where dense RFID reader environments are most likely to exist

5. Conclusion

In this chapter, we discuss hardware design and deployment issues in current passive UHF band RFID systems. Using the link budget concept, the simple method to calculate forward-

and reverse-link interrogation range is shown. Then, we consider the hardware issues on an RFID reader: phase diversity and signal combining techniques, phase noise with range correlation effect, and TX leakage reduction methods. Finally, three interference problems with an emphasis on reader-to-reader interference encountered in the deployment of RFID systems are presented.

6. References

Birari, S. M. & Iyer, S. (2005). Mitigating the Reader Collision Problem in RFID, *Proceedings of the 13th IEEE International Conference on Networks*, 16-18 Nov. 2005

Droitcour, A. D.; Lubecke, O. B., Lubecke, V. M., Lin, J. & Kovacs, G. T. A. (2004). Range correlation and I/Q performance benefits in single-chip silicon doppler radars for noncontact cardiopulmonary monitoring, *IEEE Trans. Microwave Theory Tech.*, Vol. 52, No. 3, pp.838-848, Mar. 2004, ISSN 0018-9480

Dubkin, D. M. (2008). *The RF in RFID: passive UHF RFID in practice*, Elsevier ISBN 978-0-7506-8209-1

EPCgloabl (2004). *EPC radio-frequency identity protocols class-1 generation-2 UHF RFID protocol for communications at 860MHz-960MHz version 1.0.9*, EPCglobal Standard Specification, 2004.

Finkenzeller, K. (2003). *RFID Handbook: Fundamentals and applications in contactless smart cards and identification*, John Wiley, ISBN 0-470-84402-7, Chichester

Jang, B. -J. (2008). Phase diversity and optimal I/Q signal combining methods on an UHF RFID reader's receiver, *Microwave Journal(Web Exclusive)*, Vol. 51, No. 4, April. 2008

Jang, B. -J. & Yoon, H. (2008a). Examine the effects of phase noise on RFID range, *Microwave and RF*, July. 2008, pp. 78-77, ISSN 0745-2993

Jang, B. -J. & Yoon, H. (2008b). Range correlation effect on the phase noise of an UHF RFID reader, *IEEE Microwave and Wireless Components letters*, Vol.18, No. 12, Dec. 2008, pp. 827-829, ISSN 1531-1309

Karthasu, U. & Fischer, M. (2003). Fully integrated passive UHF RFID transponder IC with 16.7-uW minimum RF input power, *IEEE J. Solid-State Circuits*, Vol. 38, No. 10, pp.1602-1608, Oct. 2003, ISSN 0018-9200

Kim, D. Y.; Yoon, H., Jang, B. -J., & Yook, J. G. (2008). Interference Analysis of UHF RFID Systems, *Progress In Electromagnetics Research B*, vol. 4, pp. 115-126, ISSN 1937-6472

Kim, D. Y.; Yoon, H., Jang, B. -J., & Yook, J. G. (2009). Effects of Reader-to-Reader Interference on the UHF RFID Interrogation Range, *IEEE Trans. Industrial Electronics*, Vol. 56, No. 7, Mar. 2004, pp.2337-2346, ISSN 0278-0046

Kim, W. -K.; Lee, M. -Q., Kim, J. -H. Lim, H. -S., Yu, J. -W., Jang, B. -J. & Park, J. -S. (2006). A passive circulator with high isolation using a directional coupler for RFID, *IEEE Microwave Symposium Digest*, pp.1177-1180, ISSN 0149-645X, June. 2006

Leong, K. S.; Ng, M. L. & Cole, P. H. (2006). Positioning Analysis of Multiple Antennas in a Dense RFID Reader Environment, *Proceedings of. International Symposium on Applications and the Internet Workshops*, pp. 23-27, ISBN 0-7695-2510-5, Jan. 2006

Nikitin, P. V. & Rao, K. V. S. (2008). Antennas and propagation in UHF RFID systems, *Proceedings of 2008 IEEE International Conference of RFID*, pp. 277-288, ISBN 978-1-4244-1711-7, Apr. 2008

Yoon, H. & Jang, B. -J. (2008). Link budget calculation for UHF RFID systems, *Microwave Journal*, Vol. 51, No. 12, Dec. 2008, pp. 78-77, ISSN 0192-6225

2

Design Considerations for the Digital Core of a C1G2 RFID Tag

Ibon Zalbide, Juan F. Sevillano and Igone Vélez
TECNUN (Universidad de Navarra) and CEIT
Spain

1. Introduction

An EPC Class 1 Gen 2 (C1G2) RFID system is composed of a reader and one or several passive tags. Passive tags obtain the required energy from the radio frequency field emitted by the reader. The forward data link (reader to tag) is embedded in this radio frequency field. The backward data link (tag to reader) is achieved by means of backscattering.
The RFID tag consists of several analog circuits and a digital core. The analog circuits perform tasks such as harvesting the energy from the electronic wave, supplying power, generating a clock signal and signal conditioning. The digital core of the tag performs data detection and implements the logical requirements of the standard.

1.1 Passive long range UHF RFID systems

Fig. 1 shows the basic architecture of a long range Ultra High Frequency (UHF) RFID tag. The antenna receives the signal emitted by the reader. The voltage multiplier rectifies the incoming signal and increments the voltage to charge the supply capacitor C_{supply}. The efficiency of this voltage conversion will depend on the architecture of the voltage multiplier. The supply capacitor is used to supply power to the rest of the tag. The analog front-end creates the signals that the rest of the tag needs to work properly, such as regulated voltages, the clock signal and the reset signal. It performs some kind of demodulation by generating an intermediate signal that can be used by the digital core to detect the received bits. The analog front-end also modulates the load impedance of the tag commanded by the digital core, to backscatter the signal emitted from the reader so that information can be transmitted backwards. The digital core handles the communication protocol and accesses the non volatile memory to retrieve and store data.
Fig. 1 also shows the basic architecture of the digital core of a passive long range UHF tag. The input signal provided by the front-end is evaluated in a symbol detector to detect incoming symbols. A command decoder determines the operation code and the arguments received and it forwards them to a control unit. In the control unit, the finite state machine defined in the standard is implemented to control the communication flow. Moreover, depending on the standard, additional features such as collision arbitration algorithms or integrity checks are performed in this unit. Usually, the number of states of the finite state machine and the integrated additional features define the complexity and functionality of the whole tag. Finally, a transmitter controls the load modulator of the front-end and backscatters the answer to the reader.

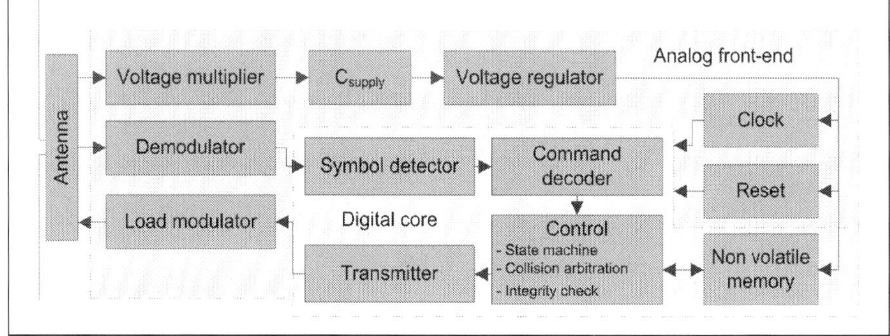

Fig. 1. Architecture of a passive RFID tag.

The C1G2 standard (EPC Global, 2005) has become the main communication protocol of passive long range UHF RFID systems. Given the success of the C1G2 standard, the ISO organization finally adopted it with minor changes as the ISO18000-6C (ISO, 2006). Nowadays, it is the dominant air interface for passive UHF RFID tags, because of its flexible functionality and the compatibility with the whole EPC network. Almost every recent research work concerning passive long range UHF RFID tags uses this communication protocol; e.g.: (Yan et al., 2006; Barnett et al. 2007; Man et al., 2007; Ricci et al., 2008; Zhang et al., 2008; Roostaie et al. 2008; Wanggen et al. 2009). Thus, this chapter is focused in the C1G2 standard. However, the concepts and ideas presented can be extended to other standards.

1.2 Communication range

The communication range of an RFID system is one of the factors that define the scope of its applications. Assuming that the reader is continuously sending a continuous wave (CW), the maximum communication distance between a passive RFID tag and the reader is mainly limited by two factors related to the tag's power consumption: the input power in the tag, and the voltage at the input of the voltage multiplier (Pardo et al., 2007; De Vita et al., 2005).

Input power in the tag

The communication range r is limited by the minimum input power. According to (Pardo et al. 2007),

$$r \leq \sqrt{\frac{P_{EIRP}\lambda^2 G X^2 \eta}{4\pi^2(R_A^2 + 4X^2) \cdot P_{TAG}}} \,, \tag{1}$$

where P_{EIRP} is the Effective Isotropic Radiated Power, λ the wavelength, G the tag antenna gain, X the reactance introduced by the load modulator, η the efficiency of the rectifier, R_A the impedance of the antenna and P_{TAG} the tag power consumption. Equation (1) shows that reducing the power consumption of the analog or digital parts increases the communication range.

Voltage at the input of the voltage multiplier

According to (Pardo et al. 2007), there is another constraint due to the minimum voltage at the input of the voltage multiplier. If the threshold voltage required to switch on the voltage multiplier is not achieved, the tag will not start working, no matter the available input power. This constraint is related to the fabrication technology and it is given by

$$r \leq \frac{0.7 Q_{MN} \lambda \sqrt{P_{EIRP} G R_A}}{4 \pi V_{tech}} , \tag{2}$$

where Q_{MN} is the quality factor of the matching network and V_{tech} is the minimum voltage at the input of the voltage multiplier. The voltage V_{tech} depends on the technology used. Q_{MN} can be expressed in terms of the equivalent input resistance of the tag as

$$Q_{MN} = \sqrt{\frac{R_P}{R_A} - 1} , \tag{3}$$

where R_P is the equivalent resistance in parallel with C_{supply} that represents the power consumption of the tag. R_P increases as power consumption decreases. Thus, as the power consumption decreases, Q_{MN} increases and a larger communication range is feasible.

Summarizing, the communication range increases when the power consumption of the tag decreases. As detailed in (Pardo et al. 2007), the most restrictive of (1) and (2) sets the actual communication range of the system. However, the relation between power reduction and range improvement is not always constant. The dependence of the communication range on the input power in the tag is stronger than on the voltage at the voltage multiplier. Thus, when the power consumption of the tag is high and (1) limits the communication range, reducing the power consumption of the tag increases notably the maximum communication distance. But when the power consumption goes down, (2) becomes the most restrictive and the range improvement slows down. At this point the technology is limiting the communication more than the power consumption.

A proper design of the tag is required to minimize the power consumption, and move from the section where the power consumption limits the communication range to the section where the technology is the limiter. This way, the maximum communication range for the selected technology can be achieved. As the digital part's power consumption can be comparable to the analog, the reduction of the digital power consumption is very important for the overall performance of the system.

The main goal of the publications focused on C1G2 digital cores is to minimize the average power consumption. Advances in the technology of semiconductors help to reduce the power consumption of integrated circuits. Designers have to work with the technology available at that time. However, there are issues where designers can focus to optimize their designs for a given technology.

The power consumption of the digital core grows with the clock frequency. Thus, designers try to reduce the clock frequency to minimize power consumption. Impinj, a C1G2 tag seller, published a white paper where this value was said to be 1.92MHz (Impinj, 2006). Even though there are some works in the literature that work at 1.92MHz, such as (Wang et al., 2007), most works propose digital cores for other clock frequencies. For example, (Hong et al., 2008) works at 4 MHz, (Man et al., 2007) at 3.3MHz, (Zhang et al. , 2008; Yan et al., 2006) at 1.28MHz and (Ricci et al., 2008) at 2MHz. A study of the constraints on the clock signal of the digital core is needed so that the clock frequency can be optimally selected.

Another approach employed by digital designers to reduce power consumption is power management. Depending on the technology, similar benefits can be obtained in a simpler way by means of clock gating. In either case, the net effect is that the digital core can not be considered to have constant power consumption, but a power consumption profile in time

(Zalbide et al., 2008). The effect of the shape of this power consumption profile in the overall performance (i.e.: communication range) needs to be studied.

1.3 Objectives of the chapter

This chapter performs an analysis of the EPC C1G2 standard to extract requirements of the tag's digital core for proper forward-link data detection and backward-link data backscattering. The EPC C1G2 standard specifies the characteristics of the waveforms employed in the forward-link communication. These characteristics of the waveforms pose requirements on the clock signal used for data detection (wander and jitter of the clock). The backscattering signal is also controlled by the digital core. Therefore, the requirements set by the standard on the backscattering signal constrain the clock signal used in the digital core as well. The chapter reviews some aspects of the operation of the digital core of the tag and presents equations and figures that can be used to select an appropriate clock signal.

On the other hand, the chapter presents methods to analyse the influence of the application of power management techniques in the communication range of the system. The results provide valuable tools to analyse different trade-offs early in the design of a RFID tag.

The chapter is organized as follows. Section 2 describes the architecture of the digital core considered for the study. Section 3 analyses the constraints on the clock signal and Section 4 studies the influence of the power management on the communication range. Finally, the main conclusions are summarized in Section 5.

2. Architecture

Fig. 2 depicts the architecture of a generic C1G2 digital core used as example for this study. The incoming symbols are detected in the symbol detector. The command decoder obtains the operation code and the arguments of the requested operation. In order to avoid a big input buffer, the variable length arguments, such as the selection mask, are processed in the command decoder. The cryptographic functions are also performed in the command decoder making all arguments transparent to the rest of modules. The control module controls the system with a finite state machine and a register bank. The collision arbitration, session management and memory lock are contained in this module. It does the necessary operations accessing the memory and the register bank. Finally, TX, the transmitter, encodes the answer with the required format. For this purpose, a backward link frequency synthesizer is included in the transmitter. The accesses to the tag memory (EEPROM) are handled by an intermediate module memory access. The power management unit, PM, controls the activity of the rest of the modules using the clock gating technique. VEEPROM is a cache memory used to reshape the power consumption profile of the tag (Zalbide, 2009). For power management, five different working states have been defined. Every working state is optimized to perform a specific operation during the communication. Table 1 shows the relationship of the working states with the modules. Each working state activates the necessary modules to fulfill its functionality and deactivates the remaining modules. Some modules, such as the memory access, are used in various different working states; others are only used in a single working state. The combination of all the working states enables the system to work properly minimizing the activity of its circuitry.

The STRTP state is the initial working state. The tag checks the kill bit, and if it is not killed, the system turns to STDBY state, waiting for the beginning of the forward link data

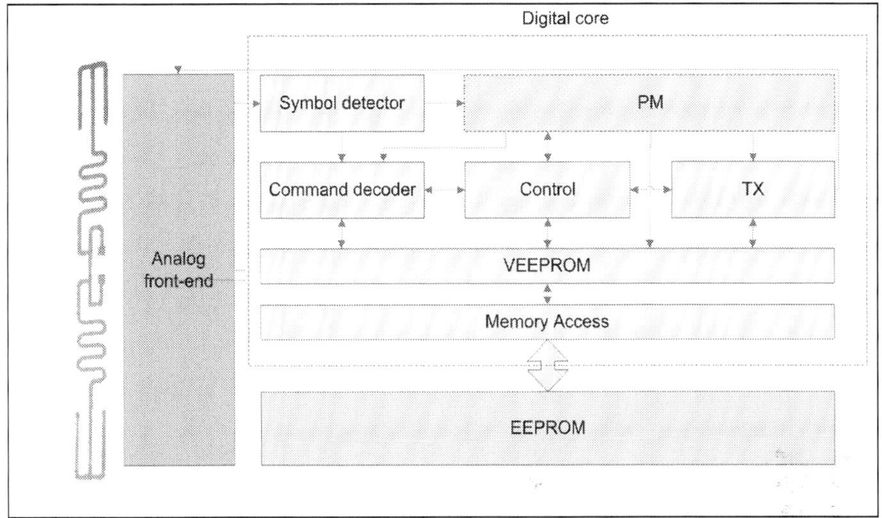

Fig. 2. Architecture of the low power C1G2 digital core.

	Working state				
	STRTP	STDBY	RX	CNTRL	TX
PM	ON	ON	ON	ON	ON
Symbol detector	OFF	ON	ON	OFF	OFF
Command decoder	OFF	OFF	ON	OFF	OFF
Control	ON	OFF	OFF	ON	OFF
Memory access	ON	OFF	ON	ON	ON
VEEPROM	ON	OFF	OFF	ON	OFF
TX	OFF	OFF	OFF	OFF	ON

Table 1. Working states of the digital core.

transmission. In this state only the symbol detector is active. When the beginning of a new message from the reader is detected, the command decoder is activated and the working state turns to RX. After receiving the whole message, the working state changes to CTRL, deactivating the command decoder and the symbol detector, and activating the control and the register bank. Finally, in the TX state the response is sent to the reader and the working state returns to STDBY.

Reading from the EEPROM is one of the most power hungry operations that the tag performs. In the design presented in this chapter, the EEPROM is read when a lot of energy is arriving to the tag. Then, the read data is stored in VEEPROM, which is less power hungry. This way, if data from the EEPROM is needed when less energy is available, they can be read from VEEPROM instead of from EEPROM. The introduction of this module allows reshaping the power distribution so that the power peaks caused by the accesses to EEPROM can be moved to less critical time intervals. In exchange, the tag spends more time initializing, as it must copy the data from the EEPROM to the VEEPROM.

3. Analysis of the clock signal requirements

As the power consumption of the digital core grows with the clock frequency, the selection of a minimum clock frequency will maximize the communication range. In the following, a detailed study of the clock signal constraints for C1G2 communication is presented. This study shows that the minimum required clock frequency depends on the characteristics of of the clock signal and the implementation of the transmitter.

The section is organized as follows. First, a model for the clock signal used by the digital core is defined. Then, the operation of the digital core is analyzed together with the specification of the standard. From this analysis, equations that constrain the clock signal parameters are obtained. These equations are computed numerically to find the regions in the clock signal parameter space where the C1G2 standard specifications are satisfied. These results facilitate the definition of the requirements for the generator of the clock signal used in the digital core.

3.1 Clock model

Ideally, the clock signal can be considered as a square wave of period T. The frequency, $f=1/T$, is assumed to be constant and invariable in time. Nevertheless, actual clock sources do not generate perfect clock signals. For instance, if we measure the average clock period over two time intervals in different days or ambient conditions, the results may be different. Moreover, the duration of the clock periods within the same time interval suffers small variations from one cycle to another. For our analysis, we will model the clock signal using two parameters:

Average period, T_a: it is the mean value of the period of the clock signal during a whole inventory round.

Random jitter, ξ_i: it is a random variable that represents the normalized deviation of the clock edges from the edges of the average period.

Thus, the duration of the ith clock period T_i is given by $T_i=T_a+\xi_i$. If the maximum random jitter of the clock signal is annotated as ξ_{max}, then for all i, $T_i \in [T_a \cdot (1-\xi_{max}), T_a \cdot (1+\xi_{max})]$

3.2 Forward link

In the forward link of C1G2 (EPC Global, 2005), a reader communicates with one or more tags by modulating a Radio Frequency (RF) carrier using Amplitude-Shift Keying (ASK) modulation with Pulse Interval Encoding (PIE). The reader transmits symbols of duration $T_S=T_H+T_L$. In each symbol, the signal has maximum amplitude during T_H seconds and minimum amplitude during T_L seconds. $T_L=PW$ for both a data-0 and a data-1. As shown in Fig. 3, in order to transmit a data-0, T_H is set so that $T_S=Tari$. In order to transmit a data-1, T_H is set so that $1.5 \cdot Tari \leq T_S \leq 2 \cdot Tari$.

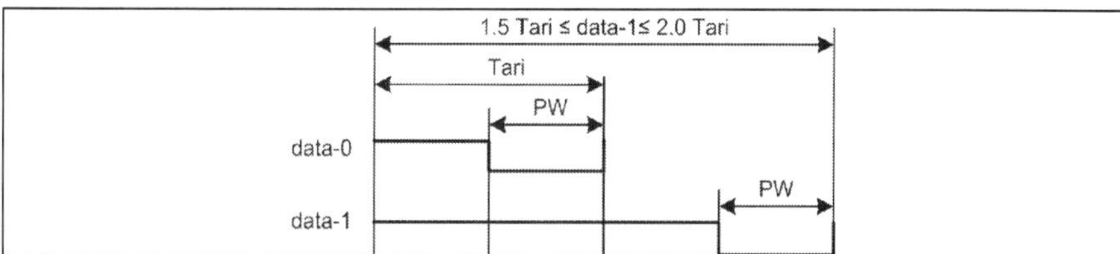

Fig. 3. PIE codification, from (EPC Global, 2005).

The forward data rate is set in the preamble of every command sent by the reader to the tag by means of symbol *RTcal*, as shown in Fig. 4. The duration of this symbol *RTcal* is equal to the duration of a data-0 plus the duration of a data-1. A tag shall measure the length of *RTcal* and compute *pivot=RTcal/2*. The tag shall interpret subsequent reader symbols shorter than *pivot* as data-0s, and subsequent reader symbols longer than *pivot* as data-1s.

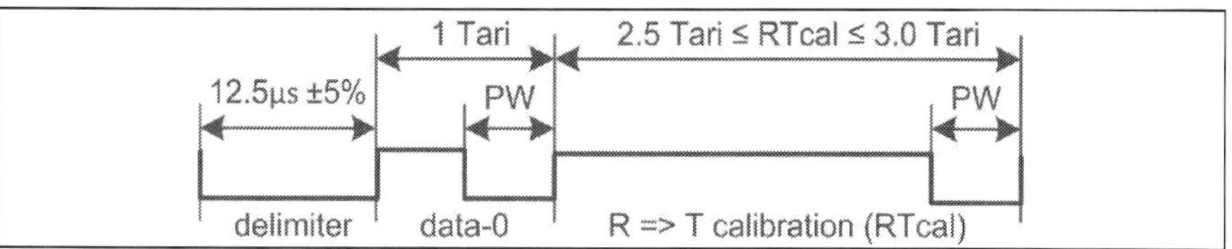

Fig. 4. Forward link calibration in the preamble, from (EPC Global, 2005).

3.2.1 Symbol detection

The front-end of the tag is assumed to have a one bit Analog to Digital Converter (ADC) to convert the envelope of the RF signal to a digital signal. The input to the digital core is assumed to have a high value during T_H and a low value during T_L. The digital core samples the input signal and identifies the incoming symbols by measuring the distance between modulated pulses. It is assumed that one sample is taken every clock cycle.

Given a generic symbol S, its duration will be annotated as t_S. The number of samples obtained when sampling S, n_S, will be in the range defined by equation (4).

$$\left\lfloor \frac{t_S}{T_a \cdot (1 + \xi_{max})} \right\rfloor \le n_S \le \left\lceil \frac{t_S}{T_a \cdot (1 - \xi_{max})} \right\rceil , \tag{4}$$

where $\lfloor \cdot \rfloor$ and $\lceil \cdot \rceil$ are the floor and the ceil functions respectively.

3.2.2 Forward link constraints

For proper operation, the digital core shall be able to detect when its input signal is in the high and in the low states. The duration in the low state, PW, is the shortest one. Therefore, the first constraint is that $n_{PW} \ge 1$. From (4), the first constraint is obtained:

$$\left\lfloor \frac{PW}{T_a \cdot (1 + \xi_{max})} \right\rfloor \ge 1 . \tag{5}$$

The second constraint comes from the fact that in order to detect the data-0 symbol properly, the number of samples obtained from a data-0 symbol has to be lower or equal to n_{pivot}: $n_{data-0} \le n_{pivot}$, where $n_{pivot} = \lfloor n_{RTcal}/2 \rfloor$. Using (4) to obtain the maximum number of samples for n_{data-0} and the minimum number of samples for n_{RTcal}, we have,

$$\left\lceil \frac{Tari}{T_a \cdot (1 - \xi_{max})} \right\rceil \le \left\lfloor \frac{1}{2} \left\lfloor \frac{RTcal}{T_a \cdot (1 + \xi_{max})} \right\rfloor \right\rfloor . \tag{6}$$

If the symbol to be detected is a data-1, then we need that $n_{data-1} > n_{pivot}$. Taking from (4) the minimum number of samples for n_{data-0} and the maximum number of samples for n_{RTcal}, we obtain the third constraint:

$$\left\lfloor \frac{RTcal - Tari}{T_a \cdot (1 + \xi_{max})} \right\rfloor \leq \left\lfloor \frac{1}{2} \left\lceil \frac{RTcal}{T_a \cdot (1 - \xi_{max})} \right\rceil \right\rfloor. \tag{7}$$

3.3 Backward link

In the backward link, a tag communicates with a reader using ASK and/or Phase-Shift Keying (PSK) backscatter modulation (EPC Global, 2005). The backward link data codification can be either FM0 baseband or Miller. Both the backward link codification and data rate are set by the reader in the last *Query* command. The backward data rate is set by means of the duration of the *TRcal* symbol in the preamble and the Divide Ratio (*DR*) specified in the payload of the last *Query* command.

A tag shall compute the backward link frequency as

$$BLF = \frac{DR}{TRcal} \tag{8}$$

and adjust its response to be inside the Frequency Tolerance (*FT*) and Frequency Variation (*FV*) limits established by the C1G2 standard (EPC Global, 2005). Additionally, the standard sets requirements on the duty cycle of the backward signal.

3.3.1 TRcal symbol detection

The first source of error in the generation of *BLF* is introduced when symbol *TRcal* is detected. The digital core measures the duration of *TRcal* as the number of entire clock cycles comprised inside the backward link calibration symbol, n_{TRcal}. The value of n_{TRcal} will be an integer in the range given by (4). The value of n_{TRcal} is used to compute the number of cycles required to synthesize one cycle of *BLF*. As n_{TRcal} is an approximate representation of the duration of *TRcal*, an error will be introduced.

3.3.2 Backward link frequency synthesis

The accuracy of the synthesized backward link signal depends on how the transmitter is implemented. In the following, we analyze three possible implementations: balanced half-T_{pri} base transmitter, unbalanced half-T_{pri} base transmitter and full T_{pri} base transmitter. A set of backward link constraints result for each of the three transmitters.

For latter use, the following definitions are performed:

- $T_{pri}=1/BLF$ is the period that the transmitter has to synthesize.
- n_{Tpri} is the number of clock cycles inside of a period of the synthesized backward link signal.
- n_H is the number of clock cycles that the transmitter maintains the output signal in high per period of the synthesized backward link signal.
- n_L is the number of clock cycles that the transmitter maintains the output signal in low per period of the synthesized backward link signal.

3.3.3 Balanced half-T_{pri} base transmitter constraints

This is the most straightforward implementation of the transmitter using a synchronous digital circuit design flow. Inside the transmitter, a counter counts $n_H=n_L$ clock cycles, and the output signal is toggled every time the counters finish. As n_H and n_L are the same, the output BLF signal stays the same number of cycles in high and in low, generating a balanced waveform. Thus, the transmitter needs to computes the number of cycles required to generate a half-T_{pri} pulse. As this value has to be an integer, rounding is performed as shown in equation (9).

$$n_H = n_L = \left\lfloor \frac{n_{TRcal}}{2 \cdot DR} + \frac{1}{2} \right\rfloor \tag{9}$$

And thus, $n_{Tpri}=2n_H$.

The average value of the synthesized backward link frequency will be $n_{Tpri}\,T_a$. Taking from equation (4) the maximum and minimum values of n_{TRcal}, we can write the following two constraints to meet the frequency tolerance requirements of the standard:

$$\frac{DR}{TRcal} \cdot (1 - FT) \leq \cfrac{1}{2 \cdot \left| \left\lceil \cfrac{\left\lceil \cfrac{TRcal}{T_a \cdot (1 - \xi_{max})} \right\rceil}{2 \cdot DR} \right\rceil + \cfrac{1}{2} \right| \cdot T_a} \tag{10}$$

$$\frac{DR}{TRcal} \cdot (1 + FT) \geq \cfrac{1}{2 \cdot \left| \left\lfloor \cfrac{\left\lfloor \cfrac{TRcal}{T_a \cdot (1 + \xi_{max})} \right\rfloor}{2 \cdot DR} \right\rfloor + \cfrac{1}{2} \right| \cdot T_a} \cdot \tag{11}$$

On the other hand, the frequency variation of the synthesized backward link signal is given by

$$FV = \left| \cfrac{\cfrac{1}{T_{pri_k}} - \cfrac{1}{\overline{T_{pri}}}}{\cfrac{1}{\overline{T_{pri}}}} \right| = \left| \cfrac{\overline{T_{pri}} - T_{pri_k}}{T_{pri_k}} \right| \cdot \tag{12}$$

Taking into account that

$$T_{pri_k} - \overline{T_{pri}} = \sum_{i=1}^{n_{Tpri}} T_a \cdot (1 + \xi_i) - n_{Tpri} \cdot T_a = \sum_{i=1}^{n_{Tpri}} \xi_i \cdot T_a \tag{13}$$

and manipulating equation (12), it can be shown that

$$FV = \cfrac{1}{\left| 1 + \cfrac{n_{Tpri}}{\sum\limits_{i=1}^{n_{Tpri}} \xi_i} \right|} \leq \cfrac{1}{\left| 1 + \cfrac{1}{\xi_{max}} \right|} \tag{14}$$

From the requirements in the standard, we find the frequency variation constraint

$$\xi_{max} < 0.0256. \tag{15}$$

This constraint is independent from the clock frequency: it only limits the maximum jitter. Finally, the duty cycle requirements are considered. The duty cycle can be expressed as

$$DC = \frac{\displaystyle\sum_{i=1}^{n_H} T_a \cdot (1+\xi_i)}{\displaystyle\sum_{i=1}^{n_H} T_a \cdot (1+\xi_i) + \sum_{i=1}^{n_L} T_a \cdot (1+\xi_i)} . \tag{16}$$

Working on (16), the following equation is obtained,

$$DC = \frac{1}{1 + \dfrac{n_L + \displaystyle\sum_{i=1}^{n_L} \xi_i}{n_H + \displaystyle\sum_{i=1}^{n_H} \xi_i}} . \tag{17}$$

Introducing the worst case jitter values in (17), the minimum and maximum duty cycles are obtained. Taking the requirements from the standard we have

$$\min(DC) = \frac{1}{1 + \dfrac{n_L \cdot (1+\xi_{max})}{n_H \cdot (1-\xi_{max})}} \geq 0.45 \tag{18}$$

$$\max(DC) = \frac{1}{1 + \dfrac{n_L \cdot (1-\xi_{max})}{n_H \cdot (1+\xi_{max})}} \leq 0.55 . \tag{19}$$

In this type of transmitter $n_H = n_L$, and equations (18) and (19) yield the same duty cycle constraint:

$$\xi_{max} \leq 0.1. \tag{20}$$

3.3.4 Unbalanced half-T_{pri} base transmitter constraints

In this case, we also perform a synchronous digital circuit design flow, but we first compute the value of n_{Tpri} as

$$n_{Tpri} = \left\lfloor \frac{n_{TRcal}}{DR} + \frac{1}{2} \right\rfloor . \tag{21}$$

And then, the values of n_H and n_L are selected as,

$$n_H = \left\lfloor n_{Tpri}/2 \right\rfloor \tag{22}$$

$$n_L = n_{Tpri} - n_H . \tag{23}$$

The counter in the transmitter counts n_H clock cycles while the output is set to high, and n_L clock cycles while the output signal is set to low.

Proceeding in a similar way to the former transmitter, we find the two frequency tolerance constraints to be

$$\frac{DR}{TRcal} \cdot (1 - FT) \leq \frac{1}{\left[\left\lceil \frac{TRcal}{\frac{T_a \cdot (1 - \xi_{max})}{DR}} \right\rceil + \frac{1}{2} \right] \cdot T_a} \tag{24}$$

$$\frac{DR}{TRcal} \cdot (1 + FT) \geq \frac{1}{\left[\left\lfloor \frac{TRcal}{\frac{T_a \cdot (1 + \xi_{max})}{DR}} \right\rfloor + \frac{1}{2} \right] \cdot T_a}. \tag{25}$$

The frequency variation constraint is the same as for the former transmitter and is given by (20).

In this transmitter, the values of n_H and n_L are different. If we replace equations (22) and (23) in equations (18) and (19), we obtain the two duty cycle constraints.

The backward link signal synthesized with this transmitter has a more accurate frequency. Nevertheless, the duty cycle is worse than in the former transmitter, because the number of cycles that the output signal is set to high and the number of cycles that the output signal is set to low can be different. This generates an unbalanced output waveform.

3.3.5 Full-T_{pri} base transmitter constraints

This approach can be found in (Ricci et al., 2008). Part of the backward link signal synthesis is performed out of the digital circuit synchronous domain of the transmitter as shown in Fig. 5. The transmitter controls a multiplexer, which sets the output BLF signal to '1', '0', 'clk' or '**not** clk'. With this technique, the time granularity needed by the transmitter is T_{pri} instead of $T_{pri}/2$, because the availability of 'clk' and '**not** clk' makes it possible to toggle the input to the load modulator two times per clock cycle. Therefore, the values of n_H and n_L can take values with a precision of a half period:

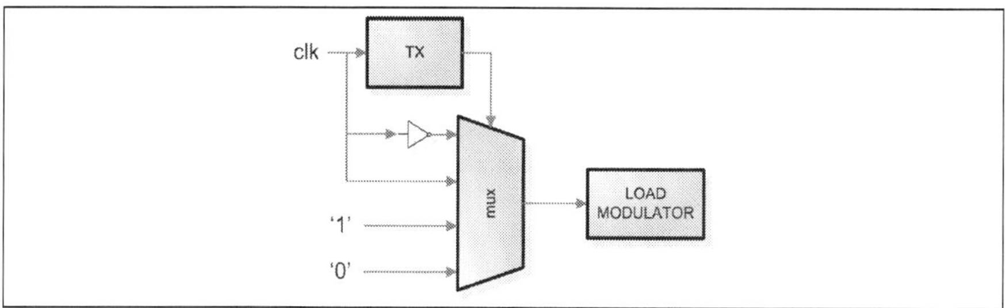

Fig. 5. Full-T_{pri} base transmitter.

$$n_H = n_L = n_{Tpri}/2 \tag{26}$$

where n_{TPri} is computed using equation (21).

The frequency tolerance constraints for this transmitter are the same as for the former unbalanced half-T_{pri} base transmitter and they are given by equations (24) and (25). The frequency variation constraint is equation (20), as for the two former transmitters.

In order to analyze the duty cycle, we define $n_H{}^{(i)}$ as the number of complete clock cycles that the signal is in high:

$$n_H^{(i)} = \lfloor n_H \rfloor \tag{27}$$

and $n_H{}^{(f)}$ as a variable that takes the value one when the signal has to be in high for half a clock cycle and cero when not; i.e.:

$$n_H^{(f)} = n_H \bmod 2 . \tag{28}$$

Using these definitions, the duty cycle for this transmitter is given by

$$DC = \frac{\sum_{i=1}^{n_H^{(i)}} T_a \cdot (1+\xi_i) + n_H^{(f)} \cdot T_a \cdot (\frac{1}{2}+\xi_i)}{\sum_{i=1}^{n_{Tpri}} T_a \cdot (1+\xi_i)} . \tag{29}$$

Introducing the worst case jitter values, the minimum and maximum duty cycles are obtained. Then, taking into account the requirements from the standard, we obtain the two duty cycle constraints for this transmitter:

$$\min(DC) = \frac{n_H^{(i)} + \dfrac{n_H^{(f)}}{2} - (n_H^{(i)} + n_H^{(f)}) \cdot \xi_{\max}}{n_{T_{pri}} \cdot (1+\xi_{\max})} \geq 0.45 \tag{30}$$

$$\max(DC) = \frac{n_H^{(i)} + \dfrac{n_H^{(f)}}{2} + (n_H^{(i)} + n_H^{(f)}) \cdot \xi_{\max}}{n_{T_{pri}} \cdot (1-\xi_{\max})} \leq 0.55 . \tag{31}$$

The accuracy of the backward link signal synthesized with this transmitter is the same as for the former transmitter, but this transmitter has no negative effect on the duty cycle, as the synthesized output signal is balanced.

3.4 Results

In order to comply with all the C1G2 specifications, the clock signal has to fulfil all the presented constraints. As some of these constraints depend on the implemented transmitter type, in the following, the clock constraints are evaluated separately for the three transmitters. The results have been obtained sweeping the range of possible values of all the parameters. *Tari*, *RTcal* and *TRcal* have been swept with a resolution of 1µs for both values of *DR*. The resolution in $1/T_a$ is of 1 kHz and of 0.1% in $\xi_{\max}$.

Fig. 6, Fig. 7 and Fig. 8 show the main constraints for a C1G2 digital core with a balanced half-T_{pri} base transmitter, an unbalanced half-T_{pri} base transmitter and a full-T_{pri} base transmitter, respectively. The results are presented in a two dimensional plot, where the horizontal axis represents $1/T_a$ and the vertical axis represents ξ_{max}. The forward link curve separates the $(1/T_a, \xi_{max})$ combinations that violate any of the forward link constraints from the $(1/T_a, \xi_{max})$ combinations that satisfy all of them. For the backward link, the constraints have been plotted separately, so that we can better see their effect in the clock source requirements. Any combination $(1/T_a, \xi_{max})$ inside the filled area fully complies with all the C1G2 clock requirements. Given a value of ξ_{max}, several ranges of compliant values of $1/T_a$ are found. The clock source implemented in the design has to generate a clock signal whose frequency is inside this range and its jitter is lower than the maximum allowed for the selected range.

If we analyse Fig. 6, we can observe that, for a digital core with a balanced half-T_{pri} base transmitter, it is possible to satisfy the C1G2 specifications with a clock frequency as low as 2.5 MHz. Nevertheless, in order to work in this region, the clock source needs to be very accurate and stable. We propose to work in the range (3.2 MHz-4.3 MHz) with looser requirements for the clock source stability and allowing a maximum jitter of 1%.

An unbalanced half-T_{pri} base transmitter allows synthesizing a more accurate BLF than with the balanced half-T_{pri} base transmitter. However, we can observe in Fig. 7 that this gain in accuracy has a negative effect in the duty cycle. As the duty cycle constraints are really restrictive in this case, the minimum clock frequency actually required is much higher than in the previous case. In fact, the clock frequency for such a design has to be higher than 6.4 MHz.

Fig. 8 shows that a C1G2 digital core with a full-T_{pri} base transmitter obtains the best results related to the clock constraints. A wide secure operating region is found at $1/T_a = 1.9$ MHz with $\xi_{max}=0.5\%$. Moreover, with an accurate enough clock source, it is possible to satisfy the C1G2 clock signal constraints with a clock frequency as low as 1.30 MHz.

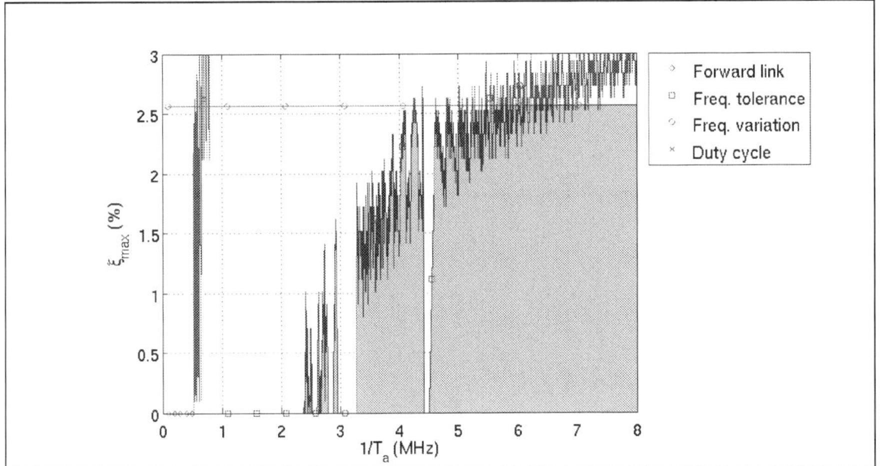

Fig. 6. Clock frequency constraints for C1G2 digital core with a balanced half-T_{pri} base transmitter.

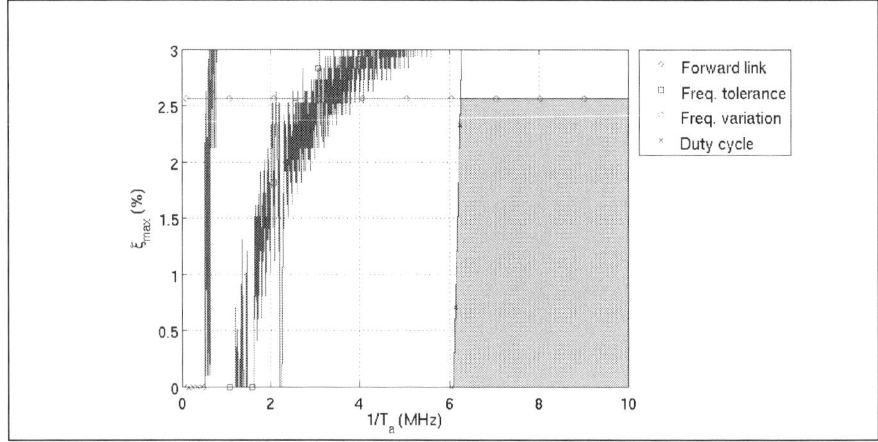

Fig. 7. Clock frequency constraints for C1G2 digital core with an unbalanced half-T_{pri} base transmitter.

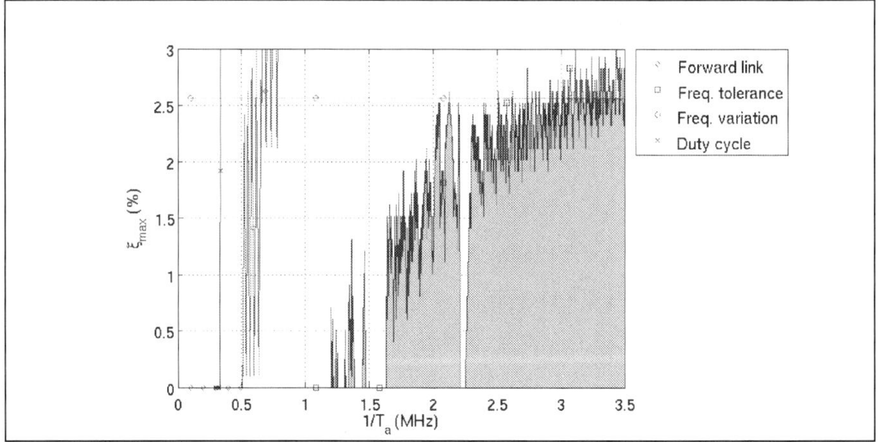

Fig. 8. Clock frequency constraints for C1G2 digital core with a full-Tpri base transmitter.

4. Energetic study

As explained in Section 1.2, the communication range of the system is strongly related to the power consumption of the tag. However, these power constraints are obtained assuming that the reader transmits a constant amount of power and that the tag also consumes power uniformly. None of these assumptions is true when the communication between reader and tag starts. The signal emitted from the reader is modulated, so that there is no continuous energy input at the tag. Moreover, the power consumption of the tag usually changes during the communication process. Thus, it is necessary to perform an energetic study to analyze the real behaviour of passive tags, and to understand the real limitations of the system.

The C1G2 communication protocol specifies that the forward link communication shall be ASK with a modulation depth of 90%, and the backward communication can use ASK or PSK backscattering. During the forward link the RF envelope is modulated with pulses of duration PW as shown in Fig. 9. High PW favours a clear communication. Low PW, instead, minimizes the time periods with no input power. The C1G2 standard defines the limits of acceptable PW values.

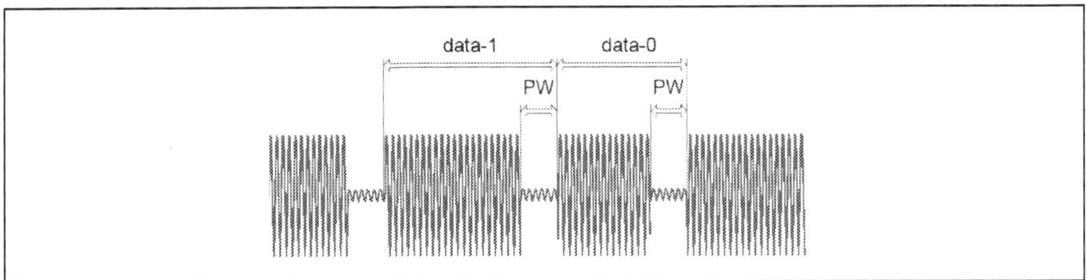

Fig. 9. Modulated RF signal during downlink communication.

Assuming that the power received during PW is negligible, the supply capacitor will supply the energy required by the rest of the tag. This will produce an energy discharge during PW. A similar effect occurs when the tag replies to the reader backscattering the received signal. The modulation in the backscattered signal is produced switching the reflection coefficient of the antenna between two states to differentiate a '0' from a '1'. Thus, the tag can communicate with the reader, but cannot receive all the energy of the input signal. The energetic discharge of the supply capacitor causes a drop in the supply voltage. This voltage drop is related with the discharged energy amount and with the value of the supply capacitor, C_{supply}. As the circuitry of the tag requires a minimum supply voltage to work, energetic constraints can be obtained for the value of C_{supply}. Moreover, the C1G2 standard specifies the minimum charge time of the tag, which also limits the value of C_{supply}.

This section is organized as follows. First, we present the models employed to analyze the energetic behaviour of the tag. An expression is obtained for the constraint on the maximum value of C_{supply} and expressions that can be used to evaluate the constraint on the minimum value of C_{supply} are presented. Next, we describe the methodology to evaluate the constraint on the minimum C_{supply}. Finally, a case study is presented as example.

4.1 Tag model

In order to perform the energetic analysis, a simplified model of the tag is defined. The model is divided into three sub models, each of one representing a specific state of the RFID communication. The first model represents the behaviour of the tag during the charge of the supply capacitor. In this model, it is assumed that the front-end of the tag includes power on reset (POR) circuitry. This POR block is usually included in RFID front-ends in order to switch on the tag only after the supply capacitor has been charged. This way, the tag consumes almost no power during the charge period allowing a faster charge and avoiding uncontrolled activity in the tag due to low supply voltage. The second model describes the energetic behaviour of the tag when the supply capacitor is charged, all the circuits are working and a continuous power is arriving to the antenna. Finally, the third model describes the behaviour of the tag when the input wave is modulated, and times of period with no input power are present.

In order to calculate the available power in the tag, the Friis equation is used to estimate the power available in the antenna, and the power conversion efficiency factor of the tag is applied. The power available in the tag is given by

$$P_{IN} = P_{EIRP} \cdot (\frac{\lambda}{4\pi r})^2 \cdot G_{TAG} \cdot \eta \qquad (32)$$

where P_{EIRP} is the equivalent isotropic radiated power emitted from the reader, λ is the wavelength of the operation frequency, r is the communication range, G_{TAG} is the gain of the tag antenna and η is the power conversion efficiency of the tag.

The characterization of the front-end includes the power conversion efficiency η, the power consumption of the tag P_{TAG}, the required minimum supply voltage V_{min} and the maximum allowed supply voltage V_{max}. The front-end creates a regulated voltage at V_{min} to supply the rest of the blocks of the tag. Moreover, the supply voltage is limited to V_{max}, so that the technology does not break.

4.1.1 Charge of C_supply

When the reader starts emitting power for the first time, the tag begins accumulating energy in the supply capacitor C_{supply}. During this process, all the blocks of the tag remain switched off, so that all the incoming energy is stored in C_{supply}. The model for this behaviour is shown in Fig. 10. It consists on a power source connected to C_{supply}. This model is a simplified description of the behaviour of the tag during the charge process until the supply capacitor reaches the supply voltage V_{min}. At this point, the POR switches all the modules of the tag on, and the behaviour of the tag changes.

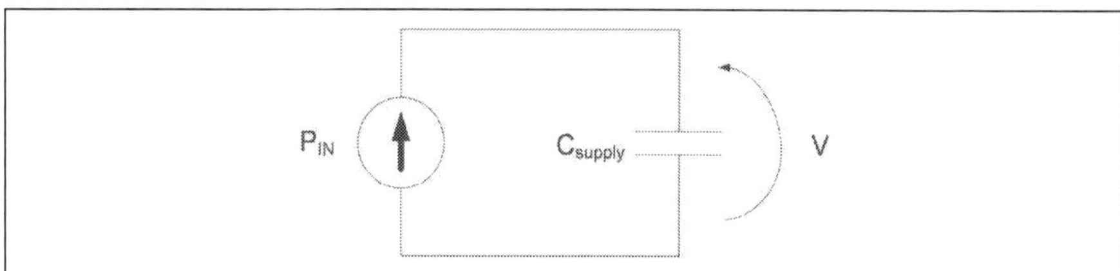

Fig. 10. Tag model during charge process of the supply capacitor.

The energy stored in C_{supply}, is related with the voltage V as follows:

$$E = \int_0^Q \frac{q}{C_{supply}} dq = \frac{1}{2} \frac{Q^2}{C_{supply}} = \frac{1}{2} \cdot C_{supply} \cdot V^2 . \qquad (33)$$

Given that the energy received during the time period t is equal to $P_{IN} \cdot t$, from (33), the charge time required for a specific C_{supply} to reach V_{min} is obtained as

$$t_{charge} = \frac{C_{supply} \cdot V_{min}^2}{2 \cdot P_{IN}} . \qquad (34)$$

Introducing the maximum charge time specified in the standard in (34), and isolating C_{supply}, the maximum C_{supply} constraint is obtained,

$$C_{max} = \frac{2 \cdot P_{IN} \cdot 1.5 \cdot 10^{-3}}{V_{min}^2} \, . \tag{35}$$

This constraint depends on the available input power, and thus, it depends on the communication range between reader and tag.

4.1.2 Tag working with input power

Once V_{min} has been reached, more blocks in the tag are active and, thus, the power consumption of the tag increases. As the supply voltage of the different blocks is regulated, the power consumption of the tag does not change with the supply voltage at C_{supply}. Thus, the power consumption of the tag is inserted in the model as a power source in the opposite direction to the input power source. Fig. 11 shows the model of the tag with all the blocks switched on and receiving constant power from the reader.

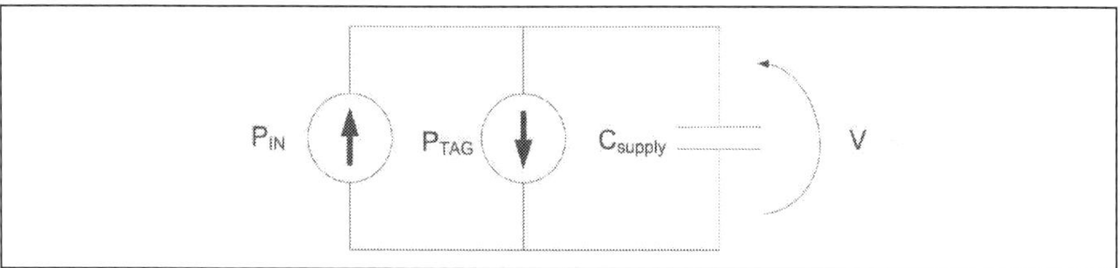

Fig. 11. Tag model receiving CW from the reader.

In this case, depending on the communication range, P_{IN} may be greater or lower than P_{TAG}. If $P_{IN} > P_{TAG}$, V_{supply} still receives some energy and keeps on charging. If $P_{IN} < P_{TAG}$, instead, C_{supply} has to provide the difference, and will discharge slowly. The maximum stable communication range is the one where $P_{IN} = P_{TAG}$. Eventually, this is the power limitation seen in Section 1.2.

Given that at t_0 the supply voltage had a value of V_0, the energy accumulated in the capacitor at this moment was

$$E_{t_0} = \frac{1}{2} \cdot C_{supply} \cdot V_{t_0}^2 \, . \tag{36}$$

Considering a stable situation receiving continuous power from the reader between t_0 and t_1, the available energy in C_{supply} at t_1 is

$$E_{t_1} = E_{t_0} + (P_{IN} - P_{TAG}) \cdot (t_1 - t_0) \, . \tag{37}$$

From (33), (36) and (37), the supply voltage at t_1 is obtained,

$$V_{t_1} = \sqrt{V_{t_0}^2 + \frac{2 \cdot (P_{IN} - P_{TAG}) \cdot (t_1 - t_0)}{C_{supply}}} \, . \tag{38}$$

4.1.3 Tag working without input power

When the real communication starts, the reader modulates the RF wave. During the modulation, periods of time with no input power exist. The same happens during the backscattering of the answer. For these periods of time, the model shown in Fig. 12 is used. Both, the forward and backward modulation cause a voltage drop in C_{supply}.

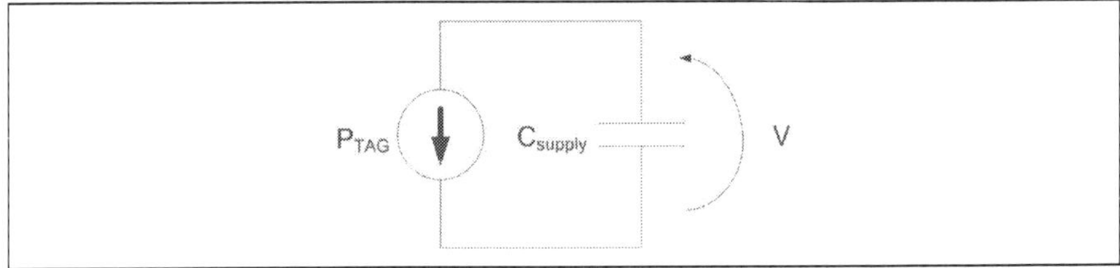

Fig. 12. Tag model receiving no power from the reader.

In this case, considering a stable situation when no input energy is received from the reader between t_1 and t_2, the energy at t_2 is given by

$$E_{t_2} = E_{t_1} - P_{TAG} \cdot (t_2 - t_1).$$

(39)

And the supply voltage at t_2 is given by

$$V_{t_2} = \sqrt{V_{t_1}^2 - \frac{2 \cdot P_{TAG} \cdot (t_2 - t_1)}{C_{supply}}}.$$

(40)

4.2 Methodology for the energetic study

When a communication link is established, the RF wave is modulated in order to transport information, and the input power available in the tag is variable. This can cause a drop in the supply voltage. If the supply voltage drops below a minimum voltage value, V_{min}, the tag operation can fail.

Analysing (38) and (40), we can observe that the supply voltage at a certain instant depends both on C_{supply} and on the power consumption of the tag. When power management techniques are applied to a design, this power consumption can be very variable in time. Additionally, different commands and forward and backward frequencies produce different profiles of power consumption in the time. There will be a certain command and certain forward and backward frequencies that yield the maximum power consumption of the tag, P_{max}. Given a communication distance (or a specific input power in the antenna), a minimum value of the supply capacitor, C_{min}, can be obtained that ensures that the supply voltage does not drop below V_{min} even if P_{max} is consumed.

4.2.1 Estimation of power consumption profile

Given a system $S = \{m_0, m_1, m_2, ..., m_{N-1}\}$ composed by N modules, the power consumption without power-management is given by

$$P_{TOTAL} = \sum_{m_i \in S} P_{m_i}.$$

(41)

However, when power management is implemented, different working states are defined depending on the activity of each module. In this case, a better characterization of the power consumption is obtained determining the average power consumption of the design in every working state (WS) as

$$P_{WS} = \sum_{m_i \in WS} P_{m_i} + \sum_{m_j \notin WS} P_{leakage_{m_j}} + P_{m_{PM}}, \tag{42}$$

where m_{PM} is the new power management module introduced to generate the clock gating control signals for the modules. The average power consumption of the different modules of the design can be determined making them work independently at full load.

Combining the power consumption during each working state and the time spent in each of them, the profiles of the power consumption for all possible commands and configurations are obtained.

4.2.2 Selection of the optimum value of C_{supply}

Equation (38) can be used to analyse the voltage drop during the charge and equation (40) during the discharge. The power profile that causes the highest voltage drop is considered to be the worst power profile. The search for the worst case shall consider the different commands and configurations. Given a value of C_{supply}, there is a maximum value of the communication range where the supply voltage does not fall below V_{min} in the worse case. Thus, for each value of C_{supply} the maximum communication range can be obtained.

On the other hand, for each communication range, the constraint defined in (35) sets an upper limit on the value of the supply capacitor, C_{max}. Taking into account the C_{min} constraint caused by the modulation of the RF wave and the C_{max} constraint set by the maximum charge time established by the C1G2 standard, the optimum value of C_{supply} can be selected.

4.3 Case study

In the following example, a front-end which consumes 25 µW and has an efficiency of $\eta = 30\%$ is assumed. The voltage limiter of the front-end is assumed to be set to 2.0 V, and the lower limit of the voltage, V_{min} to 1.2 V. P_{EIRP} is set to 2 W, which is the maximum power emission allowed in Europe for RFID communication at the operation frequency of 868 MHz, and G_{TAG} is set to 1 assuming an ideal isotropic antenna. The clock signal has been set to 1.5 MHz and the typical PVT operating conditions have been used.

Using the procedure described in Section 4.2.1, the power consumption in the five working states defined in Table 1 can be obtained. As the activity of the input signal depends on the forward link frequency determined by *Tari*, the actual power consumption of the design in RX working state also depends on the value of *Tari*. Similarly, the activity of the output signal depends on the *BLF* employed. Thus, when presenting any power consumption result of C1G2 digital cores, the configuration of the forward and backward links has to be specified. There is a lack of information about these parameters in the literature, where the results obtained are presented without further specifications.

If we know the power consumption of the tag in each working state for all values of Tari and *BLF*, the power consumption profile of any command can be generated. As an example, Fig. 13 (a) shows the power consumption profile of a *Read* command with *Tari* = 25 µs and *BLF* = 40 kHz. The *Read* command in this example requires that the whole EPC bank (96 bits) is read. It can be observed that the power consumption changes from one working state to another.

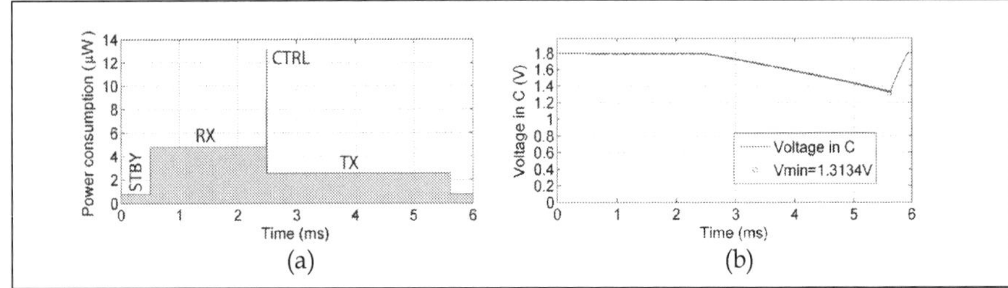

Fig. 13. Power distribution and voltage drop of a *Read* command for *Tari* = 25 µs and *BLF* = 40 kHz.

4.3.1 Worst command

In order to determine the worst command, the power distribution of every command in the C1G2 standard must be obtained, as well as the voltage drop caused by them in the supply capacitor. As an example, for *Tari* = 25 µs and BLF = 40 kHz, Fig. 13 (b) shows the voltage drop caused by the *Read* command. In this case, the time that the tag has to recover energy per received symbol is greater than the recovery time per transmitted cycle. Thus, even though the power consumption in RX is higher than in TX, the voltage drop caused by the reception of the command is smaller than the voltage drop caused by the transmission of the answer.

In our example, the final voltage in the supply capacitor is 1.3134 V. This is the greatest voltage drop in the supply capacitor produced by any EPC command for the design considered in our case study. This is due to the fact that the *read* command of the whole EPC bank is one of the longest commands received by the tag and requires the transmission of the largest amount of data. Thus, in order to characterize our digital core, the average power consumption for a *Read* command requesting the data of the biggest memory bank shall be used.

4.3.2 Worst configuration

The effects of employing different forward and backward data rates shall also be studied. The worst case *Read* command will be used to study the effect of the backward link configuration. However, in this command, the voltage drop caused by the forward link is mostly covered by the voltage drop caused by the backward link. Thus, results for the command with longest forward link communication, *Select*, will be also presented to observe the effects of *Tari*.

Fig. 14 presents the power distribution and supply voltage drop of the worst case *Read* command for *Tari* = 25 µs and *BLF* = 640 kHz. If we compare these results with the ones shown in Fig. 13, we can observe that the power consumption during the TX working state reduces as lower *BLFs* are employed. However, reducing *BLF* makes the communication slower and requires that that tag stays more time in the TX working state. At the end, this produces a bigger voltage drop in the supply capacitor. In order to maximize the communication range of C1G2 RFID systems, a high *BLF* configuration is suggested. However, in order to characterize our digital core, results with the lowest BLF shall be used, i.e.: BLF = 40 kHz.

Fig. 15 shows the power distribution and supply voltage drop of the *Select* command for *Tari* = 6.25 µs and *BLF* = 240 kHz. Fig. 16 presents the power distribution and supply voltage drop of the same command for *Tari* = 25 µs and *BLF* = 240 kHz. Comparing both configurations, it can be observed that for higher values of *Tari* the power consumption in the RX working state is reduced and, thus, the discharge of the supply capacitor slows down. However, as the forward link frequency is reduced, the time required to transmit the same number of symbols increases. Due to this fact, the supply voltage drop at the end of the operation is greater with a high *Tari*. Thus, the communication range of a C1G2 RFID system may be increased by configuring low *Tari* values. In order to ensure the correct operation of the tag in any case, the characterization of the digital core has to be done with the worst case forward link configuration, which is *Tari* = 25 µs.

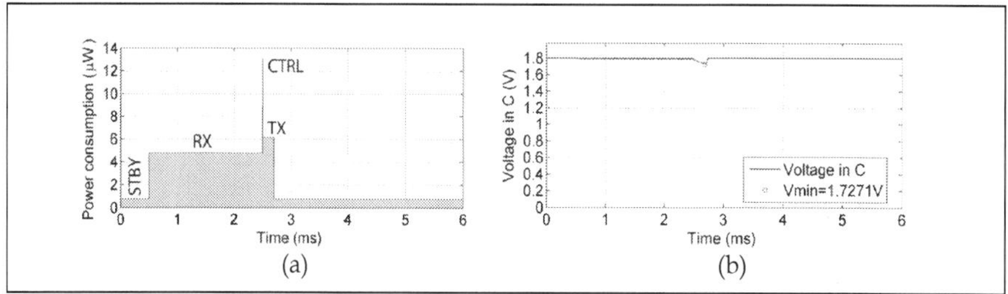

Fig. 14. Power distribution and supply voltage drop of the *Read* command for *Tari* = 25 µs and *BLF* = 640 kHz.

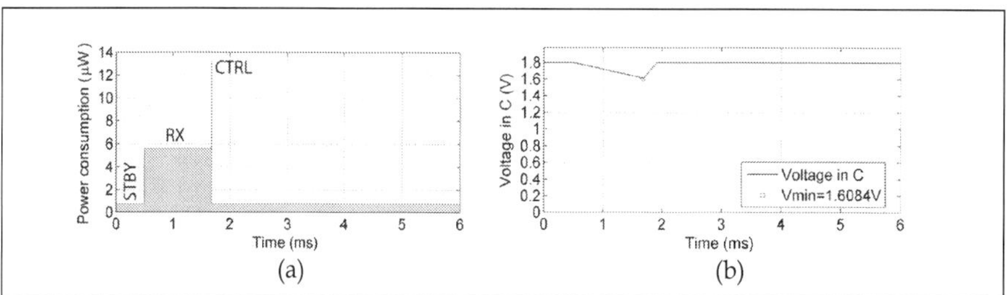

Fig. 15. Power distribution and supply voltage drop of the *Select* command for *Tari* = 6.25 µs and *BLF* = 240 kHz.

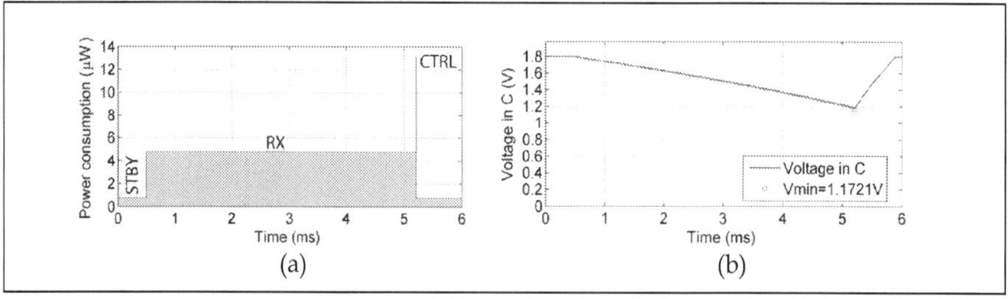

Fig. 16. Power distribution and supply voltage drop of the *Select* command for *Tari* = 25 µs and *BLF* = 240 kHz.

4.3.3 Energetic constraints

For the worst command and worst configuration, the energetic constraints have been calculated using the model described in Section 4.1 and the methodology described in Section 4.2.2. Fig. 17 shows the obtained results for our example. On the one hand, for operation points below the C_{min} constraint line, the supply capacitor is high enough to keep the supply voltage over V_{min}. On the other hand, C_{max} establishes the maximum value of the supply capacitor in order to fulfil the charge time specification. The area that is below both constraint lines is the operative region where the C1G2 standard is completely fulfilled.

In this case, with values of C_{supply} below 0.35 nF, the supply voltage always drops below V_{min}. The maximum communication range is achieved for values of C_{supply} above 90 nF. However, with such a capacity, the C_{max} constraint is violated. C_{supply} =30 nF is the point where both constraints cross. This is the value of C_{min} that maximizes communication distance fulfilling all the energetic constraints.

It can be observed in Fig. 17 that the dimensioning of C_{supply} has a relevant impact on the actual communication range. In this case, the system would work properly with a supply capacitor of 1 nF, but the communication range would be limited by the energetic constraint to 2.8 m. Increasing C_{supply}, the communication range is increased to 3.3 m.

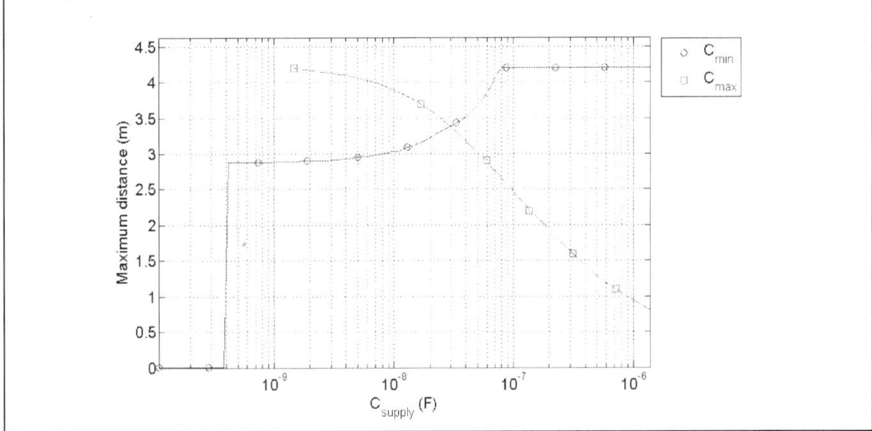

Fig. 17. Maximum distance achievable for different values of C_{supply}.

5. Conclusion

A communication link standard poses many constraints to the clock of the digital core of a tag for proper forward-link data detection and backward-link data backscattering. For the C1G2 standard, the backward link requirements are the ones that set the most restrictive constraints. There are several frequency bands where the C1G2 specifications are fulfilled. Depending on the characteristics of the clock source, such as the average period and the maximum jitter, and on the type of transmitter, the most suitable operation point can be selected using the results presented in Section 3.

In the literature, the average power consumption is usually presented to characterize a tag. However, we have seen in Section 4 that this value is not enough. Due to the energetic behaviour of the tag, the communication range is also limited by the value of C_{supply}. Moreover, this constraint depends on the profile of the power consumption, which changes from one command to another and from one communication mode to another. In order to obtain a complete characterization of the proposed design, a procedure as the one shown in Section 4.3 shall be followed.

6. References

Barnett R., Balachandran G., Lazar S., Kramer B., Konnail G., Rajasekhar S. & Drobny V. (2007). A passive UHF RFID transponder for EPC Gen 2 with -14dBm sensitivity in 0.13 μm CMOS, *Digest of Technical Papers of the IEEE International Solid-State Circuits Conference*, pp. 582-623, ISBN: 978-1-4244-0853-5, San Francisco (USA), February 2007, IEEE, Piscataway (USA)

De Vita G. & Annaccone G. (2005). Design criteria for the RF section of UHF and microwave passive RFID transponders, *IEEE Transactions on Microwave Theory and Techniques*, Vol. 53, No. 9, September 2005, pp 2978-2990, ISSN: 0018-9480

EPC Global (2005). Specification for RFID air interface. EPC Global Class-1 Gen-2 UHF RFID Version 1.0.9

Hong Y., Chan C. F., Guo J., Ng J. S., Shi W., Leung L. K., Leung K. N., Choy C. S. & Pun K. P. (2008). Design of Passive UHF RFID Tag in 130nm CMOS Technology, *Proceedings of IEEE Asia Pacific Conference on Circuits and Systems*, pp. 1371-1374, ISBN: 978-1-4244-2341-5, Macao, December 2008, IEEE, Piscataway (USA)

Impinj (2006). Gen 2 Tag Clock Rate - What You Need To Know

ISO (2006). ISO18000-6C, Information technology Radio frequency identification for item management Part 6: Parameters for air interface communications at 860 MHz to 960 MHz, Amendment 1, June 2006

Man A. S. W, Zhang E. S., Chan H.T., LauV. K.N., Tsui C.Y. & Luong H. C. (2007). Design and Implementation of a Low-power Baseband-system for RFID Tag, *Proceedings of International Symposium on Circuits and Systems*, pp. 1585-1588, ISBN: 1-4244-0920-9, New Orleans (USA), May 2007, IEEE, Piscataway (USA)

Pardo D., Vaz A., Gil S., Gomez J., Ubarretxena A., Puente D., Morales-Ramos R., García-Alonso A. & Berenguer R. (2007). Design Criteria for Full Passive Long Range UHF RFID Sensor for Human Body Temperature Monitoring, *Proceedings of IEEE International Conference on RFID*, pp. 141-148, ISBN: 1-4244-1013-4, Grapevine, March 2007, IEEE, Piscataway (USA)

Ricci A., Grisanti M., De Munari I. & Ciapolini P. (2008). Design of a 2 W RFID Baseband Processor Featuring an AES Cryptography Primitive, *Proceedings of the 15th IEEE International Conference on Electronics, Circuits and Systems, ICECS*, pp. 376-379, ISBN: 978-1-4244-2181, St. Julien's, September 2008, IEEE, Piscataway (USA)

Roostaie V., Naja V., Mohammadi S. & Fotowat-Ahmady A. (2008). A low power baseband processor for a dual mode UHF EPC Gen 2 RFID tag, *Proceedings of International Conference on Design and Technology of Integrated Systems in Nanoscale Era, DTIS*, pp. 1-5, ISBN: 978-1-4244-1576-2, Tozeur, March 2008, IEEE, Piscataway (USA)

Wang J., Li H. & Yu F. (2007). Design of Secure and Low-cost RFID Tag Baseband, *Proceedings of International Conference on Wireless Communications, Networking and Mobile Computing*, pp. 2066-2069, ISBN: 978-1-4244-1311-9, Shangai, September 2007, IEEE, Piscataway (USA)

Wanggen S., Yiqi Z., Xiaoming L., Xianghua W., Zhao J. & Dan W. (2009). Design of an ultra-low-power digital processor for passive UHF RFID tags, *Journal of Semiconductors*, Vol. 30, No. 4, April 2009, pp. 045004-1-045004-4, ISSN: 16744926

Yan, H, Jianyun, H., Qiang, L. & Hao. M. (2006). Design of low-power baseband-processor for RFID tag, *Proceedings of the International Symposium on Applications and the Internet Workshops*, pp. 60-63, ISBN: 0-7695-2510-5, Phoenix, January 2006, IEEE Computer Society, Los Alamitos (USA)

Zalbide I., Vicario J. & Vélez I. (2008). Power and energy optimization of the digital core of a Gen2 long range full passive RFID sensor tag, *Proceedings of IEEE International Conference on RFID (Frequency Identification)*, pp. 125-133, ISBN: 978-1-4244-1712-4, Las Vegas (USA), April 2008, IEEE, Piscataway (USA)

Zalbide I. (2009). *Design of a digital core for a C1G2 RFID sensor tag*, PhD. Thesis, Universidad de Navarra

Zhang Q., Li Y. & Wu N. (2008). A Novel Low-Power Digital Baseband Circuit for UHF RFID Tag with Sensors, *Proceedings of Solid-State and Integrated-Circuit Technology*, pp. 2128-2131, ISBN: 978-1-4244-2185-5, Beijing (China), October 2008, IEEE, Piscataway (USA)

3

Design of Space-Filling Antennas for Passive UHF RFID Tags

Benjamin D. Braaten[1], Gregory J. Owen[2] and Robert M. Nelson[3]
[1]*North Dakota State University*
[2]*Sebesta Blomberg and Associates*
[3]*University of Wisconsin – Stout*
United States

1. Introduction

Every year researchers and engineers are finding new and useful applications for Radio Frequency Identification (RFID) systems (Finkenzeller, 2003). Because of the growing use of RFID systems, many different areas of research have also been developed to improve the performance of such systems. A few of these areas of research include novel antenna designs (Rao et at., 2005; Calabrese & Marrocco, 2008; Amin et al. 2009), analysis on the backscatter properties of RFID tags (Yen et al., 2007; Feng et al., 2006), mutual coupling between RFID tags (Li et al., 2008; Owen et al., 2009) and the deployment of RFID systems to complex and extreme environments (Qing & Chen, 2007; Sanford, 2008). There are many aspects to each area of research in RFID. A major topic in many of these areas involves research on the antenna design for RFID tags.

This chapter will focus on the design of efficient space-filling antennas for passive UHF RFID tags. First, an introduction to RFID systems is presented. This is done by describing the major components in a RFID system and how they communicate. Then, the particular backscattering properties of a passive tag are described from a unique electric field integral equation standpoint and from an overall systems perspective (i.e., using the Friis transmission equation). This discussion will then be followed by a section describing a practical design process of various space-filling antennas for passive tags. Finally, a summary of future work and a conclusion about the chapter is presented.

2. An introduction to RFID systems

The two main components of a RFID system are the readers and the tags. An overview of a RFID system is shown in Fig. 1. A reader consists of an antenna, transmitter/receiver and typically an interface with a PC (or other device for viewing information) while a tag has an antenna and an integrated circuit (IC) connected to the antenna (Fig. 2). The reader is a device that transmits electromagnetic energy and timing information into the space around itself to determine if any tags are in the region. This region around the reader is sometimes called the interrogation zone (Finkenzeller, 2003). If a tag is in the interrogation zone (i.e., or interrogated by the reader), the tag will use the IC connected to the antenna to establish communications with the reader and transmit the appropriate information. The max possible distance that a tag can be interrogated by the reader is referred to as the max read range of the tag.

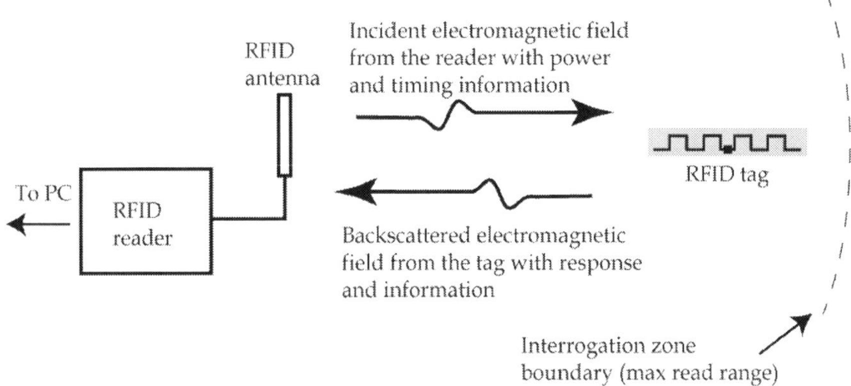

Fig. 1. Overview of a RFID system.

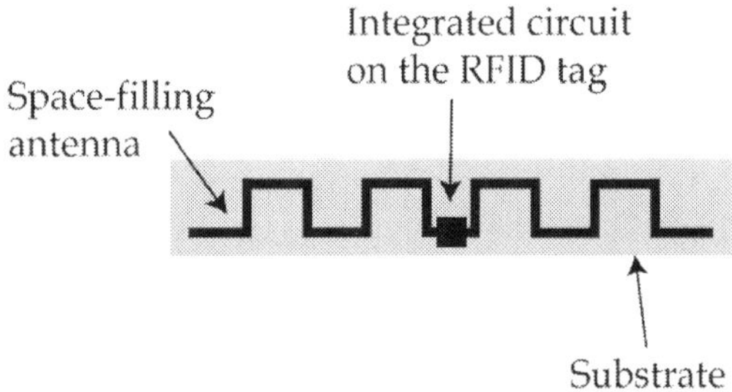

Fig. 2. A Passive RFID tag.

In general, RFID systems can be placed into three major categories: active, semi-passive and passive (Finkenzeller, 2003). An active tag has an onboard battery and can communicate with the reader using only the power from the battery on the tag. A semi-passive tag (or battery assisted tag) is awakened by the incident electromagnetic field from the reader antenna and uses the onboard battery to communicate with the reader. This greatly increases the read range of the tag. Finally, a passive tag does not have an onboard battery. The incoming electromagnetic field from the reader induces a port voltage on the tag antenna while a power harvesting circuit in the IC loads the antenna and uses the port voltage and current to provide power to the digital portion of the IC. This power is then used by the IC to identify itself and communicate back to the reader. An IC on a passive tag is usually referred to as a passive IC. A passive IC communicates with the reader by changing the input impedance of the power harvesting circuit. The impedance is changed between two different values to represent a logic 0 and 1 in a digital signal. Changing this input impedance results in two different scattered fields from the tag. These scattered fields (backscattered fields) are received by the reader and the digital information is processed. Understanding how these backscattered fields radiate in the region around the RFID tag is important. Because of this and the fact that this chapter focuses on antenna design, the next few sections will present various methods for understanding the backscattered field from the RFID tag.

3. Backscattering properties of a passive RFID tag

Typically, the performance of a RFID system is described using the Friis transmission formula (Rao et al, 2005; Marrocco, 2003). This is a very useful approach to present the performance of a RFID system, and will be done later, but many of the details associated with the tag are not easily extracted from such a presentation. For example, the current distribution on the tag antenna during an interrogation may be of interest or information on the mutual coupling between multiple tags may be a concern. One method to describe other aspects of a passive RFID tag is to derive expressions for the electric field in the region around the tag antenna in terms of the current distribution on the antenna. Once the electric field in the region around the tag is known, many other aspects associated with the RFID tag can be explored. In the following section, the backscattered field from a thin-wire dipole is derived in terms of the load impedance of the antenna. This will show how matching the load impedance with the antenna will result in a much lower scattered field when compared to the case when the terminals are shorted-circuited (Braaten et al., 2006).

3.1 Backscattering from a thin-wire dipole

First, consider the thin-wire dipole shown in Fig. 3 (a) immersed in free-space. E_{inc} represents the incoming wave from the reader and Z_L represents the input impedance of the passive IC. For this discussion, it is assumed that the length L of the antenna is $\lambda/2$ where λ is the wavelength of E_{inc} (i.e., the wavelength of the frequency at which the reader is transmitting at) and that the tag is in the far-field of the reader. This simplifies E_{inc} to a constant value. E_{inc} is travelling in the $-y$ direction and has a $\hat{z}$- component. As E_{inc} impinges on the thin-wire dipole, a current is induced. This induced current on the thin-wire dipole is assumed to be (Stutzman & Thiele, 1998)

$$I(z) = I_m \sin\left[\beta\left(\tfrac{L}{2} - |z|\right)\right] \qquad (1)$$

where I_m is the maximum current along the antenna, β is the free-space phase constant and $|z| \leq L/2$. This assumption is valid as long as Z_L is chosen in a manner that preserves the sinusoidal current distribution on the thin-wire dipole. One example that would preserve the sinusoidal current distribution would be a 50 Ω load connected to a half-wavelength dipole (Braaten et al., 2006). Next, using (1) in the induced emf method, an expression for the open circuit voltage V_{oc} at the port of the dipole can be written in the following manner (Balanis, 2005; Stutzman & Thiele 1998):

$$V_{oc} = -\tfrac{1}{I(0)} \int_{-L/2}^{L/2} I_m \sin\left[\beta\left(\tfrac{L}{2} - |z'|\right)\right] E_{inc}\, dz' \qquad (2)$$

where $I(0)$ is the current at the terminals of the dipole. Subsequently, assuming E_{inc} is a constant value and evaluating (2) results in the following expression (Braaten et al., 2006):

$$V_{oc} = -\frac{2E_{inc}}{\beta}\tan\left(\frac{\beta L}{4}\right). \tag{3}$$

Equation (3) is a simple expression for the open circuit voltage of the dipole and it is clear how the incident field from the reader can be used to control the induced voltage.

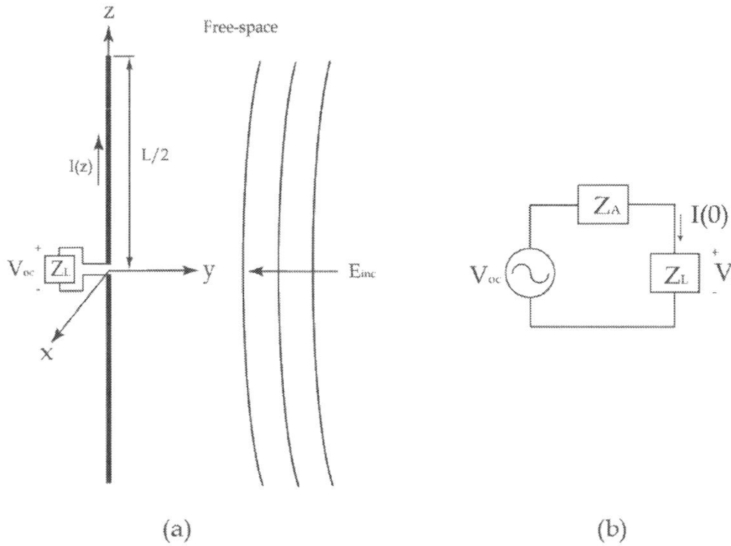

Fig. 3. (a) A thin-wire dipole in free-space; (b) the equivalent circuit of the receiving dipole

Next, consider the equivalent circuit of the receiving dipole shown in Fig. 3 (b) where Z_L and Z_A are the load and antenna impedance values, respectively. Using voltage division, the load voltage can be written as

$$V_L = \frac{Z_L}{Z_L+Z_A}V_{oc} = Z_L I(0). \tag{4}$$

Then, substituting (1) and (3) into (4) and solving for I_m gives:

$$I_m = -\frac{E_{inc}}{(Z_L+Z_A)\beta\cos^2\left(\frac{\beta L}{4}\right)}. \tag{5}$$

Next, an expression for the electric field from the dipole can be written in terms of I_m in the following manner (Balanis, 2005):

$$E_z(\bar{r}) = -j30I_m\left[\frac{e^{-j\beta r_1}}{r_1} + \frac{e^{-j\beta r_2}}{r_2} - \frac{2\cos(\beta L/2)e^{-j\beta r_0}}{r_0}\right] \tag{6}$$

where $\beta = 2\pi/\lambda$, $r_0 = |\bar{r}|$, $r_1 = |\bar{r}_+ - \bar{r}|$, $r_2 = |\bar{r}_- - \bar{r}|$ and $\bar{r}_\pm = \pm\frac{L}{2}\hat{z}$. Substituting (5) into (6) results in the following expression for the electric field:

$$E_{z,s}(\bar{r}) = \frac{j30E_{inc}}{(Z_L+Z_A)\beta\cos^2\left(\frac{\beta L}{4}\right)}\left[\frac{e^{-j\beta r_1}}{r_1} + \frac{e^{-j\beta r_2}}{r_2} - \frac{2\cos(\beta L/2)e^{-j\beta r_0}}{r_0}\right]. \tag{7}$$

$E_{z,s}(\bar{r})$ represents the scattered field in the region around the thin-wire dipole as a result of E_{inc}. Equation (7) provides insight as to how the scattered field is closely related to the load impedance. Next, using the notation $E_{inc} = E_{reader}$ and $Z_L = Z_{tag}$, (7) can be written in a more descriptive form:

$$E_{z,s}(\bar{r}) = \frac{j30E_{reader}}{(Z_{tag}+Z_A)\beta \cos^2\left(\frac{\beta L}{4}\right)} \left[\frac{e^{-j\beta r_1}}{r_1} + \frac{e^{-j\beta r_2}}{r_2} - \frac{2\cos(\beta L/2)e^{-j\beta r_0}}{r_0}\right] \tag{8}$$

where E_{reader} represents the incident field from the reader and Z_{tag} represents the input impedance of the passive IC. From (8), it is clear that by changing Z_{tag} the tag is able to change the scattered field. In an RFID system, this change in scattered field propagates back to the receiving antenna at the reader and the digital information is processed. Noticing how this scattered field is changed by the passive IC is important for successful communication between the reader and the RFID tag.

Next, several load impedances were defined ($Z_{tag} = 0\,\Omega, 25\,\Omega, 50\,\Omega$ and $75\,\Omega$) and the resulting scattered field from these different loads were calculated using (8). The scattered field was calculated 1 m from the middle of the dipole in the y-direction with a $1V/m$ incident field from the reader. The different magnitudes of the scattered fields computed by (8) are shown in Fig. 4. Two characteristics stand out from these computations. First, notice that the largest scattered field is for the short-circuit case. This indicates that more of the energy incident on the dipole is being radiated back into the space around the dipole and not being used by the passive IC. This also improves the chances of the reader picking up the backscattered field from the tag. Second, the lowest scattered field is for the $75\,\Omega$ load. This is because a $75\,\Omega$ load has the closest match with the input impedance of the antenna. This also indicates that less energy is being scattered into the region around the dipole and more of it is being used by the load.

The results in Fig. 4 directly show how the field in the region around the dipole can be changed by using different loads at the port of the dipole. Equation (8) could also be used to study the mutual coupling between RFID tags. This coupling information may be important for applications that require many RFID tags to be located in close proximity to one another. In the next section, the far-field characteristics of a RFID system are presented using the Friis transmission equation (Stutzman & Thiele, 1998). This will provide insight as to what characteristics of an RFID tag determine the max read range.

3.2 Describing the wireless RFID system using the Friis transmission equation

The RFID system described in Fig. 1 essentially consists of two transceivers (the reader and passive RFID tag). This type of communication system can be described using the Friis transmission equation (Stutzman & Thiele, 1998):

$$P_{tg} = P_{rd}\frac{G_{rd}G_{tg}\lambda^2}{(4\pi R)^2}q \tag{9}$$

where P_{rd} is the power transmitted by the reader, P_{tg} is the power received by the passive tag, G_{rd} is the gain of the transmitting antenna on the reader, G_{tg} is the gain of the space-filling antenna on the tag, λ is the free-space wavelength of the transmitting frequency by the reader, R is the distance between the antenna on the reader and the antenna on the tag and q is the impedance mismatch factor ($0 \leq q \leq 1$) between the passive IC and the antenna on the tag.

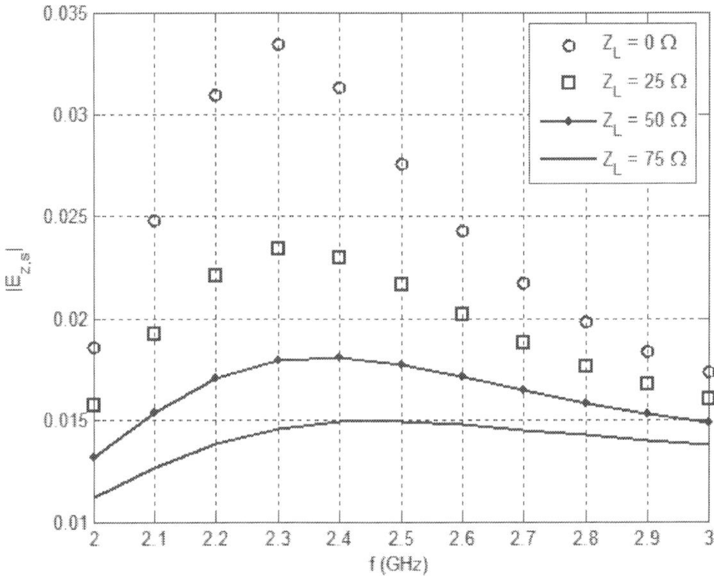

Fig. 4. Backscattered field as a result of changing the load impedance (L = 6cm, a = 1mm).

Equation (9) assumes a polarization match between the antenna used by the reader and the antenna on the tag and that the tag is in the far-field of the reader. Several comments can be made about (9). First, the power at the tag depends very much on G_{tg} (the gain of the space-filling antenna on the RFID tag). A larger gain would mean more power for the passive IC on the tag. Second, using a longer wavelength would improve the power at the tag. Third, a good match between the passive IC and antenna is essential for improving power delivery to the passive IC on the tag. Fourth, the power available to the tag reduces by the distance squared as the tag and reader antenna are moved apart.

An alternate expression for (9) can also be derived. Solving for R in (9) results in the following expression (Braaten et al., 2008; Rao et al., 2005):

$$R = \frac{\lambda}{4\pi} \sqrt{\frac{q G_{rd} G_{tg} P_{rd}}{P_{tg}}}. \tag{10}$$

Next, if the threshold power required to activate the passive IC and communicate with the reader is denoted as P_{th}, then a maximum read range r_{max} can be derived from (10) with a small substitution:

$$r_{max} = \frac{\lambda}{4\pi} \sqrt{\frac{q G_{rd} G_{tg} P_{rd}}{P_{th}}}. \tag{11}$$

Equation (11) is a very useful expression for predicting the max read range of a passive RFID tag. Normally, P_{th} of the passive IC used on the tag is known while P_{rd} and G_{rd} are fixed. This leaves the two variables G_{tg} and q available for a designer. Typically, a tag is designed to make the most of r_{max}. One method of maximizing r_{max} is to design a tag with a good value of q and a sizeable G_{tg}.

In summary, (11) effectively shows which antenna properties on the RFID tag will determine the max read range of the tag. In the next section, the concept of effective apertures is presented and the maximum power available for the passive IC on the RFID tag is discussed. It will be shown that even with a perfect conjugate match (i.e., $q = 1$) between the antenna on the RFID tag and the input impedance of the passive IC, only half of the available power incident on the tag will be available to the passive IC. The other half of the power is scattered into the region around the tag.

3.3 Effective apertures and power delivery on the RFID tag

Several computations can be carried out to determine the maximum available power from the RFID reader. First, the power density S of the incident wave from the reader has to be considered. From this incident power, the tag has to extract as much power as possible. This extracted power is proportional to S by some factor. This proportionality factor is denoted as the maximum effective aperture A_e and is defined in the following manner (Stutzman & Thiele, 1998):

$$P_{tg} = A_e S \tag{12}$$

where again P_{tg} is the power received by the passive tag. A_e can be derived in terms of the load impedance and S. To derive this expression, first consider the equivalent circuit of the RFID antenna in Fig. 3 (b). The current flowing through the load impedance can be written as:

$$I(0) = \frac{V_{oc}}{Z_L + Z_A} \tag{13}$$

where $Z_L = R_L + jX_L$ and $Z_A = R_A + jX_A$. The real power used by the load resistance is $P_{tg} = I^2(0)R_L$. Next, solving for $I(0)$ in the previous expression and substituting into (13) results in the following expression for P_{tg}:

$$P_{tg} = R_L \left(\frac{V_{oc}}{Z_L + Z_A}\right)^2. \tag{14}$$

Then, assuming $Z_L = Z_A^*$ for max power absorption by the load, (14) reduces to the following:

$$P_{tg} = \frac{V_{oc}^2}{4R_L}. \tag{15}$$

Equating (12) to (15) and solving for the effective aperture gives:

$$A_e = \frac{V_{oc}^2}{2R_L S}. \tag{16}$$

The effective aperture can also be thought of as the effective area of the antenna. The units for (16) are m^2, which is also another useful way of thinking about the behavior of an antenna on a RFID tag. By reducing the load resistance, the effective area of the RFID antenna can be improved and hence more power can be delivered to the passive IC.

Also notice that the resistance in the antenna is also absorbing power. This can be seen when considering the equivalent circuit in Fig. 3 (b). Again, assuming $Z_L = Z_A^*$, the power absorbed by the antenna is the same as the power absorbed by the tag as described in (15). This means that only half of the power delivered to the antenna is absorbed by the passive IC. The rest of the power is scattered into the region around the RFID tag. This scattering loss characteristic is referred to as the scattering aperture A_s of the antenna (Balanis, 2005) and can be calculated using

$$P_A = A_s S \tag{17}$$

where P_A is the power absorbed by the antenna (not the passive IC).

The previous discussion was meant to outline many of the important properties of RFID systems. Several of these concepts will be used in the next sections to design compact space-filling antennas for passive UHF RFID tags with very useful max read range values.

4. Designing space-filling antennas

Space-filling antennas are very useful for designing printed dipoles that resonate in a very small space. Because of these characteristics, space-filling (or meander-line) antennas have been very popular designs for passive RFID tags. It turns out that by using the periodic nature of space-filling antennas, a designer can design compact space-filling antennas with good matching properties and large gains (>2.0 dBi).

4.1 Design of a compact and efficient space-filling antenna

The design process presented here starts with the meander-line antenna shown in Fig. 5 (a) (Braaten et al., 2008). Each dipole arm of the meander-line antenna has N elements. The first step is to determine the electrical length of each dipole arm, which is the sum of the electrical lengths of each meander-line segment in Fig. 5 (b). The electrical length between nodes e_{m-1} and a_m is denoted as $L^e_{e_{m-1},a_m}$. Similarly, the electrical length between the remaining nodes in Fig. 5 (b) are written as $L^e_{a_m,b_m}$, $L^e_{b_m,c_m}$, $L^e_{c_m,d_m}$ and $L^e_{d_m,e_m}$. Thus, the entire electrical length of the mth segment is $L^e_m = L^e_{e_{m-1},a_m} + L^e_{a_m,b_m} + L^e_{b_m,c_m} + L^e_{c_m,d_m} + L^e_{d_m,e_m}$. Therefore, the total electrical length of each pole is

$$L^e_p = \sum_{n=1}^{N} L^e_n. \tag{18}$$

Notice the expression in (18) does not have an assumption on the symmetry of each meander-line element in Fig. 5 (b). To simplify the design process, the following symmetry assumptions will be enforced on each meander-line segment: $H^e_m = L^e_{e_{m-1},a_m} = L^e_{a_m,b_m} = L^e_{c_m,d_m} = L^e_{d_m,e_m}$. This then simplifies (18) down to

$$\tilde{L}^e_p = \sum_{n=1}^{N} H^e_n + L^e_{b_n,c_n}. \tag{19}$$

The next step in the design process is to define the meander-line antenna in Fig. 5 (a) with 2-3 meander-line segments on the dielectric substrate the RFID tag will be attached to. Then, add (or subtract) meander-line elements symmetrically to each dipole arm until a maximum gain is reached. This process can easily be performed in many different numerical electromagnetics software. Once the desired gain is achieved, several pivot points need to be defined on the meander-line antenna. Then, each pole of the meander-line antenna is moved around these pivot points to reduce the overall size of the antenna. This movement is shown in Fig. 6. This reduces the very long meander-line antenna down to a much smaller space-filling antenna. It has been noticed that significant gains can still be achieved by pivoting the meander-line antenna in this manner. Once the antenna fits into the desired space, an inductive loop may need to be added to conjugate match the input impedance of the antenna with the input impedance of the passive IC.

4.2 Design examples
To illustrate the design process, a space-filling antenna was designed for a passive tag on a single dielectric substrate and a passive tag with a dielectric superstrate. The space-filling antenna was modeled in Momentum (Agilent Technologies, 2009).

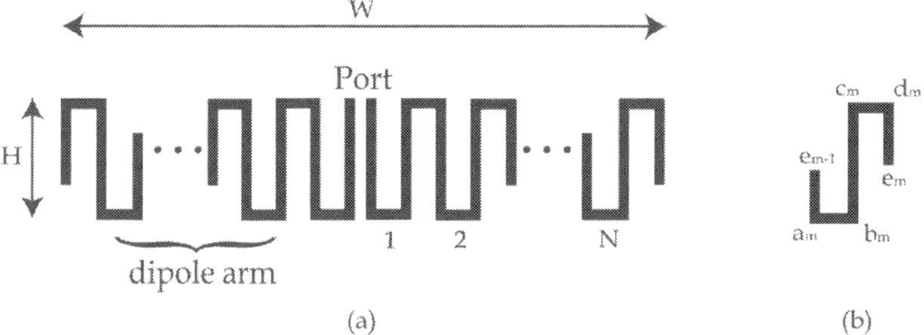

Fig. 5. (a) Meander-line antenna with N-elements on each arm; (b) the mth meander-line element.

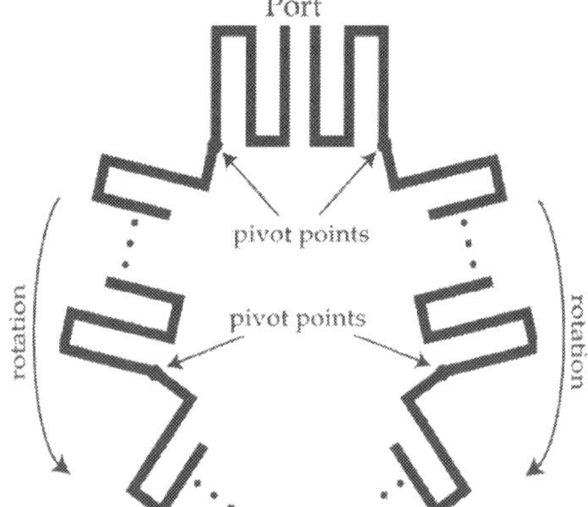

Fig. 6. Rotation of the meander-line antenna into a compact space-filling antenna.

4.3 Space-filling antenna on a single dielectric substrate
As described in the previous section, the first step was to define the size of each meander-line section. For this example, $H_m^e = L_{e_{m-1},a_m}^e = L_{a_m,b_m}^e = L_{c_m,d_m}^e = L_{d_m,e_m}^e = 5$ mm, $L_{b_m,c_m}^e = 10$ mm and the trace width was 1 mm. The substrate had a thickness of $d_1 = 1.58$ mm, permittivity of $\varepsilon_1 = 2.2$ and was lossless. The center frequency was 920 MHz. Next, each

meander-line section was connected in series to form a dipole. Meander-line sections were added until a max gain of 4.6 dBi was observed. This resulted in the design shown in Fig. 7 with N = 13 on each dipole arm. The input impedance of the meander-line antenna at 920 MHz was Z_{in} = 9.455+j23.967 Ω. Next, several pivot points were defined along each dipole arm of the meander-line antenna and an inductive matching loop was defined at the port of the antenna. The space-filling antenna in Fig. 8 was a result of this next step. The gain of the space-filling antenna was 2.0 dBi and the input impedance at 920 MHz was Z_{in} = 32.9+j168 Ω, which matches well with the input impedance of a passive IC (Rao et al., 2005).

Fig. 7. Meander-line antenna on a single substrate (ε_1= 2.2, d_1 = 1.58 mm, G = 4.6 dBi, W = 263.8mm and H = 11mm).

Fig. 8. Space-filling antenna on a single substrate (ε_1= 2.2, d_1 = 1.58 mm, G = 2.0 dBi, W = 73 mm and H = 46.1 mm).

4.4 Space-filling antenna with a single dielectric superstrate

Next, a lossless superstrate with a thickness of d_2=1.58 mm and a permittivity of ε_2 = 2.2 (the substrate properties were the same as in the previous example, d_1=1.58 mm and ε_1 = 2.2) was defined above the meander-line antenna. The meander-line sections used in the previous subsection were used in the meander-line design with the superstrate. Meander-line sections were added until a max gain of 4.81 dBi was observed. This resulted in the design shown in Fig. 9 with N = 12 on each dipole arm (a similar value of N can be found in the examples presented by Braaten et al, (2008)). The input impedance of the meander-line antenna at 920 MHz was Z_{in} = 8.256-j2.281 Ω. Just as in the previous section, several pivot points were defined along each dipole arm of the meander-line antenna and an inductive matching loop was defined at the port of the antenna. The space-filling antenna in Fig. 10 was a result of these steps. The gain of the space-filling antenna was 2.0 dBi and the input impedance at 920 MHz was Z_{in} = 26.080+j178.592 Ω. The antenna in Fig. 10 shows that with proper design, a space-filling antenna with a superstrate can have the same properties as a space-filling antenna without a superstrate.

Fig. 9. Meander-line antenna on a single substrate (ε_1 = 2.2, d_1 = 1.58 mm, ε_2 = 2.2, d_2 = 1.58 mm, G = 4.81 dBi, W = 243.8mm and H = 11mm).

Fig. 10. Meander-line antenna on a single substrate (ε_1 = 2.2, d_1 = 1.58 mm, ε_2 = 2.2, d_2 = 1.58 mm, G = 2.0 dBi, W = 73 mm and H = 46.1 mm).

4.5 Other space-filling antenna designs

Finally, it should be mentioned that space-filling antenna designs can be found in other work (Braaten et al., 2008). These designs involve a meander-line antenna on a substrate with a permittivity of 4.25 and a superstrate with a permittivity of 4.0. Several different designs are presented with various meander-line sections and trace widths. The passive RFID prototype tag presented by Braaten et al. (2008) had a max read range of 3.2 m. This was accomplished using the space-filling design techniques presented in the paper and the steps described in the subsections of this chapter. Also, several metamaterial-based (Eleftheriades & Balmain, 2005) designs have been published (Braaten et al., 2009). These designs use metamateral-inspired elements to reduce the overall size of a space-filling antenna on a passive RFID tag.

5. Future work

A new and rapidly emerging field is the study of metamaterials (Eleftheriades & Balmain, 2005; Marques et al. 2008) to improve the performance of printed antennas, filters, lenses and shielding. Initial studies have shown that the overall size of printed antennas can be reduced while preserving many of the properties of a much larger printed antenna (Lee et al. 2005; Lee et al., 2006; Mirza et al., 2008; Ziolkowski & Lin, 2008). One of the drawbacks of such metamaterial-based antennas is the complicated ground structures, vias and materials needed to reduce the overall size of the antenna. In many instances, these structures are much too complex for use on a passive RFID tag. One possibility that may avoid the need for complex structures is to use coplanar waveguide structures (CPW). These structures are usually printed on a single conducting plane with an ungrounded substrate. Some researchers have successfully applied metamaterial ideas to achieve much smaller CPW filters (Mao et al., 2007). It is anticipated, that these same principles could be extended to the use of planar antennas on passive UHF RFID tags. In particular, these ideas may be applicable for reducing the space-filling antennas presented in this chapter.

6. Conclusion

In this chapter, an introduction to RFID systems has been presented. The major components of a RFID system were defined and the role of each one was discussed. Then, the characteristics of a RFID system were described using electric field integral equations, the Friis transmission equation and effective apertures. The electric field integral equations showed how the backscattered fields were directly related to the load at the port of the RFID antenna. The Friis transmission line equation was used to clearly show what determines the max read range of a passive RFID tag. The discussion on antenna apertures revealed how in the best possible situation only half of the incident power from the reader can be used by the passive IC. The rest of the power is scattered into the region around the tag. Next, a design methodology for producing compact space-filling antennas was presented. This section was immediately followed by two examples showing this design process. The result of this chapter is an understanding of the fundamental and important concepts of RFID systems and a structured design process for producing compact, useful space-filling antennas for many different applications.

7. References

Advanced Design System, Agilent Technologies, www.agilent.com.

Amin, Y.; Batao, S.; Hallstedt, J.; Prokkola, S.; Tenhunen, H.; & Zhen, L.-R., "Design and characterization of efficient flexible UHF RFID tag antennas," *3rd European Conference on Antennas and Propagation*, pp. 2784-2786, March, 2009, Berlin, Germany.

Balanis, C.A. (2005). *Antenna Theory: Analysis and Design*, 3rd ed., Harper and Row, Publishers, New York.

Braaten, B.D.; Feng, Y. & Nelson, R.M., "High-frequency RFID tags: an analytical and numerical approach for determining the induced currents and scattered fields," *IEEE International Symposium on Electromagnetic Compatibility*, pp. 58-62, August 2006, Portland, OR.

Braaten, B. D.; Owen, G. J.; Vaselaar, D.; Nelson, R. M.; Bauer-Reich, C.; Glower, J.; Morlock, B; Reich, M.; & Reinholz, A., "A printed Rampart-line antenna with a dielectric superstrate for UHF RFID applications," in *Proceedings of the IEEE International Conference on RFID*, pp. 74-80, April, 2008, Las Vegas, NV.

Braaten, B. D.; Scheeler, R. P.; Reich, M.; Nelson, R. M.; Bauer-Reich, C.; Glower, J. & Owen, G. J., "Compact metamaterial-based UHF RFID antennas: deformed omega and split-ring resonator structures," *Accepted for publication in the ACES Special Journal Issue on Computational and Experimental Techniques for RFID Systems and Applications*, 2009.

Calabrese, C. & Marrocco, G., "Meander-slot antennas for sensor-RFID tags," *IEEE Antennas and Wireless Propagation Letters.*, vol. 7, pp. 5-8, 2008.

Cooney, E. (2005). *RFID+:The Complete Review of Radio Frequency Identification*, Cengage Delmar Learning.

Curty, J.-P.; Declercq, M.; Dehollain C. & and Joehl, N. (2006). *Design and Optimization of Passive UHF RFID Systems*, Springer-Verlag New York, LLC.

Eleftheriates, G. V. & Balmain, K. G. (2005) *Negative-Refraction Metamaterials: Fundamentals Principles and Applications*, John Wiley and Sons, Hoboken, New Jersey.

Feng, Y.; Braaten, B.D. & Nelson, R.M., "Analytical expressions for small loop antennas-with application to EMC and RFID systems," *IEEE International Symposium on Electromagnetic Compatibility*, pp. 63-68, August 2006, Portland, OR.

Finkenzeller, K. (2003). *RFID Handbook:Fundamentals and Applications in Contactless Smart Cards and Identification*, John Wiley and Sons, West Sussex, England.

Hunt, V.D.; Puglia, A. & Puglia, M. (2007). *RFID-A Guide to Radio Frequency Identification*, John Wiley and Sons, Inc.

Lee, C.-J.; Leong, K. M. K. H. & Itoh, T, "Design of a resonant small antenna using composite right/left-handed transmission line," *IEEE Antennas and Propagation Society International Symposium*, July, 2005, Washington, DC.

Lee, C.-J.; Leong, K. M. K. H. & Itoh, T, "Composite right/left-handed transmission line based compact resonant antennas for RF module integration," *IEEE Transactions on Antennas and Propagation*, vol. 54, no. 8, August, 2006, pp. 2283-2291.

Li, H.-J.; Lin, H.-H. & Wu, H.-H., "Effect of antenna mutual coupling on the UHF passive RFID tag detection," *IEEE Antennas and Propagation Society International Symposium*, July, 2008, San Diego, CA.

Mao, S.-G.; Chueh Y.-Z. & Wu, M.-S., "Asymmetric dual-passband coplanar waveguide filters using periodic composite right/left-handed and quarter-wavelength stubs," *IEEE Microwave and Wireless Components Letters*, vol. 17, no. 6, June, 2007, pp. 418-420.

Marques, R.; Martin, F. & Sorolla, M. (2008). *Metamaterials with Negative Parameters: Theory, Design and Microwave Applications*, John Wiley and Sons, Inc., Hoboken, New Jersey.

Mirza, I.O.; Shi, S.; Fazi, C.; and Prather, D. W., "Stacked patch antenna miniaturization using metamaterials," *IEEE Antennas and Propagation Society International Symposium*, July, 2008, San Diego, CA.

Owen, G. J.; Braaten, B. D.; Nelson, R. M.; Vaselaar, D.; Bauer-Reich, C.; Glower, J.; Reich, M.; & Reinholz, A., "On the Effect of Mutual Coupling on LF and UHF Tags Implemented in Dual Frequency RFID Applications," *Proceedings of the 2009 IEEE International Symposium on Antennas and Propagation*, June, 2009, Charleston, SC.

Paret, D.; Riesco R.; & Riesco, R. (2005). *RFID and Contactless Smart Card Applications*, John Wiley and Sons, Inc..

Qing, X. & Chen, Z.N., "Proximity effects of metallic environments on high frequency RFID reader antenna: Study and Applications," *IEEE Transactions on Antennas and Propagation*, vol. 55, no. 11, November, 2007, pp. 3105-3111.

Rao, K. V. S.; Nikitin, P.V.; & Lam, S.F., "Antenna Design for UHF RFID Tags: A Review and a Practical Application," *IEEE Transactions on Antennas and Propagation*, vol. 53, no. 12, December, 2005, pp. 3870-3876.

Sanford, J. R., "A novel RFID reader antenna for extreme environments," *Proceedings of the IEEE International Symposium on Antennas and Propagation*, July, 2008, San Diego, CA.

Stutzman, W.L. & Thiele, G.A. (1998). *Antenna Theory and Design*, 2nd ed., John Wiley and Sons, Inc., New York.

Yen, C.-C.; Gutierrez, A.E.; Veeramani, D.; & Van der Weide, D., "Radar Cross-Section Analysis for Backscattering RFID Tags," *IEEE Antennas and Wireless Propagation Letters*, vol. 6, 2007, pp. 279-281.

Ziolkowski, R. W. & Lin, C.-C., "Metamaterial-inspired magnetic-based UHF and VHF antennas," *IEEE Antennas and Propagation Society International Symposium*, July, 2008, San Diego, CA.

4

Design of Passive UHF RFID Tag Antennas Using Metamaterial-Based Structures and Techniques

Benjamin D. Braaten[1] and Robert P. Scheeler[2]
[1]North Dakota State University
[2]University of Colorado – Boulder
United States

1. Introduction

Metamaterials (Marques et al., 2008; Eleftheriades & Balmain, 2005; Herraiz-Martinez et al. 2009) are one of the many new technologies being adopted to improve the performance of radio frequency identification (RFID) systems (Finkenzeller, 2003; Stupf et al., 2007). In particular, metamaterial-based antenna designs are being used more frequently to improve the read range and reduce the size of passive UHF RFID tags. This chapter will introduce the concept of RFID systems and the relevant parameters for proper antenna design. Then, expressions for the phase constants, propagation constants and the characteristic (or Bloch) impedance of a wave propagating down an infinite transmission line (TL) will be derived. These expressions will then be used to introduce the concept of LH-propagation. Subsequently, the design of several metamaterial-based antennas for passive UHF RFID tags will be summarized followed by a section on the conclusion and future applications of metamaterial-based antennas to RFID systems.

2. An introduction to passive UHF RFID systems using the Friis transmission equation

The RFID system in Fig. 1 consists of a reader and several RFID tags in the space around the reader. A transmit and receive antenna is connected to the reader and each tag has a single antenna used for both transmitting and receiving. Digital circuitry that communicates with the reader is attached to the antenna on the RFID tag. This digital circuitry is often denoted as the passive IC on the RFID tag. To communicate with the tags, the reader sends out an electromagnetic field using the transmitting antenna. This electromagnetic field has power and timing information that will be used by the tag. If a tag is close enough to the reader, the tag will harvest some of the incoming energy from the electromagnetic field to power the digital circuitry in the passive IC. If it is appropriate, the passive IC will communicate with the reader using backscattered waves. By changing the input impedance of the passive IC connected to the tag antenna, the tag is able to create two different backscattered waves in the direction of the reader. One backscattered wave corresponds to a logic 0 and the other backscattered wave corresponds to a logic 1. By using timing information with the two backscattered waves, the tag is able to transmit a digital signal back to the reader. This backscattered wave with the digital information is then received by the receive antenna connected to the reader and processed.

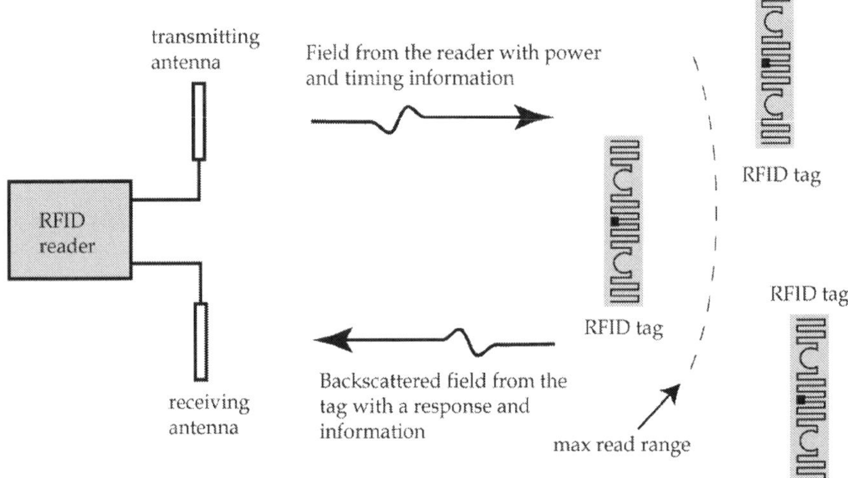

Fig. 1. Overview of a passive RFID system.

There are several different methods to describe the characteristics of a passive UHF RFID system (Braaten et al., 2006; Finkenzeller, 2003). The Friis transmission equation (Stutzman & Thiele, 1998) will be used here to show the relevant properties of an antenna on a RFID tag for achieving a maximum read range.

In general, the gain of the antennas, transmit power, frequency and sensitivity of the receiver determine the distance and rate at which a communication system can transfer digital information wirelessly. A convenient expression for describing the characteristics of such wireless systems is the Friis transmission equation (Stutzman & Thiele, 1998):

$$P_r = P_t \frac{G_r G_t \lambda^2}{(4\pi R)^2} q \tag{1}$$

where P_t is the transmitted power, P_r is the receive power, G_t is the gain of the transmitting antenna, G_r is the gain of the receiving antenna, λ is the free-space wavelength of the transmitting frequency, R is the distance between the antenna on the transmitter and the antenna on the receiver and q is the impedance mismatch factor ($0 \leq q \leq 1$) between the receiver and the receiving antenna. Equation (1) assumes a polarization match between the transmitting antenna and receiving antenna and that the receiving antenna is in the far-field of the transmitting antenna.

Equation (1) can be adopted to describe the performance of the RFID system on Fig. 1. In a RFID system, the transmitter is the reader and the receiver is the tag. The reader is connected to a transmitting antenna with a fixed gain and the tag has a receiving antenna with a fix gain. Rewriting (1) with a few substitutions results in the following expression:

$$P_{tg} = P_{rd} \frac{G_{rd}G_{tg}\lambda^2}{(4\pi R)^2} q \tag{2}$$

where P_{rd} is the power transmitted by the reader, P_{tg} is the power received by the passive tag, G_{rd} is the gain of the transmitting antenna on the reader, G_{tg} is the gain of the space-filling antenna on the tag, λ is the free-space wavelength of the transmitting frequency by the reader, R is the distance between the antenna on the reader and the antenna on the tag and q is the impedance mismatch factor ($0 \leq q \leq 1$) between the passive IC and the antenna on the tag. Next, solving for R in (2) results in the following expression (Braaten et al., 2008; Rao et al., 2005):

$$R = \frac{\lambda}{4\pi} \sqrt{\frac{qG_{rd}G_{tg}P_{rd}}{P_{tg}}}. \tag{3}$$

Equation (3) represents the distance needed to observe a particular value of P_{tg} for some fixed transmit power by the reader, fixed transmit gain and a fixed gain for the antenna on the RFID tag. Therefore, if the threshold power required to activate the passive IC and communicate with the reader is denoted as P_{th}, then a maximum read range r_{max} can be derived from (3) with a simple substitution:

$$r_{max} = \frac{\lambda}{4\pi} \sqrt{\frac{qG_{rd}G_{tg}P_{rd}}{P_{th}}}. \tag{4}$$

Equation (4) is a very useful expression and often common method for predicting the max read range of a passive RFID tag (Vaselaar, 2008; Rao et al, 2007). Typically, in a passive UHF RFID system it is very desirable to achieve the longest possible read range. Usually, P_{th} is fixed by the manufacture of the passive IC, while P_{rd}, G_{rd} and λ are fixed by the laws of the country the RFID system may be operating in. This leaves G_{tg} and q available for the design of the antenna on the passive RFID tag to maximize the read range.

3. Introduction to Left-handed propagation

Many different methods exist for improving the read range and reducing the size of a passive RFID tag. One such method is to incorporate metamaterial concepts into the design of the antenna on the RFID tag (Braaten et al., 2009a). In the next section, the concept of metamaterials is introduced by deriving expressions for the propagation constant, phase velocity and Bloch impedance of a LH-wave propagating down an infinite transmission line (Gil et al., 2007; Ryu et al., 2008). But first, a few comments on the terminology of LH-propagation are in order.

The terms RH- and LH-propagation refers to the direction of the wave vector k. In a traditional RH-TL, the electric field is curled into the magnetic field using the right hand. The field components for the RH-case are shown in Fig. 2 (a). After the curl, the thumb is pointing in the direction of the Poynting vector S and k . In a LH-TL, k is pointing in the opposite direction as S. This case is shown in Fig. 2 (b). This then requires curling the electric field into the magnetic field using the left hand. Then the direction of the thumb is pointing in the direction of k but in the opposite direction as S. Notice in both cases that S is always pointing in the same direction which indicates that the power is always flowing in the same direction (i.e., power is flowing to the load). This is the case regardless if the TL supports RH- or LH-propagation.

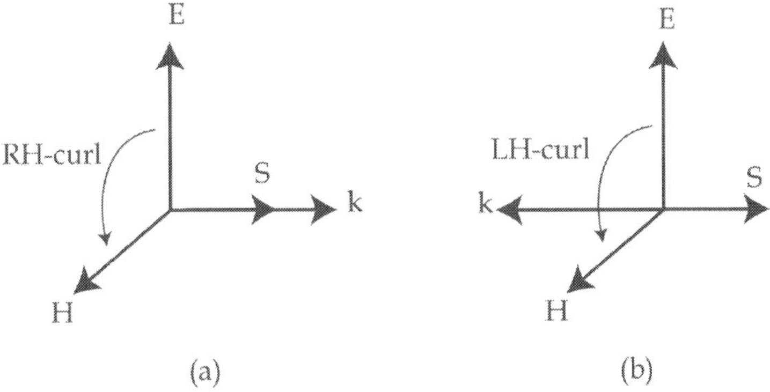

Fig. 2. (a) RH-propagation terminology; (b) LH-propagation terminology.

3.1 Deriving the Bloch impedance from an infinite periodic structure with loads that are in parallel.

First consider the equivalent circuit of the infinite TL in Fig. 3. Fig. 3 shows a periodically loaded TL with admittances jb. Using ABCD matrices (Pozar, 2005), the voltage and current at either side of the nth unit cell is

$$\begin{bmatrix} V_n \\ I_n \end{bmatrix} = \begin{bmatrix} A & B \\ C & D \end{bmatrix} \begin{bmatrix} V_{n+1} \\ I_{n+1} \end{bmatrix}. \tag{5}$$

Then using Table 4.1 in Pozar (2005), the normalized form of the ABCD matrix can be written as:

$$\begin{bmatrix} A & B \\ C & D \end{bmatrix} = \begin{bmatrix} \cos\frac{\theta}{2} & j\sin\frac{\theta}{2} \\ j\sin\frac{\theta}{2} & \cos\frac{\theta}{2} \end{bmatrix} \begin{bmatrix} 1 & 0 \\ jb & 1 \end{bmatrix} \begin{bmatrix} \cos\frac{\theta}{2} & j\sin\frac{\theta}{2} \\ j\sin\frac{\theta}{2} & \cos\frac{\theta}{2} \end{bmatrix}$$

$$= \begin{bmatrix} \cos\theta - \frac{b}{2}\sin\theta & j\left(\sin\theta + \frac{b}{2}\cos\theta - \frac{b}{2}\right) \\ j\left(\sin\theta + \frac{b}{2}\cos\theta + \frac{b}{2}\right) & \cos\theta - \frac{b}{2}\sin\theta \end{bmatrix} \tag{6}$$

where $\theta = kd$ and k is the propagation constant of the unloaded line. Next, the voltage and current for a wave propagating in the +z-direction can be written as:

$$V(z) = V(0)e^{-\gamma z} \tag{7}$$

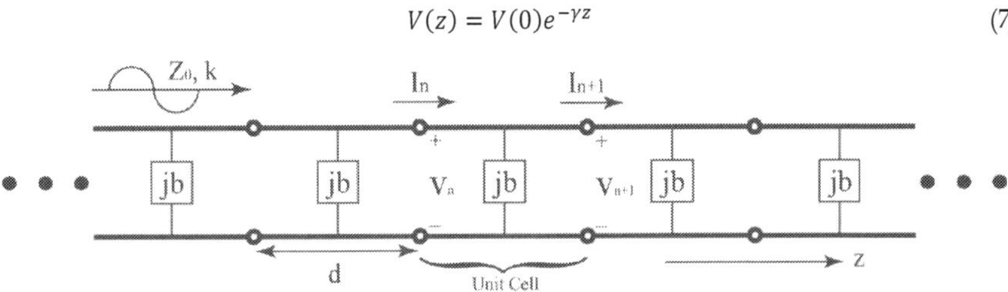

Fig. 3. Infinite periodic transmission line with parallel loads.

and

$$I(z) = I(0)e^{-\gamma z}. \tag{8}$$

Since the structure in Fig. 3 is ininitely long, V and I at the $(n+1)^{th}$ terminal differ from the nth terminal by a factor of $e^{-\gamma d}$, where d is the length of each unit cell along the TL. This then gives

$$V_{n+1} = V_n e^{-\gamma d} \tag{9}$$

and

$$I_{n+1} = I_n e^{-\gamma d}. \tag{10}$$

Solving for the voltage and current at the nth terminal gives:

$$V_n = V_{n+1} e^{\gamma d} \tag{11}$$

and

$$I_n = I_{n+1} e^{\gamma d}. \tag{12}$$

Thus,

$$\begin{bmatrix} V_n \\ I_n \end{bmatrix} = \begin{bmatrix} A & B \\ C & D \end{bmatrix} \begin{bmatrix} V_{n+1} \\ I_{n+1} \end{bmatrix} = \begin{bmatrix} V_{n+1} e^{\gamma d} \\ I_{n+1} e^{\gamma d} \end{bmatrix}. \tag{13}$$

Subtracting the matrix on the right of (13) from the middle matrices in (13) gives:

$$\begin{bmatrix} A - e^{\gamma d} & B \\ C & D - e^{\gamma d} \end{bmatrix} \begin{bmatrix} V_{n+1} \\ I_{n+1} \end{bmatrix} = 0. \tag{14}$$

Next, the determinant of (14) must vanish for a nontrivial solution (Pozar, 2005), or

$$(A - e^{\gamma d})(D - e^{\gamma d}) - BC = AD - De^{\gamma d} - Ae^{\gamma d} + e^{2\gamma d} - BC = 0. \tag{15}$$

Factoring (15) gives

$$e^{2\gamma d} - e^{\gamma d}(A + D) + 1 = 0. \tag{16}$$

Since (6) is a normalized matrix, $AD - BC = 1$. This reduces (16) to

$$A + D = e^{-\gamma d} + e^{\gamma d}. \tag{17}$$

Then using $\cosh(\gamma d) = \dfrac{e^{-\gamma d} + e^{\gamma d}}{2}$ results in

$$\cosh(\gamma d) = \frac{A+D}{2} \tag{18}$$

where A and D are taken from the matrix in (6). Since $\gamma = \alpha + j\beta$, the following expression can also be written

$$\cosh(\gamma d) = \cos\theta - \frac{b}{2}\sin\theta. \tag{19}$$

Notice that (19) is written in terms of the propagation constants and the length of the TL. Next, an expression for the characteristic impedance for the wave along the TL is derived. This impedance is sometimes called the Bloch impedance Z_B. To derive Z_B, first start with

$$Z_B = Z_0 \frac{V_{n+1}}{I_{n+1}} \tag{20}$$

which is the characteristic impedance of the nth unit cell in Fig. 3. Next, solving for V_{n+1} in (13) gives:

$$V_{n+1} = \frac{-BI_{n+1}}{A - e^{\gamma d}}. \tag{21}$$

Substituing (21) into (20) results in

$$Z_B = \frac{-BZ_0}{A - e^{\gamma d}}. \tag{22}$$

Next, solving for the root in (16) results in the following expresssion for $e^{\gamma d}$:

$$e^{\gamma d} = \frac{(A+D) \pm \sqrt{(A+D)^2 - 4}}{2} \tag{23}$$

Next, substituing (23) into (22) gives

$$Z_B^{\pm} = \frac{-2BZ_0}{2A - A - D \mp \sqrt{(A+D)^2 - 4}}. \tag{24}$$

For reciprocal networks, $A = D$. Thus, (24) reduces to

$$Z_B^{\pm} = \frac{\pm BZ_0}{A^2 - 1}. \tag{25}$$

Equation (25) is the characteristic impedance, or Bloch impedance, of the infinite periodic TL in Fig. 3. Therefore, once the admittance jb is known, A and B can be taken from (6) to evaluate the Bloch impedance along the TL. In the the next section, derivations for a similar expression to (25) are presented for a general infinite periodic TL.

3.2 Deriving the Bloch impedance from an infinite periodic structure with loads that are in series and parallel.

Again, the first step in the derivation of the Bloch impedance for the the infinite TL in Fig. 4 is to write the ABCD matrix. Using Table 4.1 in Pozar (2005), the following expressions can be written for A, B, C and D:

$$A = 1 + \frac{Z_s}{Z_p} \tag{26}$$

$$B = Z_s \left(2 + \frac{Z_s}{Z_p} \right) \tag{27}$$

$$C = \frac{1}{Z_p} \tag{28}$$

$$D = 1 + \frac{Z_s}{Z_p} = A. \tag{29}$$

Also, note again that $AD - BC = 1$ and that the voltage and current for a wave propagating in the +z-direction can be written as:

$$V(z) = V(0)e^{-\gamma z} \tag{30}$$

and

$$I(z) = I(0)e^{-\gamma z}. \tag{31}$$

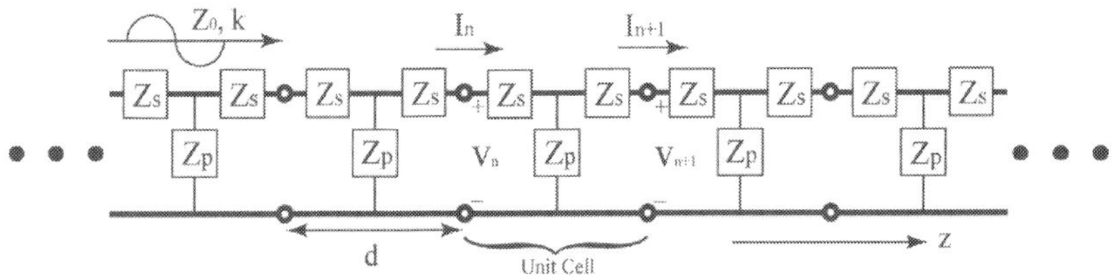

Fig. 4. Generalized infinite periodic transmission line with series and parallel loads.

Since the structure in Fig. 4 is ininitely long, V and I at the $(n + 1)^{th}$ terminal differ from the nth terminal by a factor of $e^{-\gamma d}$ where d is the length of each unit cell along the TL. This then gives

$$V_{n+1} = V_n e^{-\gamma d} \tag{32}$$

and

$$I_{n+1} = I_n e^{-\gamma d}. \tag{33}$$

Solving for the voltage and current at the nth terminal gives:

$$V_n = V_{n+1} e^{\gamma d} \tag{34}$$

and

$$I_n = I_{n+1} e^{\gamma d}. \tag{35}$$

Thus,

$$\begin{bmatrix} V_n \\ I_n \end{bmatrix} = \begin{bmatrix} A & B \\ C & D \end{bmatrix} \begin{bmatrix} V_{n+1} \\ I_{n+1} \end{bmatrix} = \begin{bmatrix} V_{n+1}e^{\gamma d} \\ I_{n+1}e^{\gamma d} \end{bmatrix}. \tag{36}$$

Subtracting the matrix on the right of (36) from the middle matrices in (36) gives:

$$\begin{bmatrix} A - e^{\gamma d} & B \\ C & D - e^{\gamma d} \end{bmatrix} \begin{bmatrix} V_{n+1} \\ I_{n+1} \end{bmatrix} = 0. \tag{37}$$

Next, the determinant of (37) must vanish for a nontrivial solution (Pozar, 2005), or

$$(A - e^{\gamma d})(D - e^{\gamma d}) - BC = AD - De^{\gamma d} - Ae^{\gamma d} + e^{2\gamma d} - BC = 0. \tag{38}$$

Factoring (38) gives

$$e^{2\gamma d} - e^{\gamma d}(A + D) + 1 = 0. \tag{39}$$

Since $AD - BC = 1$, (39) reduces to

$$A + D = e^{-\gamma d} + e^{\gamma d}. \tag{40}$$

Then using $\cosh(\gamma d) = \frac{e^{-\gamma d} + e^{\gamma d}}{2}$ results in

$$\cosh(\gamma d) = \frac{A+D}{2} = \cosh(\alpha d)\cos(\beta d) + j\sinh(\alpha d)\sin(\beta d) = \frac{1 + \frac{Z_s}{Z_p} + 1 + \frac{Z_s}{Z_p}}{2} \tag{41}$$

where A and D are taken from (26) and (29), respectively, and the trigonometric expresssion are from the normalized matrix (6). This implies

$$\cosh(\alpha d)\cos(\beta d) + j\sinh(\alpha d)\sin(\beta d) = 1 + \frac{Z_s}{Z_p}. \tag{42}$$

Now, for the propagation mode we have $\alpha = 0$ and $\beta \neq 0$. Substituting α and β into (42) gives

$$\cosh(\beta d) = 1 + \frac{Z_s}{Z_p}. \tag{43}$$

Now for the Bloch impedance,

$$Z_B = Z_0 \frac{V_{n+1}}{I_{n+1}}. \tag{44}$$

Substituing (37) into (44) and solving for Z_B gives

$$Z_B = \frac{-BZ_0}{A - e^{\gamma d}}. \tag{45}$$

Next, solving for the root in (39) results in the following expresssion for $e^{\gamma d}$:

$$e^{\gamma d} = \frac{(A+D) \pm \sqrt{(A+D)^2 - 4}}{2} \tag{46}$$

Substituting (46) into (45), using $A = D$ and factoring results in the following expresssion for Z_B (Marques et al., 2008):

$$Z_B = \sqrt{Z_s(Z_s + 2Z_p)} \tag{47}$$

which is an expression for the Bloch impedance in terms of the series and parallel loads along the TL. Next, Z_s and Z_p will defined in a manner to support both RH- and LH-propagation. Expressions will also be derived for Z_B in both instances.

First consider the RH-TL. In a RH-TL the series impedance is $Z_s = j\omega L/2$ and the parallel (or shunt) impedance is $Z_p = -j/\omega C$ in Fig. 4 (Marques et al., 2008; Eleftheriades & Balmain, 2005). This then reduces (43) to

$$\cosh(\beta_R d) = 1 + \frac{\omega^2 LC}{2}. \tag{48}$$

Note that the subscript R will be added to the variables to denote the RH-propagation. Similarly, a subscript L will be added to the variables to denote LH-propagation. Also, for the Bloch impedance,

$$Z_{BR} = \sqrt{Z_s(Z_s + 2Z_p)} = \sqrt{\frac{L}{C}\left(1 - \frac{\omega^2}{\omega_{CR}^2}\right)} \tag{49}$$

where $\omega_{CR}^2 = \left(\frac{2}{\sqrt{LC}}\right)^2$.

Next, in a LH-TL the series impedance is $Z_s = -j/2\omega C$ and the parallel (or shunt) impedance is $Z_p = j\omega L$ in Fig. 4 (Marques et al., 2008; Eleftheriades & Balmain, 2005). This then reduces (43) to

$$\cosh(\beta_L d) = 1 + \frac{1}{2LC\omega^2}. \tag{50}$$

Similarly for the Bloch impedance,

$$Z_{BL} = \sqrt{\frac{L}{C}\left(1 - \frac{\omega_{CL}^2}{\omega^2}\right)} \tag{51}$$

where $\omega_{CL}^2 = \left(\frac{2}{\sqrt{LC}}\right)^2$.

For the previous analysis it was assumed that $\lambda_d \gg d$ where λ_d is the internal wavelength and d is the segment length in Fig. 4. To ensure this inequality, the segment length d must be reduced. This translates to smaller values of L and C. Doing so increases the cutoff frequency ω_{CR}, thus the following expressions are only valid for frequencies that satisfy the inequality $\omega \ll \omega_{CR}$. This implies that $\frac{\omega}{\omega_{CR}} \ll 1$. This inequality simplifies (49) to

$$Z_{BR} \approx \sqrt{\frac{L}{C}(1-0)} = \sqrt{\frac{L}{C}}. \tag{52}$$

Also, for the LH-TL, (51) reduces to

$$Z_{BL} = \sqrt{\frac{L}{C}(1-0)} = Z_{BR} \tag{53}$$

because $\omega \gg \frac{\omega_{LC}^2}{\omega^2} \approx 0$.

The previous steps illustrate the process of deriving the Bloch impedance values for a RH- and LH-TL. In the next section, the derivation of the expressions for the propagation constants and phase velocities along a RH- and LH-TL will be presented.

3.3 Deriving the propagation constants and phase velocity expressions from an infinite periodic structure with loads that are in series and parallel.

First, taking the Taylor series expansion of (48) and truncating after the second term gives

$$\cos(\beta_R d) \approx 1 - \frac{(\beta_R d)^2}{2!} = 1 - \frac{\omega^2 LC}{2}. \tag{54}$$

Solving for the phase constant in (54) gives

$$\beta_R d = \omega\sqrt{LC}. \tag{55}$$

Then for the phase $V_{\phi R}$ and group V_{gR} velocity,

$$V_{\phi R} = \frac{\omega}{\beta_R} = \frac{d}{\sqrt{LC}} > 0 \tag{56}$$

and

$$V_{gR} = \left(\frac{\partial \beta_R}{\partial \omega}\right)^{-1} = V_{\phi R} > 0. \tag{57}$$

Next, using the Taylor series expansion of (50) and truncating after the second term gives

$$\cos(\beta_L d) \approx 1 - \frac{(\beta_L d)^2}{2!} = 1 - \frac{1}{2\omega^2 LC}. \tag{58}$$

Then solving for the phase constant in (58) and choosing the negative sign gives

$$\beta_L d = -\frac{1}{\omega\sqrt{LC}}. \tag{59}$$

Then for the phase $V_{\phi L}$ and group V_{gL} velocity,

$$V_{\phi L} = \frac{\omega}{\beta_L} = -\omega^2 d \sqrt{LC} < 0 \tag{60}$$

and

$$V_{gL} = \left(\frac{\partial \beta_L}{\partial \omega}\right)^{-1} = -V_{\phi L} > 0. \tag{61}$$

Note the inequalities in (60) and (61). In particular, notice the sign change in the phase velocity, but the group velocities remain positive. Looking at the summary in Table 1, it is clear which expressions in the LH-TL change sign for the LH-propagation. The RH-wave has a positive phase constant and phase velocity while a LH-wave have a negative phase constant and phase velocity. Both LH- and RH-waves have the same Bloch impedance (i.e., characterstic impedance) and positive group velocity. The sign of the group velocities were chosen to be both positive, which was done to agree with the definition of power flow (it is assumed that power flows from the source on the left to the load on the right of the TL in Fig. 4). Then in both instances, the group velocity is delivering power in the correct direction.

RH-TL	LH-TL
$Z_{BR} = \sqrt{\dfrac{L}{C}}$	$Z_{BL} = Z_{BR}$
$\beta_R d = \omega\sqrt{LC} > 0$	$\beta_L d = -\dfrac{1}{\omega\sqrt{LC}} < 0$
$V_{\phi R} = \dfrac{d}{\sqrt{LC}} > 0$	$V_{\phi L} = -\omega^2 d \sqrt{LC} < 0$
$V_{gR} = V_{\phi R} > 0.$	$V_{gL} = -V_{\phi L} > 0.$

Table 1. Summary of the derived RH- and LH-TL properties.

3.4 Dispersion diagrams for an infinite periodic structure with loads that are in series and parallel.

In this section the dispersion diagrams for the RH- and LH-TL are plotted. From Table 1, the propagation constants along a RH- and LH-TL are

$$\beta_R d = \omega\sqrt{LC} > 0 \tag{62}$$

and

$$\beta_L d = -\frac{1}{\omega\sqrt{LC}} < 0. \tag{63}$$

Solving for ω in (62) and (63) gives

$$\omega = \frac{\beta_R d}{\sqrt{LC}} \tag{64}$$

and

$$\omega = -\frac{1}{\beta_L d \sqrt{LC}}. \tag{65}$$

Plotting (64) and (65) results in the dispersion diagrams shown in Fig. 5 (normalized).

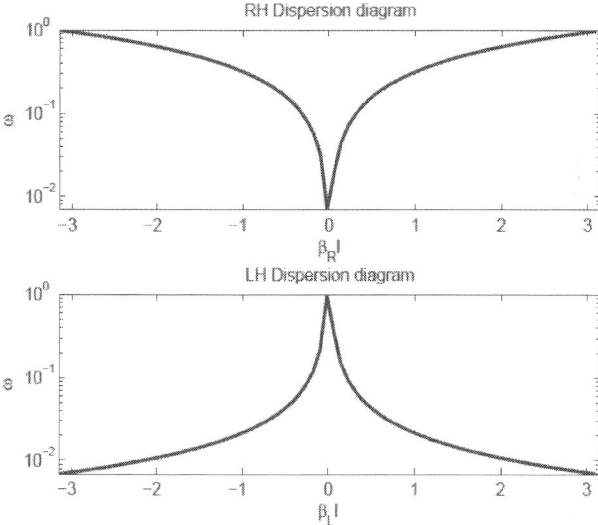

Fig. 5. RH- and LH-TL dispersion diagrams.

4. Metamaterial-based antennas for passive UHF RFID tags

As mentioned in the previous sections, one of the main advantages of using metamaterial-based elements in the design of antennas is that the resulting antenna is much smaller than traditional printed antennas (Lee et al, 2005; Lee et al., 2006; Abdalla et al., 2009; Iizuka & Hall, 2007). In this section, several of these ideas have been adopted for use on passive RFID tags. Particularly, metamaterial-based elements (Ali & Hu, 2008; Ghadarghadr et al., 2008) have been incorporated into the design of printed antennas on a single ungrounded dielectric. The first design shown in the next section uses deformed-omega elements (Mishra et al., 2008) to introduce a series inductance to the port of the antenna that causes the antenna to resonate at a much lower frequency (Braaten et at., 2009a). The second design involves using coplanar waveguide elements (CPW) to reduce the overall size of a meander-line antenna. In particular, series connected CPW inductors and capacitors found in CPW filters (Mao et al., 2007) are used to periodically load a meander-line antenna. The result is a much smaller meander-line antenna with a lower resonant frequency (Braaten et al., 2009b). Finally, the third design uses two split-ring resonators (Marques et al., 2008; Eleftheriades & Balmain, 2005) instead of a meander-line antenna to form a dipole. This type of dipole is useful for RFID tags because the input impedance is inductive above the resonant frequency.

4.1 The Meander-line antenna
The printed meander-line antenna shown in Fig. 6 (a) is very useful for achieving resonance in a very small area. This makes the meander-line antenna very popular for integration on passive UHF RFID tags (Marrocco, 2003; Rao et al., 2005). It is often desirable to describe an antenna using an equivalent circuit. To do this, first consider the meander-line section in Fig. 6 (b). Each meander-line section can be modeled as a parallel connected equivalent capacitance C_m and equivalent inductance L_m (Bancroft, 2006). The equivalent capacitance exists between the vertical segments of each meander-line section and the self-inductance is created by the horizontal segments of each meander-line section. Thus, each pole of the meander-line dipole is made up of several series connected parallel $L_m C_m$ sections. In order for the meander-line antenna to resonate, it is important to maximize the section inductance L_m.

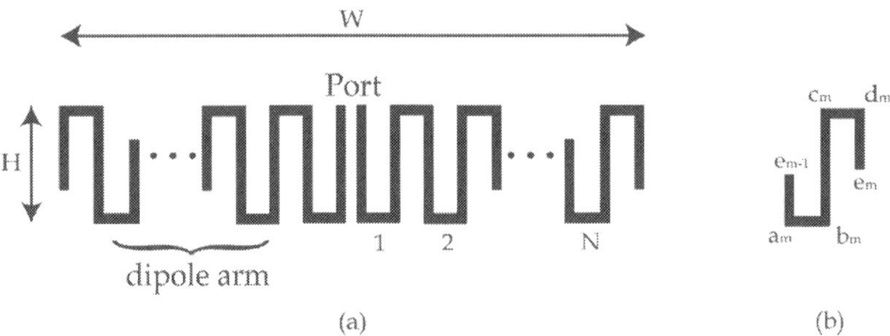

Fig. 6. (a) Meander-line antenna with N-elements on each arm; (b) the mth meander-line section.

4.2 Meander-line antenna using deformed-omega elements
One way to improve the inductance of each section is to substitute the deformed-omega element shown in Fig. 7 (a) in for several of the meander-line sections (Braaten et al., 2009a). This added inductance will cause the meander-line antenna to resonate at a much lower frequency. This allows a designer to reduce the overall size of the meander-line antenna.
The inductance of each deformed-omega element can be approximated as (Braaten et al., 2009a):

$$Z_a \approx \tfrac{1}{2} j\omega\mu_0 a \left[ln\left(\tfrac{8a}{p}\right) - 2 \right] \ \Omega \qquad (66)$$

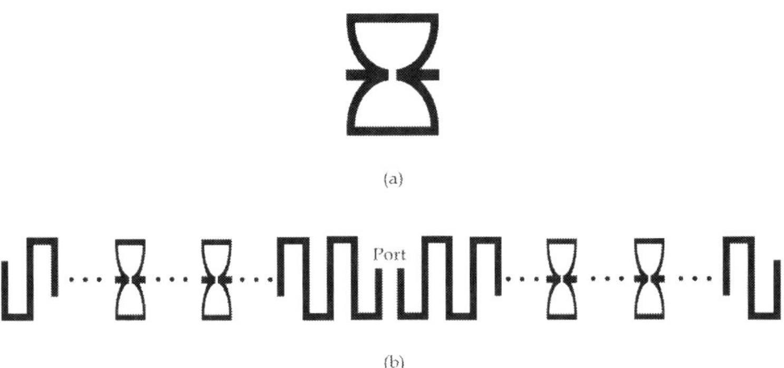

(a)

(b)

Fig. 7. (a) Deformed-omega element; (b) meander-line antenna with series connected deformed-omega elements.

where a is the width of the individual deformed-omega element and p is the trace width. Braaten et al. (2009a) has shown that the overall size of a meander-line antenna can be significantly reduced by introducing deformed-omega elements into the design. The prototype tag presented by Braaten et al. (2009a) is printed on FR-4 with a thickness of .787 mm. The overall size of the tag is 42.2 mm wide and 18.8 mm high and has a read range of 4.5 m. An image of the RFID tag is shown in Fig. 8. The size of the passive tag is being compared to a previous meander-line design by the same authors (Braaten et al., 2008).

Fig. 8. A passive RFID tag with deformed-omega elements used in the antenna design being compared to the size of a previous meander-line design.

4.3 Meander-line antenna periodically loaded with right/left-handed CPW-LC loads

There are other methods to introduce inductance or capacitance to the equivalent circuit of each meander-line section. One method is to periodically load each meander-line section with a series connected inductance (L) and capacitance (C) (Braaten et al., 2009b). It is often desirable to have the tag antenna on a single conducting layer. Thus, a CPW structure is needed to introduce the series L and C (Mao et al., 2007). An image of the series connected CPW-LC is shown in Fig. 9 (a). The load consists of an interdigitated capacitor connected to a conducting loop that introduces inductance. One method of periodically loading the meander-line antenna is shown in Fig. 9 (b). Periodic CPW-LC loads could also be introduced to the bottom of the meander-line antenna (Fig. 10).

Prototype designs using this method to load a meander-line antenna (Braaten et al. 2009b) have shown that the introduction of the periodic CPW LC-loads along the meander-line antenna reduces the overall size of the meander-line antenna by 18%. The prototype tag by Braaten et al. (2009b) was printed on 1.36 mm of FR-4 and had a max read range of 4.87 m. The overall size of the prototype tag was 14.81 mm high by 47.13 mm wide.

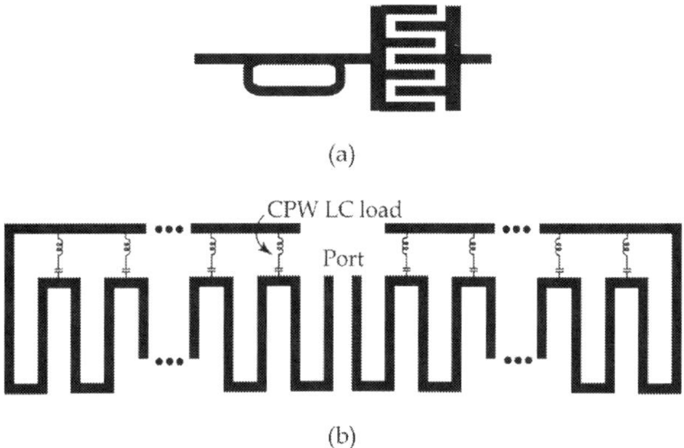

(a)

(b)

Fig. 9. (a) Series connected CPW-LC loads; (b) meander-line antenna periodically loaded with series connected LC loads

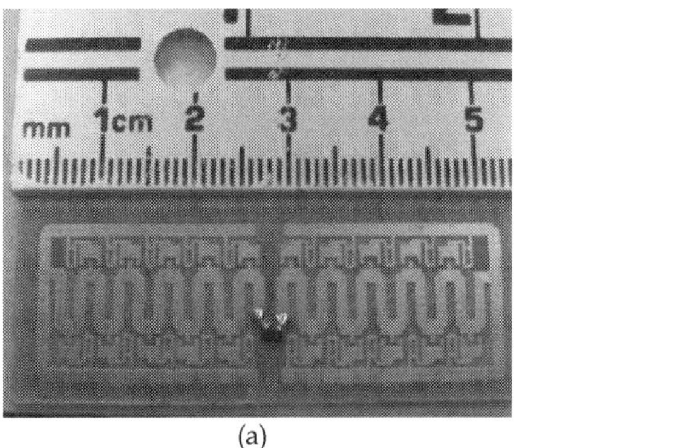

(a)

(b)

Fig. 10. (a) A meander-line antenna periodically loaded on the top and bottom with CPW-LC elements; (b) closer image of the CPW-LC loads.

4.4 Dipole antenna designed with split-ring resonators

The design of an antenna on a RFID tag does not always have to involve a meander-line antenna. As long as the antenna can be designed with an input impedance that is inductive, is compact in size and has a usable gain, then other elements may work. One element that can be used is the split-ring resonator (SRR) (Eleftheriates & Balmain, 2005) shown in Fig. 11. The equivalent circuit of a SRR is the same as each meander-line section shown in Fig. 6 (b). The particularly useful characteristic of a SRR is that this element is inductive above resonance (Dacuna & Pous, 2007). Therefore, a dipole could be made using two SRR as long as the dipole is driven above the resonance frequency.

Prototype RFID tags have been manufactured using a single SRR (Dacuna & Pous, 2007) and two SRRs (Braaten et al., 2009a) as a dipole. In both cases, the SRR were driven above resonance to achieve an inductive input impedance. Max read ranges of 6.5 m have been reported (Dacuna & Pous, 2007).

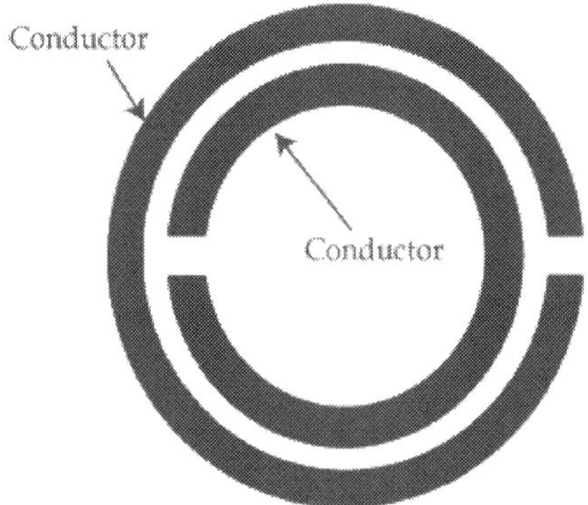

Fig. 11. Split-ring resonator element.

5. Conclusion

The first topic in this chapter was an introduction to RFID systems. This was followed immediately with a discussion on metamaterials and LH-propagation. Expressions for the propagation constants, phase velocity and Bloch impedance were derived and discussed. Next, several metamaterial-based antenna designs for passive RFID tags were presented. The designs offered showed that by incorporating elements found in metamaterials in the design of the antenna on a RFID tag, the antenna could be made to resonate at a much smaller dimension. The result is a compact passive RFID tag with very useful max read range values.

6. Future work

One common characteristic is shared among every antenna design in this chapter. Every design is based around an antenna with RH-propagation. An area that could be investigated would be to achieve LH-propagation along a RFID antenna on a passive RFID tag. An antenna that achieves LH-propagation may have the added advantage of being much smaller than traditional meander-line antennas but many questions on the far-field characteristics (i.e., backscattering properties) of the antenna still need to answered.

7. References

Abdalla, M. A. Y.; Phang, K. & Eleftheriades, G. V., "A planar electronically steerable patch array using tunable PRI/NRI phase shifters," *IEEE Transactions on Microwave Theory and Techniques*, vol. 57, vo. 3, pp. 531-541, March, 2009.

Ali, A. & Hu, Z., "Metamaterial resonator based wave propagation notch for ultrawideband filter applications," *IEEE Antennas and Wireless Propagation Letters*, vol. 7, pp. 210-112, 2008.

Bancroft, R. (2006). *Microstrip and Printed Antenna Design*, Scitech Publishing, Inc., Raleigh, North Carolina.

Braaten, B.D.; Feng, Y. & Nelson, R.M., "High-frequency RFID tags: an analytical and numerical approach for determining the induced currents and scattered fields," *IEEE International Symposium on Electromagnetic Compatibility*, pp. 58-62, August 2006, Portland, OR.

Braaten, B. D.; Owen, G. J.; Vaselaar, D.; Nelson, R. M.; Bauer-Reich, C.; Glower, J.; Morlock, B; Reich, M.; & Reinholz, A., "A printed Rampart-line antenna with a dielectric superstrate for UHF RFID applications," in *Proceedings of the IEEE International Conference on RFID*, pp. 74-80, April, 2008, Las Vegas, NV.

Braaten, B. D.; Scheeler, R. P.; Reich, M.; Nelson, R. M.; Bauer-Reich, C.; Glower, J. & Owen, G. J., "Compact metamaterial-based UHF RFID antennas: deformed omega and split-ring resonator structures," *Accepted for publication in the ACES Special Journal Issue on Computational and Experimental Techniques for RFID Systems and Applications*, 2009a.

Braaten, B. D.; Reich, M. & Glower, J., "A compact meander-line UHF RFID tag antenna loaded with elements found in right/left-handed coplanar waveguide structures," *Submitted for review in the IEEE Antennas and Wireless Propagation Letters*, 2009b.

Dacuna, J. & Pous, R. "Miniaturized UHF tags based on metamaterials geometries," *Building Radio Frequency Identification for the Global Environment*, July 2007, [Online], www.bridge-project.eu.

Eleftheriates, G. V. & Balmain, K. G. (2005) *Negative-Refraction Metamaterials: Fundamentals Principles and Applications*, John Wiley and Sons, Hoboken, New Jersey.

Finkenzeller, K. (2003). *RFID Handbook: Fundamentals and Applications in Contactless Smart Cards and Identification*, John Wiley and Sons, West Sussex, England.

Ghadarghadr, S.; Ahmadi, A. & Mosellaei, H., "Negative permeability-based electrically small antennas," *IEEE Antennas and Wireless Propagation Letters*, vol. 7, pp. 13-17, 2008.

Gil. M.; Bonache, J.; Selga, J.; Garcia-Garcia, J. & Martin, F. , "Broadband resonant-type metamaterial transmission lines, " *IEEE Microwave and Wireless Components Letters,* vol. 17, no. 2, February, 2007, pp. 97-99.

Herraiz-Martinez, F. J.; Hall, P. S.; Liu, Q. & Sogovia-Vargas, D., "Tunable left-handed monopole and loop antennas," *IEEE Antennas and Propagation Society International Symposium,* July, 2009, Charleston, SC.

Iizuka, H. & Hall, P.S., "Left-handed dipole antennas and their implementations," *IEEE Transactions on Antennas and Propagation,* vol. 55, no. 5, May, 2007, pp. 1246-1253.

Lee, C.-J.; Leong, K. M. K. H. & Itoh, T. "Design of resonant small antenna using composite right/left-handed transmission line," *IEEE Antennas and Propagation Society International Symposium,* July, 2005, pp. 218-221.

Lee, C.-J.; Leong, K. M. K. H. & Itoh, T. "Composite right/left-handed transmission line based compact resonant antenna for RF module integration," *IEEE Transactions on Antennas and Propagation,* vol. 54, no. 8, August, 2006, pp. 2283-2291.

Mao, S.-G.; Chueh Y.-Z. & Wu, M.-S., "Asymmetric dual-passband coplanar waveguide filters using periodic composite right/left-handed and quarter-wavelength stubs," *IEEE Microwave and Wireless Components Letters,* vol. 17, no. 6, June, 2007, pp. 418-420.

Marques, R.; Martin, F. & Sorolla, M. (2008). *Metamaterials with Negative Parameters: Theory, Design and Microwave Applications,* John Wiley and Sons, Inc., Hoboken, New Jersey.

Marrocco, G., "Gain-optimized self-resonant meander line antennas for RFID applications," *IEEE Antennas and Wireless Propagation Letters,* vol. 2, pp. 302-305, 2003.

Mishra, D.; Poddar, D. R. & Mishra, R. K. "Deformed omega array as LHM," *Proceedings of the International Conference on Recent Advances in Microwave Theory and Applications,* pp. 159-160, Jaipur, India, November, 2008.

Pozar, D. M. (2005). *Microwave Engineering,* 3rd Ed., John Wiley and Sons, Inc., Hoboken, New Jersey.

Rao, K. V. S.; Nikitin, P. V. & Lam, S. F., "Antenna design for UHF RFID tags: a review and a practical application," *IEEE Transactions on Antennas and Propagation,* vol. 53, no. 12, pp. 3870-3876, December, 2005.

Rao, K. V. S.; Nikitin, P. V. & Lazar, S., "An overview of near field UHF RFID" *IEEE International Conference on RFID,* pp. 174-176, March, 2007.

Ryu, Y.-H.; Lee, J.-H.; Kim, J.-Y. & Tae, H.-S., "DGS dual composite right/left handed transmission lines," *IEEE Microwave and Wireless Components Letters,* vol. 18, no. 7, July, 2008, pp. 434-436.

Stupf, M.; Mittra, R.; Yeo, J. & Mosig, J. R., "Some novel design for RFID antenna and their performance enhancement with metamaterials," *Microwave and Optical Technology Letters,* vol. 49, no. 4, February, 2007, pp. 858-867.

Stutzman, W.L. & Thiele, G.A. (1998). *Antenna Theory and Design,* 2nd ed., John Wiley and Sons, Inc., New York.

Vaselaar, D., "Passive UHF RFID design and utilization in the livestock industry," Masters Paper, North Dakota State University, Fargo, ND, 2008.

Ziolkowski, R. W. & Lin, C.-C., "Metamaterial-inspired magnetic-based UHF and VHF antennas," *IEEE Antennas and Propagation Society International Symposium,* July, 2008, San Diego, CA.

5

RFID Antennas – Possibilities and Limitations

Johan Sidén and Hans-Erik Nilsson
Mid Sweden University / Sensible Solutions Sweden AB
Sweden

1. Introduction

Marking items for remote identification not only places high demands on the heart of the RFID tags, the Application Specific Circuit (ASIC), but also on the tag antennas. For long reading distances, efficient antennas are crucial and their efficiency is directly proportional to the maximum reading distance of both semi-active and passive tags.

This chapter is an in-depth investigation of the requirements for the antenna part of UHF RFID tags, with focus on antenna design, characterization and optimization from the perspectives of both costs involved and technical constraints. The main focus is devoted to antennas that could be manufactured using more or less standard manufacturing techniques available in the packaging industry. The chapter also presents some new ideas on how to utilize the antenna structure itself as a sensor for measuring different physical properties during the logistic chain.

The chapter starts by describing the most general requirements for a functional UHF antenna, explaining why designs for RFID tag antennas cannot be taken directly from traditional antennas designed for other applications since RFID chips input impedances differ significantly from traditional input impedances such as 50 Ω and 75 Ω.

Ordinary antennas, not made for RFID, are usually also designed to work at a particular location or at locations where the surroundings of the antenna are well known. This is not generally the case for RFID applications, where the antenna is largest part of the tag and where the near field region of the antenna will be defined by the object and material that the RFID tag is attached to. For cost and practical reasons, tag antennas should be two-dimensional and if they are applied to paper or soft plastics they should be non-sensitive to bending, particularly as they might be placed over an edge of an object. An antenna's resonant length is also directly dependent on its surrounding medium. If the antenna is designed to be placed in free space it will change its properties if placed next to a dielectric medium, even if this medium is electrically lossless. There is thus a demand for some kind of general tag antennas and guidelines for designing and optimizing tag antennas when their final location is known in advance. The type of antenna properties and designs to be avoided are similarly given as well as which types should be used when, for example, putting antennas onto flexible, dielectric and metallic materials. Specific solutions presented also include how to enhance the tags' commercial value by incorporating insignias such as company logos in the antenna design.

Antennas used in traditional RFID tags are mainly made out of copper or aluminum. In order to keep the antenna prices to a minimum and facilitate very large production it is desirable to fabricate the tag antennas using commercially available web shaping techniques or utilizing printing processes with electrically conductive ink. Conductance in such prints is normally based upon relatively expensive silver particles and for cost reasons, printed antennas should therefore have as small a total printed area as possible, thin trace thickness and still be robust and maintain high radiation efficiency. The chapter therefore includes a discussion and examples of the trade-off between cost and antenna performance when minimizing the amount of expensive material by thinning the printing layers.

For tagging objects containing materials that cause problems to one layer antennas and where traditional double layer PCB antennas are too expensive to use, we will also show how microstrip antennas can be manufactured using printing processes and cardboard material as substrate.

The chapter, with its regards to efficiency of RFID tag antennas, ends with a presentation of how pairs of ordinary low-cost passive RFID tags can be used as remote reading sensors and specifically moisture sensors. This is possible by arranging one antenna in such a way that it degrades in performance in proportion to the physical quantity it is designed measure and letting the other antenna serve as reference. For use as moisture sensors this is easily done by embedding one of two identical RFID tags in moisture absorbing material and leaving the other tag open. If the pair of tags are placed in a humid environment or directly exposed to wetness the embedded tag will require a higher minimum signal strength than the open tag to operate. The differences involved in reading the two tags are proportional to the level of humidity or wetness and the pair of tags in this setup can thus be used as low-cost remote sensors and similar setups can be used for remote measurements of other quantities.

2. Design of a simple one-layer RFID tag antenna

The RF front of passive RFID chips often incorporates Schottky diodes, which give rise to an input impedance far from that commonly seen in other RF systems [1]. Schottky diodes are used to rectify incoming RF signals to supply power to the RFID chip. They are also used to modulate the signal reflected by the antenna back to the reader. To obtain as high as possible feed voltage to the chip, a voltage doubler is usually implemented with the aid of two Schottky diodes. The input impedance observed by the antenna therefore has a relatively low real part and an imaginary capacitive part of some hundred Ohms. An example of Schottky diodes that could be used for this purpose involves the Agilent Technologies' HSMS-282x Series [2]. These diodes feature low series resistance, low forward voltage at all current levels, and good RF characteristics. Considering a voltage doubler built using Schottky diodes at the chip's RF input, the impedance observed by the tag antenna can be described as two parallel Schottky diodes. Through the use of the equations presented in the data sheet of the mentioned diodes, the impedance for one diode can be calculated to $72 - j \times 244$ Ω for the frequency 868 MHz, assuming room temperature, and an operating current 25 µA. Two parallel diodes thus give a chip impedance of $36 - j \times 122$ Ω. To maximize power transfer from the antenna to the chip, a conjugate impedance match is required [1], which in this case implies that the antenna should have an input impedance, Z_A, close to $36 + j \times 122$ Ω.

Looking at the Smith Chart for the most basic antenna, the half wavelength dipole, in Fig. 1 (a) and (c), one sees that this kind of antenna cannot be tuned to the desired input impedance by simply adjusting its length. The same dipole equipped with an inductive load in parallel to the antenna input terminals can however reach this order of input impedance as is illustrated in Fig. 1(b) and (d). The ordinary dipole antenna in Fig. 1 (a) has dimensions

l=163 mm, w=2 mm and g=1 mm and the inductive dipole in Fig. 1 (b) has dimensions l=185 mm, loop length ll=20, loop height lh=10 mm and line width lw=1 mm. Both Smith charts in Fig. 1 are achieved by simulating the antenna patterns in Ansoft HFSS [3] and sweeps from 500 MHz to 1500 MHz where the markers indicate the frequency and 868 MHz.

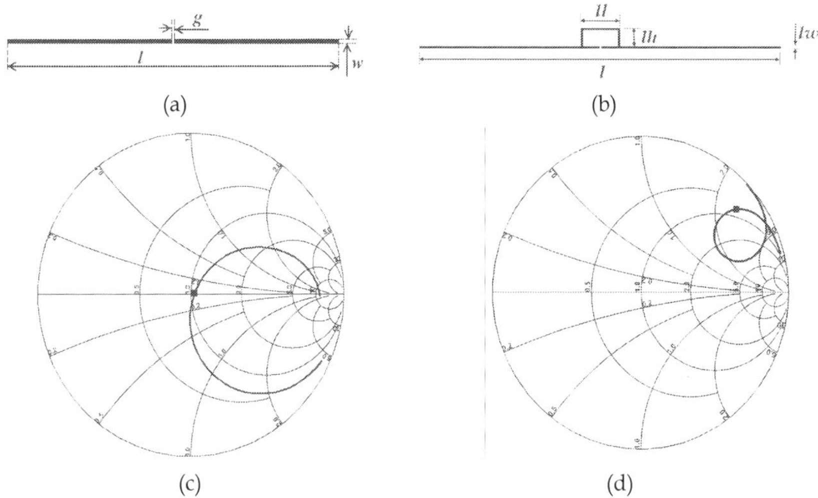

Fig. 1. An ordinary dipole (a) and (b) the same with an additional loop to make its input impedance more inductive. (c) and (d) show the Smith Charts of respectively the antennas in (a) and (b)

3. Printed antennas

Printed antennas are considered in order to reach an RFID technology fully integrated into packaging fabrication lines. The drawback with printed antennas, however, is their reduced radiation efficiency as compared to their copper counterparts as the bulk conductivity of their printed traces is lower than for solid metals. This chapter shows how careful antenna design and proper printing technologies can produce antennas which are as efficient as copper antennas and how this can be accomplished using a minimum of conductive ink.

3.1 Radiation efficiency of a printed antenna

Metallic conducting particles, which are usually in the shape of flakes or spheres, are blended into an ink vehicle and printed, for example, by means of a flexographic printing press. Conductivity is created by the overlap of the conductive particles, which is why the bulk conductivity is naturally lower than for solid metals.

Different printing technologies can be utilized to apply ink onto different substrates, for example flexography, gravure, screen-printing and Inkjet. Inkjet printing however requires expensive nano-sized particles and the printing speed of ink-jet systems is still lagging behind other techniques. Naturally, we will see technology breakthroughs in the area of ink-jet technology and this technique cannot be neglected in the future.

As mentioned, the main drawback of printed antennas is their limited conductivity when compared to fabricating antennas from solid metals. Basic laws for conductors and conductivity state that ohmic losses decrease as conductor thickness increases [4]. Even though printed ink traces are not homogenous, a similar behavior will also apply to this case. An electrical transmission line of a given length and width, and printed with a particular ink thickness, has a total resistance proportional to the length and inversely proportional to the trace width and thickness.

In the industry, flexographic printers are commonly multi-station units, containing up to ten stations in a series, where each station usually prints one color (Fig. 2). With identical printing plates and the same silver ink at several stations, a multi-station flexographic printer can also be used to print ink layers whose thickness is proportional to the number of stations used. In the specific laboratory printer setup used, the experience has been that each print pass provides a layer thickness of the order of 3-5 μm. Using such a setup, it is possible to choose layer thicknesses as multiples of 3-5 μm by choosing the number of stations to use.

Fig. 2. Image of a multi-station flexographic printer. Printed trace thickness can be adjusted by letting two or more stations print with identical printing plates.

Losses in conductor and substrate materials are very difficult to distinguish through measurements. Even though methods for this have been proposed [5], it is even more difficult when the investigated conductor is a radiating antenna [7]. The effect of limited conductivity is initially obtained by simulating the antenna in free space, i.e. with no substrate.

Fig. 3 shows the simulated input impedance locus in a Smith Chart for the antenna in Fig. 1 (b) printed with different sheet resistances and for frequencies 500-1500 MHz where 868 MHz is marked by black dots. All simulations assume an antenna without substrate. In the Smith Chart the graph representing the antenna made of a Perfect Electrical Conductor (PEC) has the widest loop, very closely followed by antennas with sheet resistances 50 and 100 mΩ/□. The antennas made with the highest sheet resistances, 1000 and 5000 mΩ/□, have narrower loops as the sheet resistance increases. A narrower loop, caused by higher ohmic losses, is an indication of the antenna receiving a lower Q-value. A lower Q-value also implies a wider bandwidth, which is observed as a shorter total distance in the Smith Chart traversed by the graph for higher sheet resistances. For higher sheet resistance the whole graph is also translated towards both higher resistance and lower reactive values, and would eventually occur in the rightmost part of the Smith Chart, i.e. the higher the sheet resistance of the antenna structure, the more similar it is to an open end transmission line.

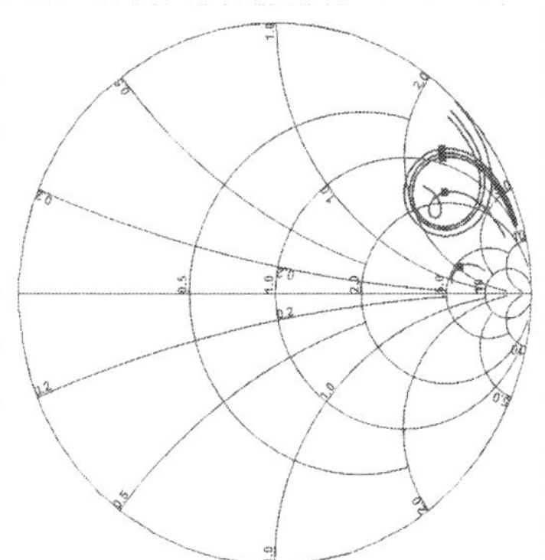

Fig. 3. Smith chart from simulations of antennas in free space and with sheet resistances 0 (PEC), 50, 100, 1000, and 5000 mΩ/□ respectively, with the higher sheet resistances corresponding to smaller loops. Frequency spans from 500 MHz at the upper end of graphs to 1.50 GHz at the lower end. 868 MHz is marked with black dots.

Since the input impedance is changed when ohmic losses are introduced, there will also be losses caused by an impedance mismatch. The efficiency of power transferred to, or received by, the antenna due to a non-perfect impedance match can according to [7] be expressed as

$$e_{mismatch} = 1 - |\Gamma|^2 = 1 - \left| \frac{Z_A - Z_C^*}{Z_A + Z_C} \right|^2 \tag{1}$$

where Γ is the voltaic reflection due to impedance mismatch, Z_A is the antenna input impedance and Z_C is the chip impedance. Maximum power transfer from the antenna to the chip occurs when Z_A and Z_C are the complex conjugates of each other, i.e. $Z_A=Z_C^*$. Table I shows the calculated radiation efficiency due to a mismatch using Equation (1) and the impedance values are presented in the same table. It is assumed that the antenna is originally designed to be fabricated as a PEC, placed in free space, and has a perfect impedance match for this PEC. The complex conjugate of the PEC antennas input impedance is therefore set as the reference chip impedance, Z_C. Mismatch due to some potential substrates is discussed in the next section.

Sheet Resistance (mΩ/□)	Input Impedance	Radiation Efficiency due to Mismatch	Radiation Efficiency due to Ohmic Loss	Total Conductor Radiation Efficiency $e_{Conductor}= e_{Mismatch} \cdot e_{Ohmic}$
PEC	$38 + j130\ \Omega$	100 %	100 %	100 %
50	$39 + j121\ \Omega$	99 %	85 %	85 %
100	$41 + j118\ \Omega$	99 %	76 %	75 %
1000	$60 + j96\ \Omega$	84 %	24 %	21 %
5000	$86 + j70\ \Omega$	43 %	4 %	2 %

Table I. Simulated Radiation Efficiency due to Limited Conductivity

From Table I it is obvious that as long as it is possible to print this antenna with sheet resistances below 100 m$\Omega/\square$, the radiation efficiency due to an impedance mismatch is almost negligible with $e_{Mismatch}$ above 99 %. It is observed that 100 m$\Omega/\square$ corresponds to a printed layer thickness of approximately 5 µm, and can be achieved by only one pass with the flexographic printer used while two passes are preferable. More important is the size of the ohmic losses which are introduced to the system due to the ink conductivity, and how this affects the total radiation efficiency. Ohmic loss is the part of the power that is absorbed in the antenna and converted into heat. The efficiency due to ohmic losses, e_{Ohmic}, is therefore defined as the quotient between radiated power, $P_{Radiated}$, and antenna input power, P_{in} [7]:

$$e_{Ohmic} = \frac{P_{Radiated}}{P_{in}} \tag{2}$$

The radiation efficiency due to ohmic losses is also retrieved from the same simulations as used to calculate the input impedance, and as presented in Table I. It is seen that ohmic losses are a much more severe contribution to loss in radiation efficiency, than that introduced by an impedance mismatch. For instance, a sheet resistance of 100 m$\Omega/\square$ gives a mismatch efficiency $e_{Mismatch}$=99 %, but the ohmic efficiency is only e_{Ohmic}=80 %. The total radiation efficiency of the printed antenna, $e_{Conductor}$, due to both impedance mismatch and ohmic losses is the product of the two loss quantities as shown in Equation (3)[7].

$$e_{conductor} = e_{Mismatch} \cdot e_{Ohmic} \tag{3}$$

The result of (3) is placed in the last column of Table I, where the print of for example 100 m$\Omega/\square$ gives a total radiation efficiency of $e_{Conductor}$ =75 %.

3.2 Radiation efficiency due to substrate loss

While the previous section characterized losses due to a printed antenna's limited conductivity, focus is now switched to potential losses introduced by common printer substrates and final objects subject to RFID tagging. Three different substrates are considered, a thin glossy paper, a plastic film, and a thick paper, all with the potential to be used as an antenna substrate in a commercial printing press. The complex permittivity of the substrate materials were measured using an Agilent 8507 Dielectric Probe Kit, together with an Agilent E5070B Vector Network Analyzer, and can be read in Table II.

Material	Thin Paper	Plastic Film	Thick Paper
Thickness (µm)	87	73	600
Relative Permittivity at 868 MHz	4.01 0 j0.29	1.89 + j0.059	2.2 + j0.14
Loss Tangent	0.07	0.031	0.064

Table II. Measured Substrate Permittivities

To only characterize the effect of substrate losses, and not the antenna conductor losses, the antenna is now regarded as a PEC in the simulations, with no ohmic losses due to antenna conductivity. Fig. 4 shows the simulated input impedance results, for the same antenna as before on the introduced substrates. The substrates behaved in a very similar manner, and the frequency 868 MHz is again marked by black dots. Referring to the dots' relative positions, the thick paper is the upper left, thin paper the middle and plastic film the lower right.

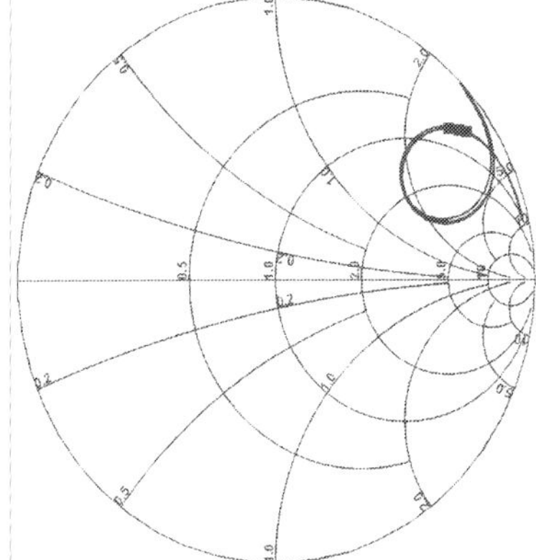

Fig. 4. Smith chart for antenna made of PEC on the substrates thin paper, thick paper and plastic film.

Substrate	Radiation Efficiency due to Mismatch	Radiation Efficiency due to Ohmic Loss	Total Radiation Efficiency $e_{Substrate} = e_{Mismatch} \cdot e_{Ohmic}$
Thin Paper	95 %	96 %	92 %
Plastic Film	99 %	99 %	98 %
Thick Paper	89 %	93 %	84 %

Table III. Radiation Efficiency due to Introduced Substrate

In a manner similar to the investigations involving finite conductivity in the previous section, Table III quantifies losses introduced by substrates in terms of the impedance mismatch and ohmic losses respectively. These losses may not be considered severe, with the worst case being the thick paper that shows a total radiation efficiency due to substrate of $e_{Substrate}$=84 %. According to Table III, the best substrate in this situation is the plastic film with an $e_{Substrate}$=98 %. As it is known from the previous section, lower ohmic losses produce wider loops in the Smith Chart but this can hardly be discerned in Fig. 4, since the substrate losses are relatively low. When comparing radiation efficiency in Table III to the Smith chart in Fig. 4, it should however be kept in mind that if a specific substrate is considered in the antenna design process, the efficiency due to a mismatch could approximate to 100 %. The ohmic losses are quite low, even though the substrates' loss tangents are significantly above zero. As seen in Table III that the plastic film, with the lowest loss tangent, also produces the lowest ohmic losses. The low ohmic losses are mostly due to the low substrate thickness relative to the wavelength, λ. For the thick paper for example the substrate thickness is about 0.0018 λ. If the substrate were to be thicker, but with the same relative permittivity, both the mismatch and ohmic losses would increase as will be shown in section 5.

3.3 Total radiation efficiency

Total radiation efficiency depends on ohmic and mismatch losses introduced by both the antenna conductor and substrate:

$$e_{Total} = e_{Conductor} \cdot e_{Substrate} \tag{4}$$

Table IV show the simulated total radiation efficiency for the introduced substrates, when printed with sheet resistances 50, 100, 1000, and 5000 mΩ/$\square$, respectively. As might be expected from the results in Table I and III, the highest radiation efficiency, 83%, is achieved for 50 mΩ/$\square$ on plastic film.

Substrate	50 mΩ/$\square$	100 mΩ/$\square$	1000 mΩ/$\square$	5000 mΩ/$\square$
Thin Paper	78 %	69 %	19 %	1.4 %
Plastic Film	83 %	74 %	20 %	1.5 %
Thick Paper	71 %	63 %	17 %	1.3 %

Table IV. Simulated Total Radiation Efficiency for different Conductivities and Substrate Materials

It has been shown that impedance mismatch is a minor contributor to the total loss for thin substrates. However, this can still be avoided by taking care in the antenna design process. If the antenna's length is changed just a couple of millimeters the impedance mismatch becomes almost negligible, while the efficiency due to ohmic losses remains almost identical, which is further discussed in section 5. These results are valid for this particular dipole and although the general trends may hold true for other designs, the relationship between conductivity and radiation efficiency is also highly dependent on the antenna design. Factors such as size, print area, slot, patch, loop, etc. will also affect the radiation efficiency.

3.4 Maximum RFID read distance

In passive RFID systems, the maximum distance at which a tag can be read is always a crucial factor. The tag's only power supply is from the interrogating radio wave, and the amplitude of this wave is strictly regulated by governmental authorities. It is well-known that the power received at a given distance from a transmitting unit is inversely proportional to the square of the transmitting distance. The amount of received power is calculated using the Friis transmission formula [4]

$$P_r = \frac{(P_t \cdot G_t) \cdot G_r \cdot \lambda^2}{(4\pi R)^2} \tag{5}$$

where R is the distance between the tag and interrogator and $P_t \cdot G_t$ the Effective Isotropic Radiated Power (EIRP) transmitted by the interrogator unit. In the US, the transmitted power allowance is 4 W EIRP, and in EU 2W ERP (= 3.28 W EIRP). G_r is the gain of the receiving tag antenna, and λ the wavelength.

As follows from Eq. (5), when the distance between the interrogator unit and the tag is doubled, power received by the tag falls by a factor of four. It also follows from the equation that in order to obtain the same power with two antennas with different total radiation efficiencies, it is required to move the less efficient antenna closer to the interrogator by a factor of the square root of the ratio of total radiation efficiency. If, for example, a printed antenna has a total radiation efficiency of 64 % compared to the reference one made of copper (100 % radiation efficiency), the operating RFID range for a tag using the printed antenna would be approximately $\sqrt{0.64} \cdot 100$ % = 80 % of that for the copper one. In this way, measured sheet resistances of printed antennas can be used to characterize antenna radiation efficiency in computer simulations that in turn can be converted to predict theoretical RFID read ranges due to the square root relationship.

The presented theories were tested in reality by printing the antenna in Fig. 1(b) with a laboratory flexographic printer that was set to print multiple passes and with differently diluted ink. The measured conductivity values for the samples printed on HP photo film were between 67 and 680 mΩ/$\square$ and RFID chips were attached and connected to the antennas with aid of manually added silver ink drops.

The maximum RFID read range was measured at an approximated (in-door) open air read range setup as illustrated in Fig. 5, where r can be varied from approximately 0 m up to 8 meters before reflections from the rightmost balcony significantly influence the measurements. The RFID system used in this experiment followed the European standard with a maximum output power of 2 W ERP. All experiments are however performed to show the relative values and normalized by the maximum reliable reading distance for a PEC dipole that was found to be about 6.4 m. The results are presented in Fig. 6 where the experimental values are slightly under the theoretical ones but have the same characteristics.

From Fig. 6 it appears that in order to achieve a decent functionality for printed RFID antennas, with the ink and printing techniques used, a sheet resistance of 100 mΩ/$\square$ or lower is desirable. The ink thicknesses corresponding to the best antenna was measured with the aid of a Mahr Millitast 1083 Digital Indicator Gage [6] to be about 10 µm. The uncertainty in ink thickness is largely due to a deviation in paper thickness of the order of several micrometers.

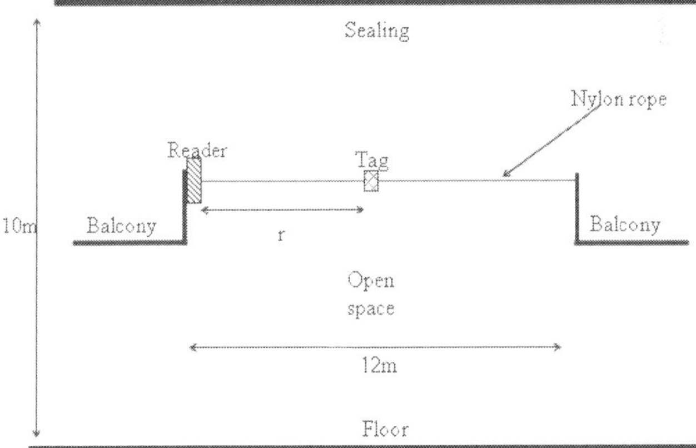

Fig. 5. Setup for determining maximum RFID read distance.

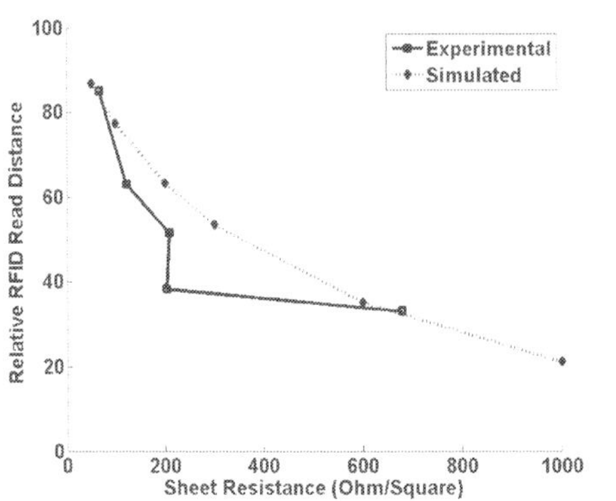

Fig. 6. Simulated and experimental values for relative RFID read range versus the sheet resistance of the printed antenna.

4. Flexible substrates and physical bending of antennas

Single layer RFID antennas on flexible substrates such as thin paper or plastics may be exposed to physical bending, especially if an RFID label is placed on a non-flat surface or over the corner of an object. This section uses numerical simulation [3] to study the impact of bending on some typical examples of RFID antennas.

The performance degradation to an RFID system caused by bending the tag antenna is investigated for the two antenna structures in Fig. 7 where the antenna in Fig. 7 (a) is relatively narrowband with a VSWR=2.0 bandwidth of 4% and the antenna in Fig. 7 (b) is a bit more wideband with bandwidth 11%. The narrowband antenna in Fig. 7 (a) is a folded dipole that is adapted to fit the impedance of passive of RFID chips by cutting a slot in the upper conductor.

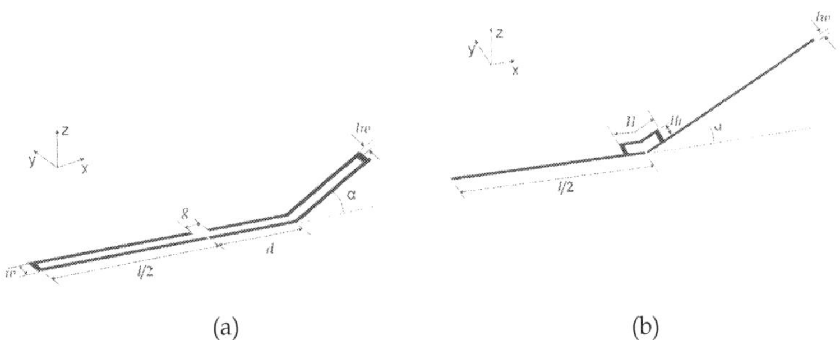

(a)   (b)

Fig. 7. An antenna will change its input impedance when distorted by bending, which is numerically simulated when bent towards the z-axis for (a) a narrowband printed folded dipole with a cut in the upper conductor for RFID impedance matching and (b) a more wideband RFID dipole.

The dipoles are fed in the middle and one of the dipole arms is bent at different angles α and at different distances d from the feed as shown in Fig. 7. The simulated folded dipole Fig. 7 (a) has a total length of l=120 mm, width w=8 mm, gap g=6 mm and line width lw=2 mm and the dipole in Fig. 7 (b) has a total length of l=185 mm, loop length ll=20, loop height lh=10 mm and line width lw=1 mm. The input return losses of the bent folded dipole for different values of α and d are shown in Fig. 8 where it can be seen how the input return loss increases as the antenna is bent at a point closer to the feed. The maximum input return loss occurs when α=90° and d=0 and one can see that the wide band dipole handles this slightly better than the narrowband dipole.

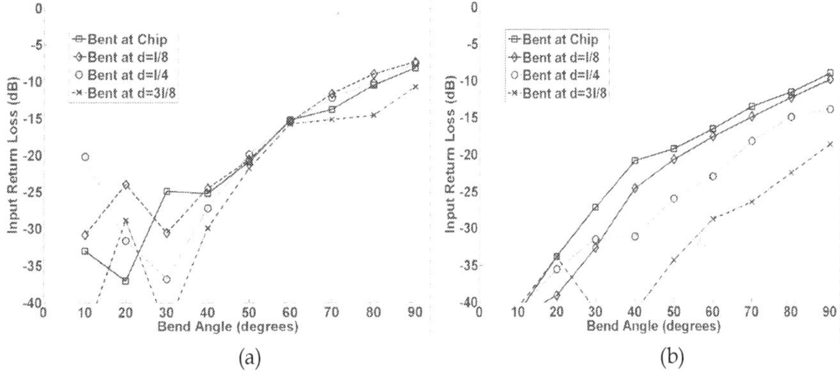

Fig. 8. Input return loss of (a) narrowband dipole and (b) wideband dipole when bent at different angles α at different positions d.

The bending of the antenna structure will no doubt result in performance degradation to an RFID system. The amount of such degradation can be expressed in terms of relative read range according to the following formula

$$RR = \sqrt{\left(1 - |\Gamma|^2\right)\frac{D'}{D}} \tag{6}$$

where RR is the relative read range as compared to a perfectly matched RFID system with the reader antenna illuminating the tag antenna in its z-direction. Γ is the reflection coefficient due to impedance mismatch which is why $1 - |\Gamma|^2$ becomes a factor that describes the relative power entering the chip. D is the tag antenna's directivity in the z-direction when flat, and D' the directivity in the z-direction when the antenna is bent. Friis' transmission formula in equation (5) tells us that the radiated power decreases in proportion to the distance squared which is why the relative read range in equation (6) becomes the square root of the mismatch and directivity factors.

The performance degradation due to antenna distortion is provided in Fig. 9 where it is observed that the operating range will be reduced to approximately 60% for both dipoles in the worst case scenario when the dipoles are bent 90° in the middle of the structure. The wideband dipole does not outperform the narrowband here as it did in Fig. 8, the reason for this is that the directivity in the antennas' z-direction changed more for the wideband antenna than for the narrowband one.

It is also observed that when bent at d=3l/8 the reduction of the operating range is almost negligible. This is an interesting observation which potentially allows for improvements to the tag construction. If the RF tag substrate is to be made less flexible in the vicinity of the dipole antenna feed point, but still allowing greater flexibility at its exterior, it could significantly reduce the tag performance degradation caused by placing it on non-flat or flexible surfaces.

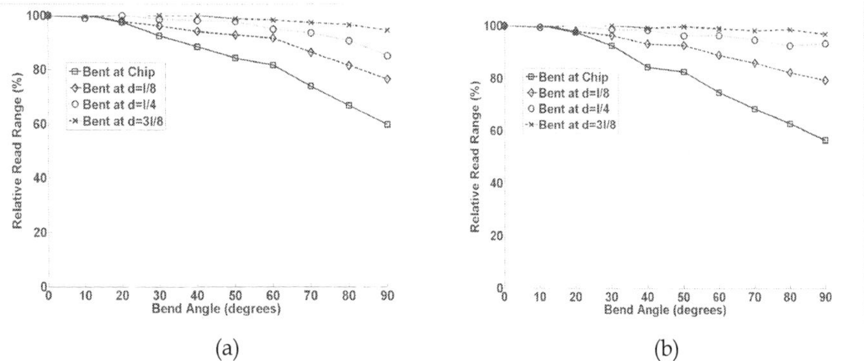

(a)   (b)

Fig. 9. Performance degradation of RFID systems for an (a) narrowband and (b) wideband dipole RFID tag antenna bent at different angles α at different distances d from the structure centre.

5. Wideband and narrowband RFID antennas behavior on background materials

As previously investigated in section 3.2, single layer RFID antennas normally have a relatively thin substrate to carry the antenna structure which does not significantly affect neither the antenna's input return loss nor its radiation efficiency. The thin carrying substrate will however place the antenna in very close proximity to other materials when applied to an object. Depending on the size of the object and the electrical properties of its material it can cause significant changes to the antenna parameters which in turn can reduce the maximum read range of the RFID tag.

Prior knowledge of the underlying material's electrical properties, losses due to impedance mismatch can to a certain degree, be compensated for in the original antenna design process by properly scaling the antenna's dimensions. Ohmic losses introduced by nearby materials can unfortunately not be compensated for by only scaling the antenna geometry but will still be present.

Pure conductive, i.e. metallic, surfaces and thick materials with very high dielectric losses, such as containers filled with water, can also be compensated for but this requires a multi-layered antenna structure which is exemplified in section 6 or the use of a spacer of a particular thickness to distance the single layer antenna from the surface.

This section shows how resonant frequency and input return loss changes when the previously used narrowband and wideband antennas in Fig. 7 were placed on a pile of paper, well representing a book, instead of the single papers investigated in section 3.2.

5.1 Changes in input return loss and radiation efficiency when put on 40 mm paper material

The paper used as substrate material has the same relative permittivity of ε_r=4.0 + j0.29 as previously but is now 40 mm thick. The simulated input return loss shown in Fig. 10 was again calculated using standard formulas for reflections due to impedance mismatch taking the input impedance of the antennas with no nearby material as the reference value.

Changes from the free space situation are obvious and it can be observed how the narrowband antenna has an input return loss of almost 0 dB when placed on the pile of paper while the wideband antenna, also out of reach, at least manages to have approximately -2 dB in the same situation. The radiation efficiency when put on the pile of paper was calculated to 26% and 65% for respectively the narrowband and wideband antenna and it was also noted that the directivity in the z-direction decreased to about 0.2 in both cases. All in all this does cause the narrowband dipole to totally malfunction but also significantly decreases the performance of the wideband dipole.

The narrowband dipole could indeed be specifically tuned to electrically fit this underlying material, but is then again only of use for that material and no other. The wideband dipole could also be geometrically adapted for optimal performance but the necessity for this is less obvious.

A dipole that inherits wideband properties is of course also beneficial for communication if a high bit-rate is desired or if a wideband-demanding modulation- and coding scheme such as a frequency hopping spread spectrum (FHSS) is to be used.

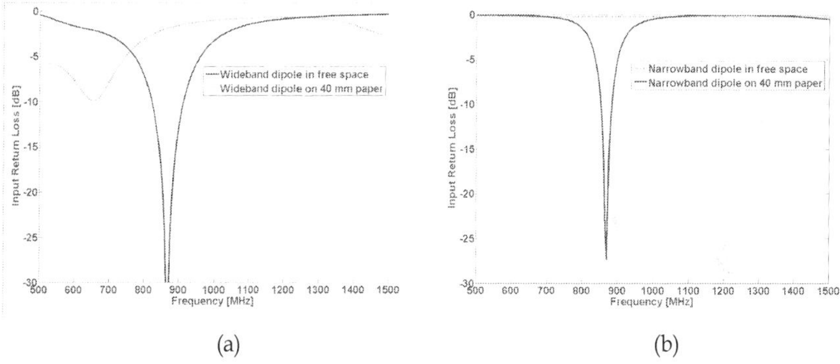

(a)   (b)

Fig. 10. (a) Input return loss for the antennas in Fig. 7 (a) and (b) when operating surrounded by free space and when put on top of 40 mm paper. While the wideband dipole does increase its input return loss significantly, the narrow dipole is more or less completely malfunctioning.

5.2 Adopting the narrowband dipole for underlying material

Resonance, very close to the original impedance point, is achieved for the dipole antenna upon the pile of paper by reducing its length from 120 mm to l=70 mm and increasing its linewidth from 2mm to lw=5 mm, and its total width w from 8 mm to w=17 mm. As mentioned, the ohmic losses cannot be compensated for by simple changes in the antenna geometry but actually increases to 42%. The antenna tuned for this specific material naturally becomes detuned again if used in free space.

Tuned narrowband antennas such as the ordinary dipole can thus be used when pre-knowledge of intended nearby materials' electrical properties is available. If a final underlying material on the other hand is unknown at the time of antenna design, narrowband antennas are consequently not a good choice and should be avoided.

6. Microstrip antennas

The microstrip antenna consists of a patch of metallization on a ground plane. These are low profile and generally have a height of only a few mm. The ground plane can consist of a patch whose size is just slightly larger than the radiating patch. Microstrip antennas have the advantage that they can be made more or less transparent to the underlying material which partly solves the problem associated with the RFID tags in an unknown environment. They are however also relatively sensitive to ohmic losses in the substrate used between the patch and the ground plane which is the reason why expensive low loss microwave substrates are often used in applications requiring microstrip antennas.

6.1 Low cost microstrip antennas for passive tags on metallic objects

The performance of a conventional tag antenna is strongly degraded by metallic objects as this affects both the impedance and radiation pattern. When a general dipole-type antenna is mounted near a metallic object, a current will be induced on both the antenna surface and the metallic surface due to the RFID reader's radiation. The metallic surface acts as an image to the original antenna, which has a negative influence on the tag antenna's scattered field and makes the tag unreadable within normal ranges [10].

Antennas that can be directly used with RFID tag for metallic objects include the planar inverted-F antenna (PIFA), U slot inverted-F tag antenna [11], [12], antennas based on slots in the metallic object itself [13] and Microstrip Antennas (MSA)[14].

As the electronic identification technology competes with rock-bottom pricing barcodes, the cost involved is the most obvious issue which is holding back the widespread adoption of RFID. Fabricating efficient MSAs from low-loss microwave substrates is therefore not possible for general RFID applications. Some low cost materials, such as foam, have been used as MSA substrate for RFID [15] but no cheaper alternative solution has been proposed for the conducting parts. Unlike a conventional dipole-type antenna, the MSA is a structure with two conducting layers acting as a resonance cavity and which is thus much more sensitive to ohmic losses in the substrate and conductors than for the simple one-layer antennas investigated previously.

This section therefore investigates the performance of silver ink printed MSAs for passive RFID operating at 868 MHz. Two compact patch structures were designed where the conducting parts of the patch antenna are flexographically printed with silver ink and common cardboard is used as the substrate. One antenna consists of a solid patch and one of very wide slots in the patch in order to minimize ink usage. The aim was to design antennas that could work as well in free space as when placed on a metallic plate.

6.2 Printed MSA design

The size of the basic MSA is quite large as it should be approximately half the wavelength divided by the square root of the substrate relative permittivity in two dimensions and have a ground plane larger than the patch. MSAs for RFID must therefore be modified. It has been demonstrated that by using a shorting plate or by cutting slots, compact MSAs can be achieved [16]. In Fig. 11.(a), a shorting plate is added to the middle of one radiating edge (the top edge). The antenna's resonance frequency is controlled by tuning the width Dp of the shorting plate, where the resonance frequency decreases with the decreasing Dp. In order to facilitate the chip mounting, a segment of microstrip line is connected to the radiating patch and a slot is cut along the feed line for the placement of an RFID chip. One end of the feed line is directly connected to the ground plane by means of a small piece of copper tape and the shorting plate is achieved by wrapping a segment of copper tape through a slot which is cut along the radiating edge, through the substrate and to the ground plane. In an industrial process line, it should be possible to substitute the small pieces of copper tape, acting as a wide via, by staples.

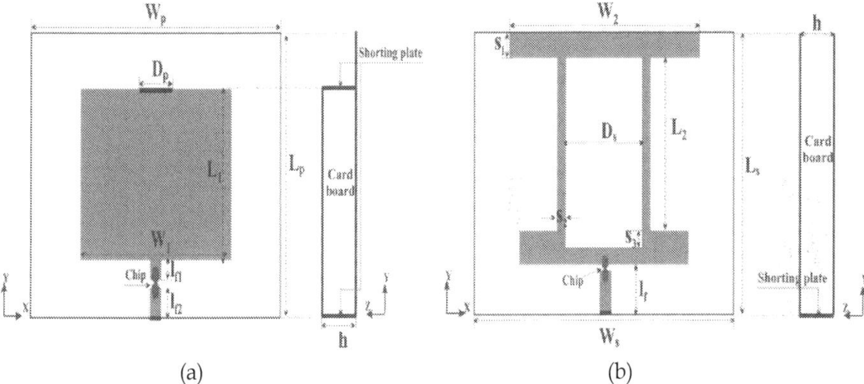

(a) (b)

Fig. 11. Geometrical Structures of (a) Shorted and (b) Slot Microstrip Patch Antenna

The shorted MSA is a compact antenna but suffers from poor gain and degradation in the radiation pattern. An alternative method of reducing the resonance frequency of the MSA is to increase the patch length of the surface current by cutting slots in the radiating patch. Generally, size reduction is achieved by adopting a C-shaped or H-shaped slot or a ring MSA [16]. In order to make the antenna more compact, both the H-shaped slot and ring are adopted in this case, resulting in the geometry shown in Fig. 11 (b). Slots are respectively cut in the middle and along the two non-radiating edges of the patch. As with the shorted MSA in Fig. 11 (a), a segment of microstrip line is used to feed the slot patch and the RFID chip is placed across the gap between the feed line and patch. The connection between the feed line and the ground plane is also achieved by the same method as that for the shorted MSA. In both cases, corrugated cardboard with a thickness of approximately 2.8mm is utilized as the substrate. The electrical properties of the cardboard material, i.e. complex permittivity parameters, could unfortunately not be measured due to lack of equipment but it would have been interesting.

The ground plane, radiating patch and feed line were flexographically printed [17] with three consecutive passes onto a plastic film with a thickness of 80µm using the silver based ink CFW-102X [18]. After being cured at 120°C for 20 minutes, an average DC sheet resistance of 70mΩ/□ was obtained. The plastic films holding the printed traces were glued to the respective sides of the cardboards. The antennas were optimized by repeated impedance and read range measurements for different dimensions with the final geometrical parameters: Wp=100 mm, Lp=Ls=75 mm, Dp=12 mm, W1=60 mm, L1=45 mm, lf1=5 mm, lf2=9 mm, Ws=110 mm, W2=80 mm, L2=38 mm, Ds=27 mm, lf=10 mm, s1=12 mm, s2=2.5 mm, s3=9 mm.

In order to evaluate the influence of the limited conductivity of the silver ink printed conductors on the antennas performance, two copper antennas were constructed with the same geometry by replacing the printed plastic film with 70 μm thick copper film. The input impedance of the printed and copper patch antennas were measured with the vector network analyzer E5070B and the results are displayed in Fig. 12. Fig. 12 (a) shows the impedance traces of the shorted patch antennas and Fig. 12 (b) shows those involving the slots. The solid lines represent the impedance traces of copper antennas and the dashed lines are those of the printed ones. The measured frequency range is from 500 to 1500 MHz and the square shaped marks represent the impedance values at 869 MHz. In the Smith Chart it can be seen that the radius of the impedance trace circles of the printed antennas in both cases are smaller than those of the copper ones, which represents more conduction loss and is similar to that observed throughout Chapter 3.

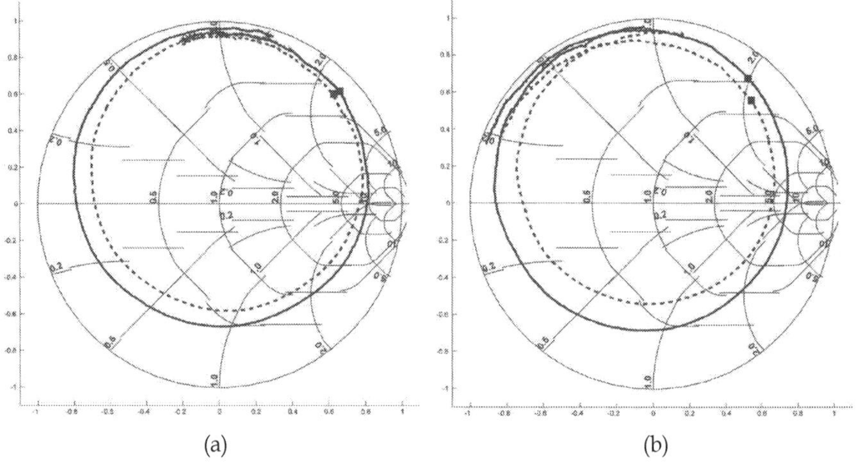

(a)   (b)

Fig. 12. Measured Input Impedance of (a) Shorted and (b) Slot Patch Antennas.

It was also observed that the radius difference is more obvious for the slot patch antenna than for the shorted one which might be because of the higher current density on traces with limited conductivity.

Alien chips with straps were placed across the slots along the respective feeds. The read range in the front direction of these four tags when there are no metallic objects nearby and when they are mounted at the centre of a metallic plate of size 305×165mm is obtained by placing the tag in the approximated open air read range in Fig. 5 with the SAMSys reader operating at 2 W ERP. The results are shown in Table XIX and observed that, without any objects nearby, the read range of the slot patch antennas are better than those of the shorted ones with the metallic plate behind the slot patch. Both antennas however perform better with the metallic plate behind, which is mainly due to no back radiation. The range enhancement is also rather better for the shorted patch antenna than for the slot one. In the Smith Chart, it can be seen that the impedance values of the slot patch antennas at 869 MHz are more close to the conjugate impedance value of the RFID chip than those of the shorted ones implying a higher input return loss for the slot patch antennas.

The read range of the copper antennas is only slightly better than the range for the printed ones and this is probably due to both lower ohmic losses and a better impedance match with the copper antennas. Since the geometrical parameters for the antenna are determined for the printed patch antennas, the input impedance will offset those of the originals when applying the same geometrical parameters to the more conductive copper antennas. This then degrades the impedance match between the antenna and the RFID chip.

Antenna type	Without metallic objects nearby	When mounted on a metallic plate with size of 305×165mm
Shorted microstrip patch antenna (copper)	2.2 m	3.3 m
Shorted microstrip patch antenna (printed)	2.0 m	2.8 m
Slot microstrip patch antenna (copper)	2.5 m	2.8 m
Slot microstrip patch antenna (printed)	2.3 m	2.4 m

Table XIX. Measured Read Range in the Front Direction of the Patch Antenna

7. Aesthetic appearance of RFID antennas

Antennas are most often designed with little or no attention to their aesthetic appearance. Typically, antennas are optimized for functionality in a specific application, or to fit within a specific footprint. RFID antennas can however be visible on packaging and products, and a graphic design could serve promotional purposes. This section shows how the adaptation of logo designs can serve as operating RFID antennas and thus add a promotional dimension to the RFID antenna design. A package, or other object, generally has a company logo or similar trademark printed on it and with the proposed concept that logo can also serve as a silver-printed RFID antenna.

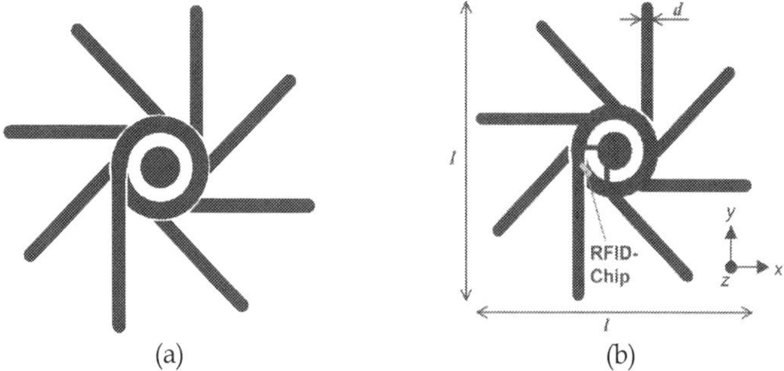

Fig. 13. (a) The Precisia LLC company logo and (b) the same logo changed into an RFID tag antenna.

7.1 Converting logotypes to RFID tag antennas

To be effective, a resonant antenna must be comparable in size to the working wavelength (usually about at least one sixth of a wavelength). For UHF RFID systems, this means about 5-15 cm in size in at least one dimension, implying that they will be very visible even when placed on relatively large objects such as containers and pallets. Typical modern RFID antennas are variations of well known structures such as dipoles, loops, slots, and meander lines. Very few RFID antenna structures are designed as an artwork even though some examples exist and include a patch antenna with the slot in the shape of an adapted University logo helping to achieve higher working bandwidth [20], and a modified meander type dipole antenna, using a text pattern [21].

As shown previously common antennas such as ordinary dipoles, folded dipoles and the like are unable to be scaled to reach impedances that fit common passive UHF RFID chips unless altered by adding or removing parts. The proposed consept include converting a logotype to an antenna where one first decides where on the geometry to place the chip and then add and remove conductive parts trying to not severely distort the original artwork. If the pre-bonded chips referred to as straps are used, the feed gap could be up to 4-5 mm. A greater challenge in converting logos to antennas is if one not only considers the input impedance but is also designing for specific directivity patterns.

As flexographic printing is the most common method for printing on cardboard material all antennas chosen for this investigation were printed using a laboratory flexographic printer by RK Print [17].

7.2 The precisia company logo as RFID antenna

Fig. 13shows the original logo by Precisia LLC [18] and its version modified to become an RFID antenna template. The original logo design contains the first letter of the company name "P" with the filled circle in the middle surrounded by filled lines forming a "star"-like pattern. The loop in the letter "P" is not completely closed and the surrounding rotated lines are not connected to the "P" (there are small gaps), except for the vertical line in the "P" itself.

The process of converting the logo into a functional RFID antenna structure was initiated by choosing a placement position for the RFID chip. With the present graphical design it was decided to make the final structure as similar to the dipole as possible. In such a case, the chip should be positioned close to the geometric centre of the structure. In this particular case, the gap in the loop of the letter "P" was chosen (see Fig. 13(b)). The central filled circle was connected to the loop by two lines, forming a shorted loop connected in parallel to the chip pads. The intention of keeping such a loop close to the chip becomes clear when looking at the altered dipole antennas used in the previous sections, as it can serve as the impedance matching quasi-lumped inductor. After trial and error with Ansoft HFSS [3] some of the remaining lines in the "star" pattern were connected to the loop. Finally, the whole antenna was scaled in the structure simulator to provide operating frequency around 867 MHz with resulting dimensions $l= 131$ mm and $d= 5$ mm.

Measured input impedances of antennas made from both copper foil and flexographic print are shown in Fig. 14.

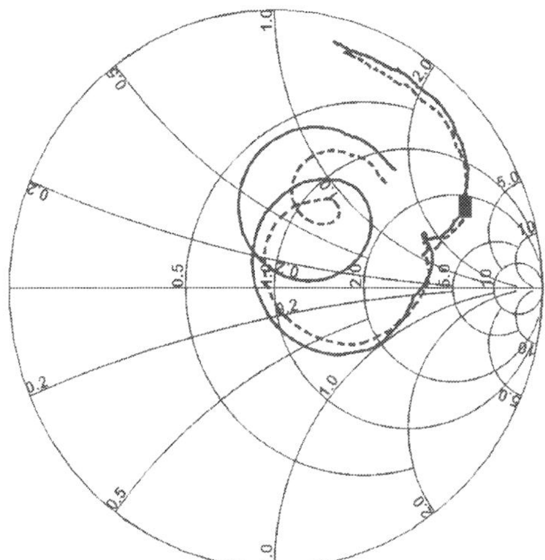

Fig. 14. Measured input impedance for the Precisia logo antenna, made from copper foil (solid line) and printed with conductive ink (dashed line). The graphs sweep from 500 to 1500 MHz and the markers on the impedance curves (the filled square) are set to 867 MHz.

The antennas were equipped with Alien Monza Gen2 chips and were measured to have maximum reliable read range of 5.16 m for the copper antenna and 4.03 m for the printed one, which can be compared 6.0 m maximum reading distance for a copper antenna very similar to the inductive dipole presented earlier.
The radiation pattern was extracted by measuring maximum read range for different angles and is shown in Fig. 15.

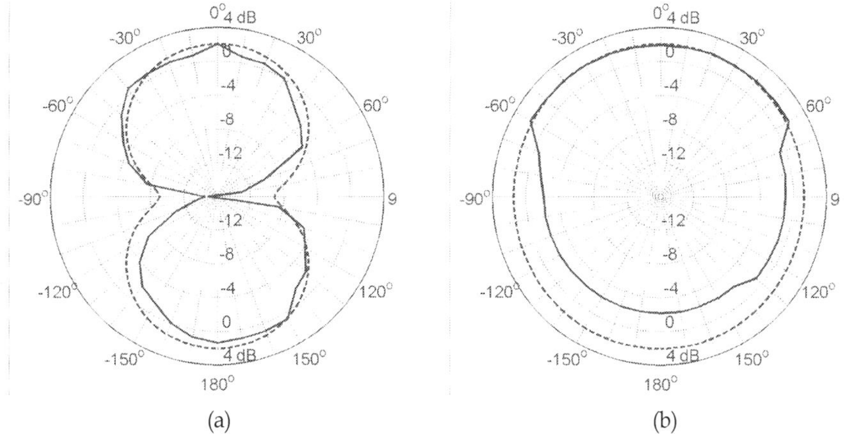

Fig. 15. Measured (solid line) and simulated (dashed line) radiation patterns for the Precisia logo antenna in (a) the xz-plane and (b) the yz-plane. Maximum directivity is about 2.0 dB, directed along the z-axis

In order to demonstrate that almost any logo could be arranged as an RFID antenna some additional logos from the Michigan region were evaluated. The reading distance of these experiments are presented in Table XIII.

Antenna pattern	Antenna printed on plastic film with silver based ink	Antenna made from copper foil
Precisia Logo	4.03 m	5.16 m
Detroit Pistons	4.81 m	6.05 m
Michigan M	5.54 m	5.79 m

Table XIII. Maximum RFID Read Range for Printed and Solid Copper Antennas

8. RFID antennas as RFID sensors

Up until today, very few RFID chips exist that include an input port also for sensor data, i.e. an extra digital or analogue input port. The analogue version could for example measure the resistance over a sensor port, allowing for simple passive sensor elements that change resistance proportional to the physical quantity of interest. One application where this would be valuable is in the remote measuring of moisture or the wetness level at the location of the tag. Moisture sensor tags with costs similar to those of ordinary ID tags could for instance be placed inside walls or floors in buildings. The humidity level inside the wall could be read by holding a handheld RFID reader at, say, one meter from the wall provided that the positions of the tags are discretely marked or mapped. By periodically reading the tags it is possible to prevent costly damage due to mould or putrefaction. The tag could be advantageously positioned directly underneath hidden water pipe connections for early leakage detection.

This section presents a concept where pairs of ordinary RFID tags are exploited for use as remote reading moisture sensors by embedding one of the two tags in a moisture absorbing material.

8.1 Open and embedded tags

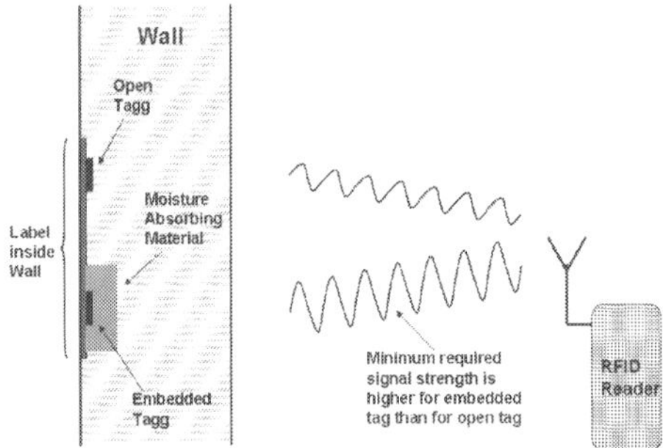

Fig. 16. Moisture sensing label incorporating two RFID tags where one of the tags is covered with a paper based moisture absorbing material. In a humid environment the embedded tag's antenna gets less efficient and needs a stronger RF signal to operate. The difference in RF power to operate is proportional to the humidity level.

Existing technologies for remote reading of humidity levels in hidden locations such as inside walls are based on different microwave technologies [22]-[23] For some locations and especially thick multi-layer walls it can however be difficult to accurately read a moisture value. Proposals for in-situ water content sensors have also been made utilizing SAW-based transponders [24].

Previous sections have shown that the performance of low cost tags, constructed with simple one-layer antennas, is very sensitive to the surrounding environment and especially to nearby metallic surfaces. Water is no exception since its relative permittivity is of the order of 80+j×80 and will directly cause ohmic losses in an antennas near-field and also change its resonance frequency. It has previously been characterized as to how this property can be used to measure wetness in soil and snow by connecting a transmission line to a buried monopole antenna [25].

The concept suggested here is based on the use of two tags on one label where one of the tags is embedded in a moisture absorbent material and the other is left open. In a humid environment the moisture concentration is higher in the absorbent material than in the surrounding environment which causes degradation to the embedded tag's antenna in terms of dielectric losses and change of input impedance as water increases both the real and imaginary parts of the paper's dielectric constant [26].

If the tags are passive, an RFID reader held at the same distance from both tags in the label must therefore emit a stronger interrogating signal in order to power up the embedded tag than the naked tag. By comparing the minimum power levels required to power up each tag it is therefore possible to determine the humidity level at the tag's location.

Moisture measurement by embedding an RFID tag could theoretically also be performed using only one tag but that would require the distance and materials between the reader and the embedded tag to always be exactly the same.

8.1.1 Characterizing pair of tags

Two passive 868 MHz Gen-2 tags from Alien using the Alien Squiggle antenna were used in experiments performed in a climate chamber. In three different setups, 10 stacked sheets of blotting papers made out of bleached kraft pulp with a basis weight of 260 g/m^2 were placed respectively in front, behind and both in front and behind one of the tags. A SAMSys RFID reader allowing control of the antenna output power was used with its antenna positioned symmetrically about 1.5 m from the pair of tags.

Results are presented in Fig. 17 (a), which can later be used as a look-up table for humidity levels when measuring labels placed at hidden locations.

The tags were also characterized for direct wetness with the result shown in Fig. 17 (b).

8.1.2 Remarks for the double tag RFID sensor approach

It is observed that the more embedded a tag is the more sensitive it is to the surrounding humidity. The larger total volume of absorbing material in the embedded tag introduces a larger total amount of water near the tag antenna. At 90% RH the half-open and totally embedded tag shows approximately the same difference when compared to the totally open reference tags. The differences in power levels for different levels of relative humidity proved to be of the order of a couple of decibels for the 80 % RH where humidity may become the source of mould. This places significant constraints on the tolerances of the readout electronics and, additionally, that the individual tag antennas within the label are not externally distorted by small variations in their respective vicinities.

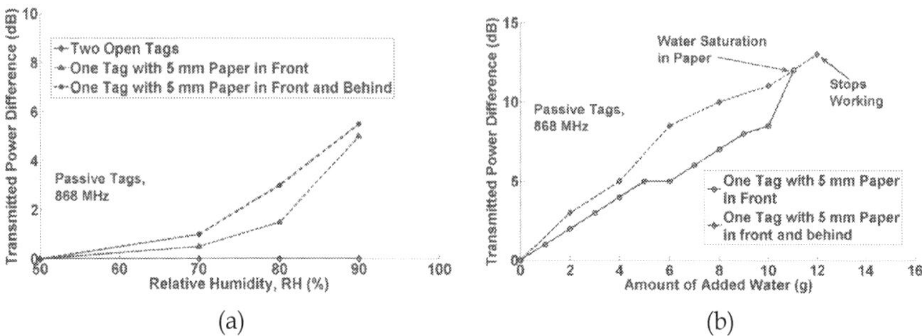

Fig. 17. Difference in minimum output power required from the RFID reader in order to read two passive tags in a label where one is covered or embedded in a moisture absorbing material.

As is shown in Fig. 17 (b), the tag embedded with 5 mm paper on each side also was not readable at all at 1.61 m when it received more than 12 grams of water. At the time of malfunction the difference in transmitted power was about 13 dB.

It is also observed that the difference in minimum transmitted power is almost linear to the amount of water added. Even though the totally embedded tag shows a greater difference than the tag covered only on one side, the difference for small amounts of water is not significant.

A more advanced label could include three or more tags where perhaps one tag could have its antenna completely dissolved if moisture exceeds a certain value. This could for example be accomplished by having a water based electrically conductive ink (i.e. silver ink). In that way one would create a memory effect that states that the relative humidity has been above a certain level and could also provide the current value.

9. References

[1] K. Finkenzeller, "RFID Handbook", ISBN 0-470-84402-7, John Wiley & Sons 2003

[2] Agilent Technologies, Inc., "HSMS-282x Series - Surface Mount RF Schottky Barrier Diodes" Data Sheet, March 2005, http://pdf1.alldatasheet.com/datasheet-pdf/view/95384/HP/HSMS282X.html, accessed 2009-09-11

[3] Ansoft High Frequency Structure Simulator (HFSS) 9.2, Ansoft Corp. www.ansoft.com, accessed 2009-09-27

[4] D. K. Cheng, "Field and Wave Electromagnetics", 2nd edition, ISBN 0-201-12819-5, Addison Wesley, UK, 1989

[5] H. Tanaka, F. Okada, "Precise Measurements of Dissipation Factor in Microwave Printed Circuit Boards", *IEEE Trans .Instrumentation and Measurement*, vol. 38, no. 2, pp. 509-514, April 1989

[6] Carl Mahr Holding GmbH, www.mahr.com, accessed 090927

[7] C.A. Balanis, "Antenna Theory: Analysis and Design, 2nd ed.", pp. 59, 61, 873, ISBN 0-471-59268-4, John Wiley & Sons 1997

[8] J. I. Kim, B.M. Lee and Y. J. Yoon, "Wideband Printed Dipole Antenna for Multiple Wireless Services", *Applied Microwave & Wireless Magazine*, pp. 70-77, September 2002

[9] Q. Xianming; Y.Ning, "A folded dipole antenna for RFID", *IEEE Int. Symp. Ant. and Prop.,* 2004, Vol. 1, pp. 97 – 100, 2004

[10] P. V. Nikitin, et al., "Power reflection coefficient analysis for complex impedances in RFID tag design," *IEEE Trans. Microw. Theory Tech.*, vol. 53, pp. 2721-2725, 2005

[11] L. Ukkonen, D. Engels, L. Sydänheimo, M. Kivikoski, "Planar wire-type inverted-F RFID tag antenna mountable on metallic objects", *IEEE Int. Symp- Antennas and Propagation 2004*, Vol. 1, pp. 101 – 104, 2004

[12] H. Kwon, B. Lee, "Compact slotted planar inverted-F RFID tag mountable on metallic objects," *Electron. Lett.*, Vol. 41, No. 24, pp. 1308-1310, 2005

[13] L. Ukkonen, M. Schaffrath, L. Sydanheimo and M. Kivikoski, "Analysis of integrated slot-type tag antennas for passive UHF RFID," *IEEE Ant. and Prop. Society Int. Symp.*, pp. 1343-1346, 2006

[14] L. Ukkonen, L. Sydanheimo, M. Kivikoski, "Effects of metallic plate size on the performance of microstrip patch-type tag antennas for passive RFID", *Antennas and Wireless Propagation Letters*, Vol. 4, pp.410 – 413, 2005

[15] H.-W. Son, G.-Y. Choi and C.-S. Pyo, "Design of wideband RFID tag antenna for metallic surfaces," *Electron. Lett.*, Vol. 42, No. 5, pp. 263-265, 2006.

[16] G. Kumar, K. P. Ray, Broadband microstrip antennas, *Artech House*, USA, pp. 205-227, 2003

[17] FlexiProof 100, RK Print Coat Instruments Ltd, *www.rkprint.com*

[18] Precisia, LLC, formerly subsidiary of the Flint Group, www.flintgrp.com, accessed 091013

[19] W. N. Caron "Antenna Impedance Matching", Amer Radio Relay League, June 1989

[20] [Y.L. Chow and C.W. Fung, "The city university logo patch antenna", *Proc. Asia-Pacific Microwave Conference, 1997. APMC '97*, Vol. 1, pp. 229 – 232, Dec. 1997

[21] [M. Keskilammi, M. Kivikoski, "Using text as a meander line for RFID Transponder Antennas", *Antennas and Wireless Propagation Letters*, Vol. 3, Issue 16, pp. 372 – 374, Dec. 2004

[22] F. Menke, R. Knochel, T. Boltze, C. Hauenschild, and W. Leschnik, "Moisture measurement in walls using microwaves", *IEEE Int. Microwave Symp.* 1995, Vol. 3, pp. 1147 – 1150, 1995

[23] T. Hauschild, and F. Menke, "Moisture measurement in masonry walls using a non-invasive reflectometer", *Electronics Letters*, Vol. 34, Issue 25, pp. 2413 – 2414, 1998

[24] L. Reindl, C.C.W. Ruppel, A. Kirmayr, N. Stockhausen, M.A. Hilhorst and J. Balendonckm, "Radio-requestable passive SAW water-content sensor", *IEEE Trans. Microwave Theory and Techniques*, Vol. 49, Issue 4, Part 2, pp. 803 – 808, April 2001

[25] A. Denoth, "The monopole-antenna: a practical snow and soil wetness sensor", *IEEE Trans. Geoscience and Remote Sensing*, Vol. 35, Issue 5, pp. 1371 – 1375, Sept. 1997

[26] L. Apekis, C. Christodoulides, P. Pissis, "Dielectric properties of paper as a function of moisture content", *Dielectric Materials, Measurements and Applications 1988, Fifth Int. Conf. on*, pp. 97 – 100, 1988

6

Antennas of RFID Tags

Ahmed M. A. Salama
College of Electronics Engineering
University of Mosul
Iraq

1. Introduction

Radio Frequency Identification (RFID) is a rapidly developing technology which uses RF signals for automatic identification of objects. RFID system generally consists of three components: 1) A small electronic data carrying device called a transponder or tag that is attached to the item to be identified, 2) A reader that communicates with the tag using radio frequency signals, 3) A host data processing system that contains the information of the identified item and distributes the information between other remote data processing systems. A typical passive RFID tag consists of an antenna and RFID chip. RFID tags can be active (with battery) or passive (without battery). In particular, passive UHF (860 ~ 960) MHz tags represent a near optimal combination of cost and performance (Hunt et al., 2007). Generally, omni directionality for the tag antenna is preferred to ensure the identification from all directions. The structure of the tag antenna should also be low cost, small in size, have good impedance matching and insensitive to the attached objects to keep performance consistent (Curty et al., 2007).

A passive RFID system operates in the following way: RFID reader transmits a modulated RF signal to the RFID tag consisting of an antenna and an integrated circuit chip. The chip receives power from the antenna and responds by varying its input impedance and thus modulating the backscattered signal. Modulation type often used in RFID is amplitude shift keying (ASK) where the chip impedance switches between two states: one is matched to the antenna (chip collects power in that state) and another one is strongly mismatched. The most important RFID system performance characteristic is tag range – the maximum distance at which RFID reader can either read or write information to the tag. Tag range is defined with respect to a certain read/write rate (percentage of successful reads/writes) which varies with a distance and depends on RFID reader characteristics and propagation environment (Nikitin & Rao, 2006).

In this chapter, the operation theory of the RFID system is described. The antenna in RFID system is discussed, and the designing considerations of the antennas for RFID applications are presented. Also the design, simulation and implementation of some commonly used antennas in the RFID system are presented and investigated. IE3D electromagnetic simulator based on Method of Moment (MoM) is used to design some of these antennas.

2. Operation theory of RFID tags

As known, passive RFID tags does not have its own power supply (i.e. battery less) ,so it depends on the received signal to power up the tag circuitry and resends the data to the reader. In this section, the operation of RFID tags is discussed and analyzed as well as the powers at the tag terminals and reader antenna are calculated.

2.1 Link budget

To calculate the power available to the reader P_r, the polarization losses will assume to be neglected and line-of-sight (LOS) communication is presented. As shown in Fig. 1, P_r is equal to $G_r P'_r$ and can be expressed as shown in equation (1) by considering the tag antenna gain G_t and the tag-reader path loss (Curty et al., 2007):

$$P_r = G_r P'_r = G_r P'_b \left(\frac{\lambda}{4\pi d} \right)^2 \tag{1}$$

$$= G_r G_t P_b \left(\frac{\lambda}{4\pi d} \right)^2 \tag{2}$$

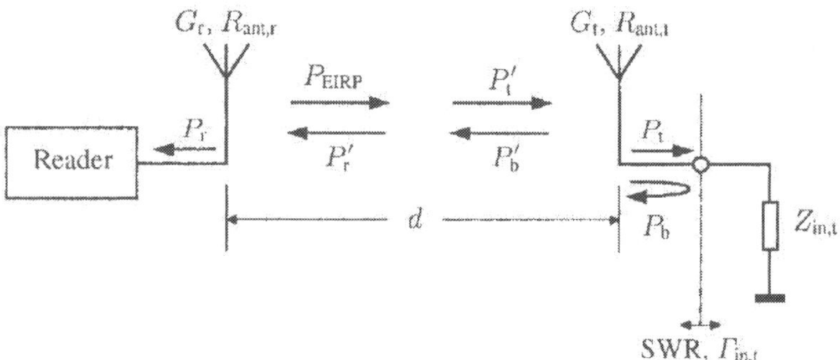

Fig. 1. Link budget calculation (Curty et al., 2007).

P'_b can be calculated using SWR between the tag antenna and the tag input impedance:

$$P_b = P_t \left(\frac{SWR - 1}{SWR + 1} \right)^2 \tag{3}$$

Or can be expressed using the reflection coefficient at the interface (Γ_{in}) as shown below:

$$P_b = P_t \left| \Gamma_{in} \right|^2 \tag{4}$$

The transmitted power (P_{EIRP}) is attenuated by reader-tag distance, and the available power at the tag is:

$$P_t G_t = P_{EIRP}\left(\frac{\lambda}{4\pi d}\right)^2 \tag{5}$$

Substituting equations (3), (4) and (5) in equation (1) will result in the link power budget equation between reader and tag.

$$P_r = G_r G_t^2 \left(\frac{\lambda}{4\pi d}\right)^4 \left(\frac{SWR-1}{SWR+1}\right)^2 P_{EIRP} \tag{6}$$

Or can be expressed in term of (Γ_{in}), so equation (2.6) will become:

$$P_r = G_r G_t^2 \left(\frac{\lambda}{4\pi d}\right)^4 |\Gamma_{in}|^2 P_{EIRP} \tag{7}$$

The received power by the reader is proportional to the $(1/d)^4$ of the distance and the matching between the tag antenna and tag RFID IC as well as (P_r) is depending on the gain of the reader and tag antennas. In other words, the *Read Range* of RFID system is proportional to the fourth root of the reader transmission power P_{EIRP}.

3. Complex conjugate concept

For the ac circuit shown in Fig. (2) which consists of fixed voltage with peak value V_s and an internal impedance $Z_s=R_s+jX_s$ and an external load $Z_L=R_L+jX_L$, the load will deliver (1/2 V_s) when $Z_L=Z_s^*$ (Zhan, 2006) .

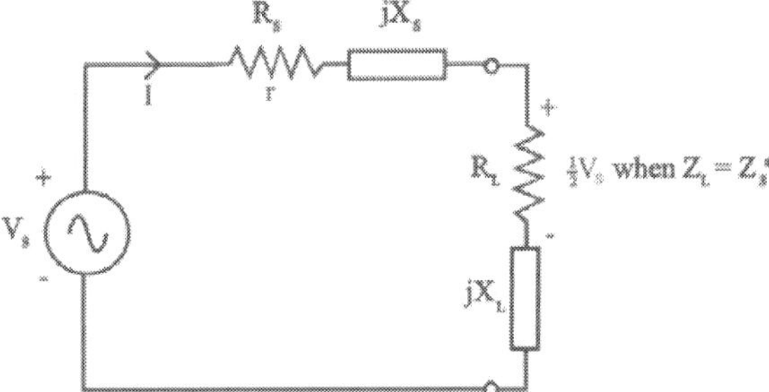

Fig. 2. Context for maximum power transfer theorem (Zhan, 2006)

The maximum power transfer theorem states that: for a linear network with fixed source impedance, the maximum power is delivered from the source to the load when the load impedance is the complex conjugate of the source impedance, that is:

$$Z_L = Z_s^* \tag{8}$$

Which means that $R_L=R_s$ and $jX_L=-jX_s$, and the circuit is said to be conjugately matched. The available source power is given by:

$$available\ source\ power = \frac{|V_s|^2}{8R_s} \tag{9}$$

As mentioned before, the RFID tag consists of an antenna and RFID integrated circuit (RFID IC) which can be illustrated by its equivalent circuits as shown below:

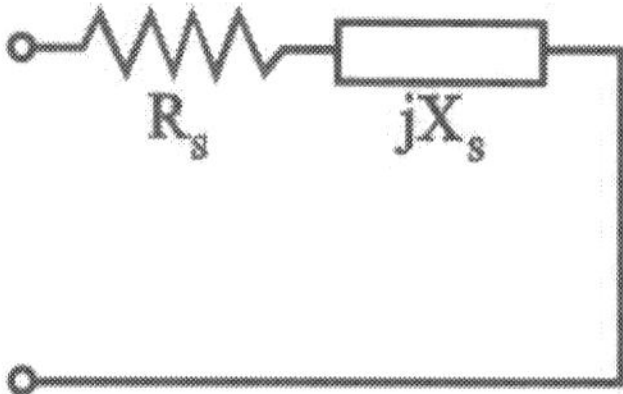

Fig. 3. The Equivalent circuit of the RFID circuit

Typically, X_s is capacitive and it comes from the rectifier capacitor which is about (1pf) this means an impedance of ($-j200\ \Omega$) at a frequency of 915 MHz, and R_s is about (10 Ω). The tag impedance will be Z_c=10-200Ω, this is an approximate value, but the exact chip impedance value can be obtained from chip manufacturer or can be measured by using network analyzer. The voltage reflection coefficient of a load Z_L on a transmission line of impedance Z_0 is defined as follow:

$$\Gamma = \frac{Z_L - Z_o}{Z_L + Z_o} \tag{10}$$

Where Z_L is the load impedance and Z_o is the line impedance. If the circuit is perfectly matched, maximum possible power will be transferred from the transmission line to the load. In the case of perfect matching between the antenna and the RFID IC there will be maximum power transfer. Also a perfect matching will result in zero voltage reflection coefficient.

Smith chart can be used for designing. If the RFID IC has input impedance of (10-j200) Ω, this value can be represented on smith chart as shown below:

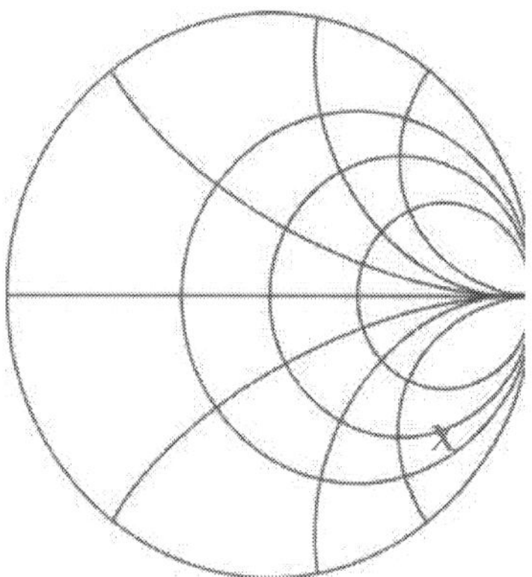

Fig. 4. Approximate position of 10Ω -j200Ω in Smith Chart

The RFID IC has capacitive impedance, so an inductive antenna with impedance of (10+j200) Ω (see Fig. 5) is required to obtain complex conjugate matching (perfect matching).

If the inductance is too low, matching networks can be used or lumped elements can be added.

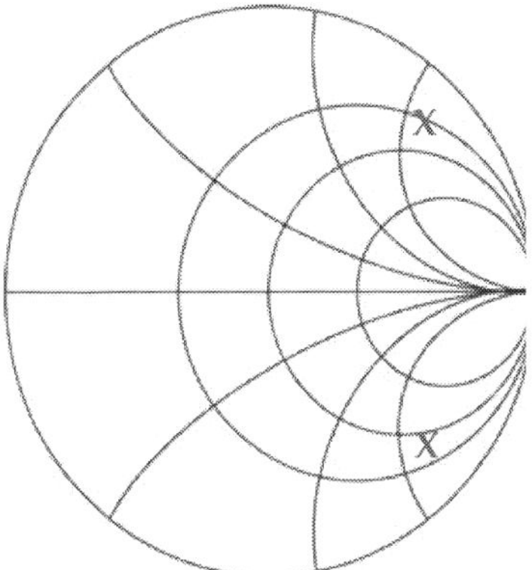

Fig. 5. Desired position of inductive antenna and capacitive chip

3. Types of RFID tag antennas

In this section, an overview of some antenna designs for passive UHF RFID tag is presented. These types are different from design to another depending on the application. There is no perfect antenna for all applications. It is the application that defines the antenna specifications. There is a high probability that many types of transponders will share the same IC but will connect to different antenna types. Patch antennas are well appropriate for metallic objects since it is possible to make use of their bodies as a ground plane (Curty et al., 2007). Inverted-F antennas are also mountable on such objects (Ukkonen et al., 2004). Other types of materials, e.g. (wood, cardboard, water, etc.), also allow differential antennas. These antennas offer the advantage of higher radiation resistance compared with single ended versions.
In the following sub-sections, some of these designs will be taken in details:

3.1 Meandered antennas
Meandered line antennas are interesting class of resonant antennas and they have been widely studied in order to reduce the physical size of the radiating elements in wire antennas like: monopole, dipole and folded dipole antennas. Increasing the total wire length in antenna of fixed axial length will lower its resonant frequency. One of the design requirements is miniaturizing the antenna, so meandering sections are added to the ordinary dipole antenna to reduce its physical size as shown below in Fig.6 (Rao et al., 2005). As the chip has a highly capacitive part in its impedance, the impedance of the designed antenna should have a highly inductive part as mentioned in the complex conjugate matching concept. To provide a better matching for the chip capacitive impedance, one meandered section was further meandered and a loading bar is added to obtain additional inductance. This antenna can be easily tuned by trimming. Lengths of meander trace and loading bar can be varied to obtain optimum reactance and resistance matching. The trimming is realized by punching holes through the antenna trace at defined locations. For

example, trimming the meander trace by Δx=5mm moves the resonant frequency up by 20 MHz as shown in Fig. 7. The gain is not significantly affected by trimming as shown in Fig.8.

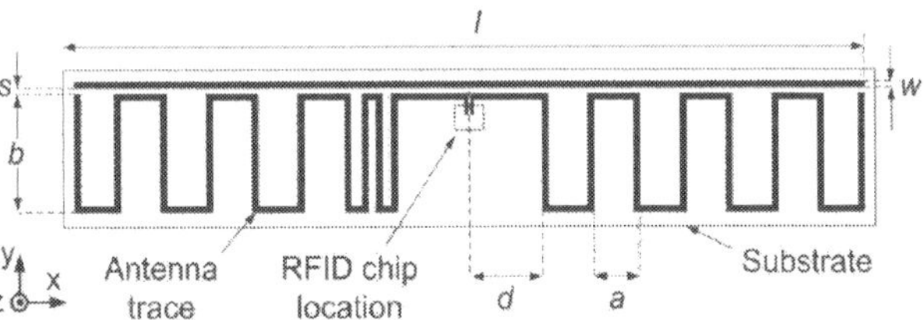

Fig. 6. Meandered line antenna

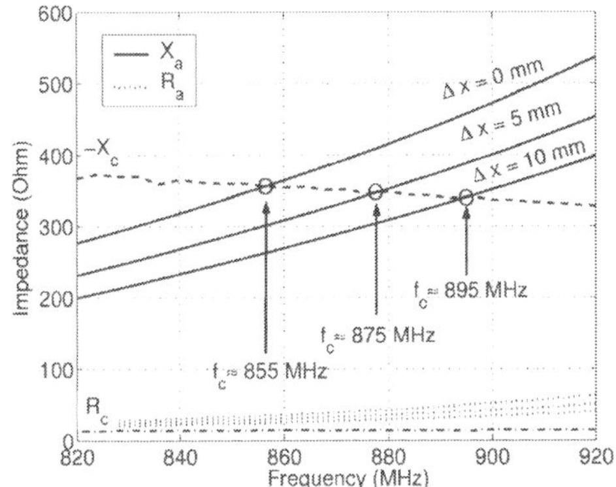

Fig. 7. Impedance of the loaded meander tag antenna (R_a, X_a) as a function of meander trace length trimming Δx

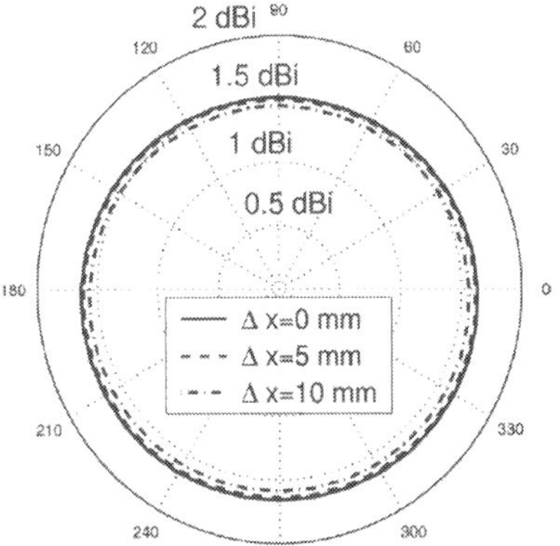

Fig. 8. Gain of the loaded meander tag antenna in yz-plane at 900 MHz as a function of meander trace length trimming Δx

3.2 Text antennas

Text can be used as a meandered line antenna (Salama & Quboa, 2008a). Using text as an antenna element in RFID tags is given with good reason; brand names or manufacturer logos can be used to form a radiating element for the RFID tag antenna which gives an additional value to the tag itself as a hi-tech advertisement. In this section the use of text as a meandered line for dipole antennas is discussed. Size reduction is compared to the ordinary dipole antenna operating at the same frequency and printed on the same substrate.

Fig.9 shows the antenna configurations of antenna No.1 and antenna No.2 where the letters of the text "**UNIVERSITY OF MOSUL**" are connected together in two different ways. In antenna No.1, the text is in contact with a straight dipole structure underneath the letters, whereas in antenna No.2, the letters are joined together from top and bottom of the letters alternatively to form a meander line structure.

Fig.10 shows the simulated return loss for the antennas No.1 and No.2. As shown in Fig.10, antenna No.2 has the better return loss. The Text antenna can be implemented and fabricated using PCB technology as shown in Fig.11.

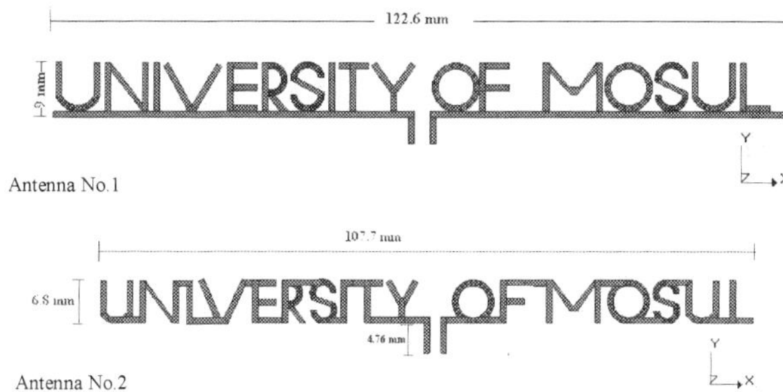

Fig. 9. Using Text as antennas for RFID tags

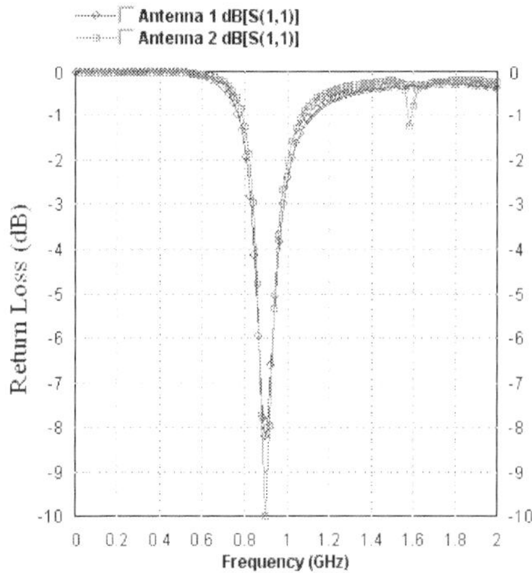

Fig. 10. The simulated return loss for the antennas No.1 and No.2.

Fig. 11. Photograph of the fabricated Text Antenna.

3.3 Fractal antennas

The interaction of electromagnetic waves with fractal geometries has been studied. Most fractal objects have self-similar shapes, which mean that some of their parts have the same shape as the whole object but at a different scale. The construction of many ideal fractal shapes is usually carried out by applying an infinite number of times (iterations) an iterative algorithms such as Iterated Function System (IFS). IFS procedure is applied to an initial structure called *initiator* to generate a structure called *generator* which replicated many times at different scales. Fractal antennas can take on various shapes and forms. For example, quarter wavelength monopole can be transformed into shorter antenna by Koch fractal. The Minkowski island fractal is used to model a loop antenna. The Sierpinski gasket can be used as a fractal monopole (Werner & Ganguly, 2003). When designing a small antenna, it is important to have a large effective length because the resonant frequency would be lower. The shape of the fractal antenna is formed by an iterative mathematical process. This process can be described by an Iterative Function System (IFS) algorithm, which is based upon a series of affine transformations which can be described by equation (11) (Baliarda et al., 2005):

$$\omega = \left[r\cos\theta - r\sin\theta \quad r\sin\theta \quad r\cos\theta \quad e \quad f \right] \tag{11}$$

Where r is a scaling factor and θ is the rotation angle, e and f are translation involved in the transformation.

Fractal antennas provide a compact, low-cost solution for a multitude of RFID applications. Because fractal antennas are small and versatile, they are ideal for creating more compact RFID equipment — both tags and readers. The compact size ultimately leads to lower cost equipment, without compromising power or read range. In this section, some fractal antennas will be described with their simulated and measured results such as: fractal dipoles and fractal loops.

3.3.1 Fractal dipole antennas

A standard Koch curve (with indentation angle of 60°) will be investigated (Salama & Quboa, 2008b), which has a scaling factor of r = 1/3 and rotation angles of θ= 0, 60, -60, and 0. There are four basic segments that form the basis of the Koch fractal antenna, which are shown in Fig. 12. The geometric construction of the standard Koch curve is fairly simple. One starts with a straight line as an initiator as shown in Fig. 12. The initiator is partitioned into three equal parts, and the segment at the middle is replaced with two others of the same length to form an equilateral triangle. This is the first iterated version of the geometry and is called the *generator* as shown in Fig. 12.

From the IFS approach, the basis of the Koch fractal curve can be written using equation (11). The fractal shape in Fig. 12 represents the first iteration of the Koch fractal curve. From there, additional iterations of the fractal can be performed by applying the IFS approach to each segment.

It is possible to design small antenna that has the same end-to-end length than their Euclidean counterparts, but much longer. When the size of an antenna is made much smaller than the operating wavelength, it becomes highly inefficient, and its radiation resistance decreases. The challenge is to design small and efficient antennas that have a fractal shape.

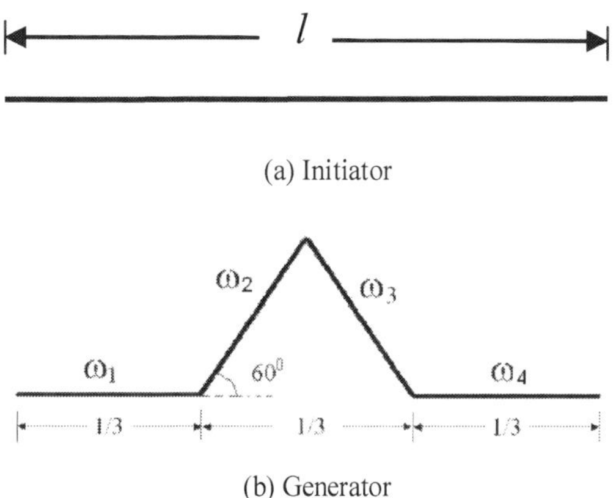

Fig. 12. Initiator and Generator of the standard Koch fractal curve.

Dipole antennas with arms consisting of Koch curves of different indentation angles and fractal iterations are investigated in this section. A standard Koch fractal dipole antenna using 3rd iteration curve with an indentation angle of 60° and with the feed located at the center of the geometry is shown in Fig.13.

Fig. 13. Standard Koch fractal dipole antenna.

Table 1 summarizes the standard Koch fractal dipole antenna properties with different fractal iterations at reference port of impedance 50Ω. These dipoles are designed at resonant frequency of 900 MHz.

The indentation angle can be used as a variable for matching the RFID antenna with specified IC impedance. Table 2 summarizes the dipole parameters with different indentation angles at 50Ω port impedance. Each dipole has an end-to-end length of 102mm.

Iter. No.	Dim. (mm)	RL (dB)	Impedance (Ω)	Gain (dBi)	Read Range (m)
K0	127.988	-27.24	54.4-j0.95	1.39	6.22
K1	108.4 X 17	-17.56	38.4+j2.5	1.16	6
K2	96.82 X 16	-12.5	32.9+j9.5	0.88	5.72
K3	91.25 X 14	-11.56	29.1-j1.4	0.72	5.55

Table 1. Effect of fractal iterations on dipole parameters.

Indent. Angle (Deg.)	f_r (GHz)	RL (dB)	Impedance (Ω)	Gain (dBi)	Read Range (m)
20	1.86	-20	60.4-j2.6	1.25	6.08
30	1.02	-22.53	46.5-j0.6	1.18	6.05
40	0.96	-19.87	41-j0.7	1.126	6
50	0.876	-14.37	35.68+j7	0.992	5.83
60	0.806	-12.2	30.36+j0.5	0.732	5.6
70	0.727	-8.99	23.83-j1.8	0.16	5.05

Table 2. Effect of indentation angle on Koch fractal dipole parameters.

Another indentation angle search between 20° and 30° is carried out for better matching. The results showed that 3rd iteration Koch fractal dipole antenna with 27.5° indentation angle has almost 50Ω impedance. This modified Koch fractal dipole antenna is shown in Fig.14. Table 3 compares the modified Koch fractal dipole (K3-27.5°) with the standard Koch fractal dipole (K3-60°) both have resonant frequency of 900 MHz at reference port 50Ω.

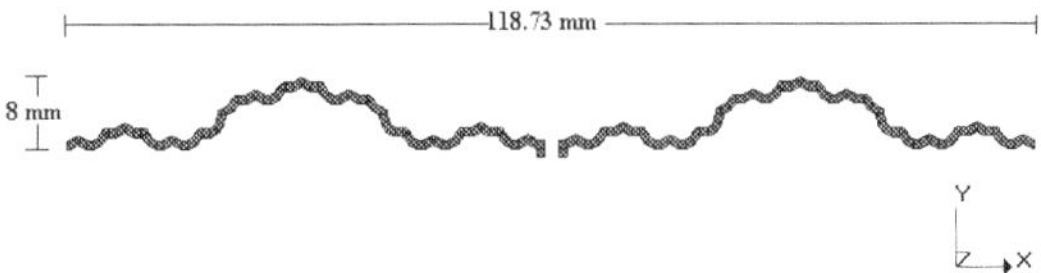

Fig 14. The modified Koch fractal dipole antenna (K3-27.5°).

Antenna type	Dim. (mm)	RL (dB)	Impedance (Ω)	Gain (dBi)	Read Range (m)
K3-60°	91.2 X 14	-11.56	29.14-j1.4	0.72	5.55
K3-27.5°	118.7 X 8	-33.6	48+j0.48	1.28	6.14

Table 3. Comparison of (K3-27.5°) parameters with (K3-60°) at reference port 50Ω.

From Table 3, it is clear that the modified Koch dipole (K3-27.5°) has better characteristics than the standard Koch fractal dipole (K3-60°) and has longer read range.

Another fractal dipole will be investigated here which is the proposed fractal dipole (Salama & Quboa, 2008b). This fractal shape is shown in Fig.15 which consists of five segments compared with standard Koch curve (60° indentation angle) which consists of four segments, but both have the same effective length.

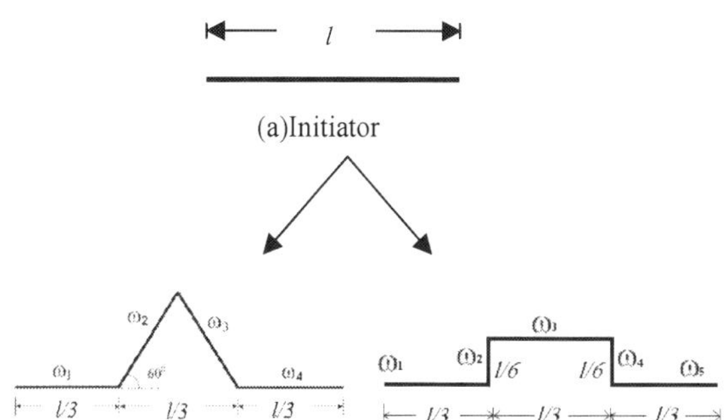

Fig. 15. First iteration of the fractal curves.

Additional iterations can be performed by applying the IFS to each segment to obtain the proposed fractal dipole antenna (P3) which is designed based on the 3rd iteration of the proposed fractal curve at a resonant frequency of 900 MHz and 50Ω reference impedance port as shown in Fig. 16 below.

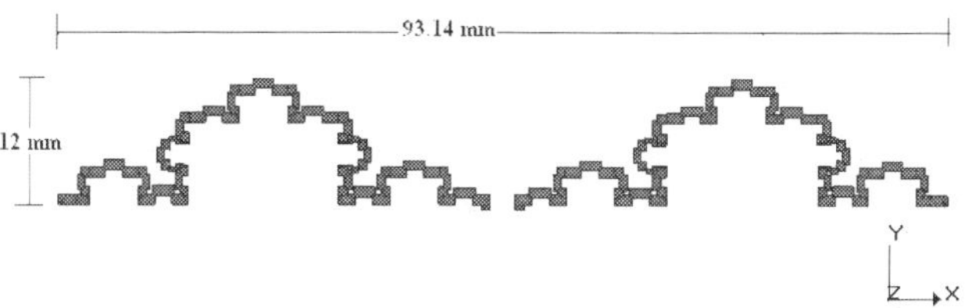

Fig. 16. The proposed fractal dipole antenna (P3) (Salama & Quboa, 2008b).

Table 4 summarizes the simulated results of P3 as well as those of the standard Koch fractal dipole antenna (K3-60°).

Antenna type	Dim. (mm)	RL (dB)	impedance (Ω)	Gain (dBi)	Read Range (m)
K3-60°	91.2 X 14	-11.56	29.14-j1.4	0.72	5.55
P3	93.1 X 12	-14.07	33.7+j3	0.57	5.55

Table 4. The simulated results of P3 compared with (K3-60°)

These fractal dipole antennas can be fabricated by using PCB technology as shown in Fig.17 and Fig.18 respectively. A suitable 50Ω coaxial cable and connector should be connected to that fabricated antennas. In order to obtain balanced currents, Bazooka balun may used. The performance of the fabricated antennas is verified by measurements. Radiation pattern and gain can be measured in anechoic chamber to obtain accurate results. The measured radiation pattern for (K3-27.5°) and (P3) fractal dipole antennas also shown in Fig.19 which is in good agreement with the simulated results.

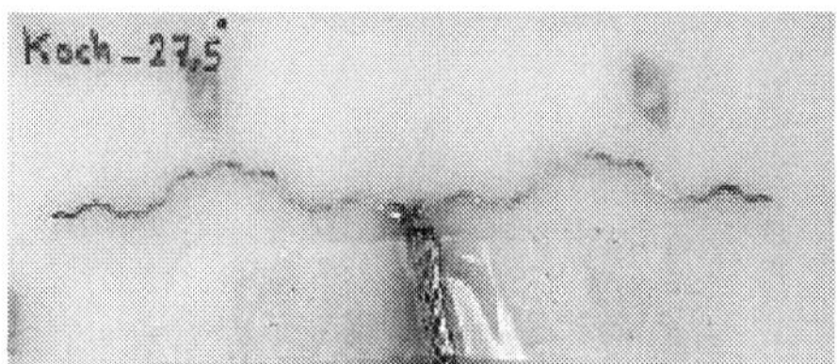

Fig. 17. Photograph of the fabricated K3-27.5° antenna.

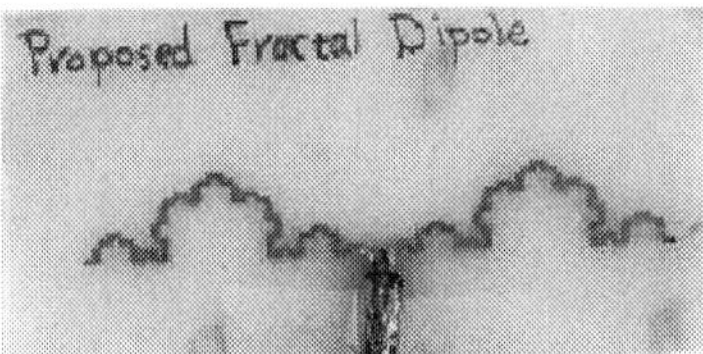

Fig. 18. Photograph of the fabricated (P3) antenna

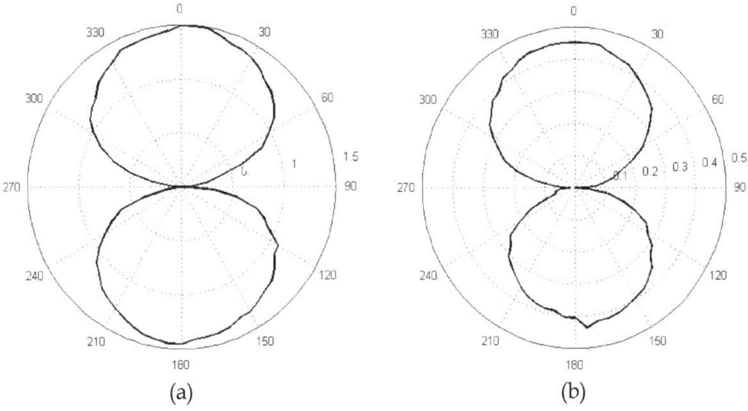

(a)   (b)

Fig. 19. Measured radiation pattern of (a) (K3-27.5°) antenna and (b) (P3) antenna

3.3.2 Fractal loop antennas

In this section, the design and performance of Two fractal loop antennas designed for passive UHF RFID tags at 900 MHz will be investigated; the first one based on the 2nd iteration of the Koch fractal curve and the other based on the 2nd iteration of the new proposed fractal curve with line width (1mm) for both as shown in Fig. 20 (Salama & Quboa, 2008c).

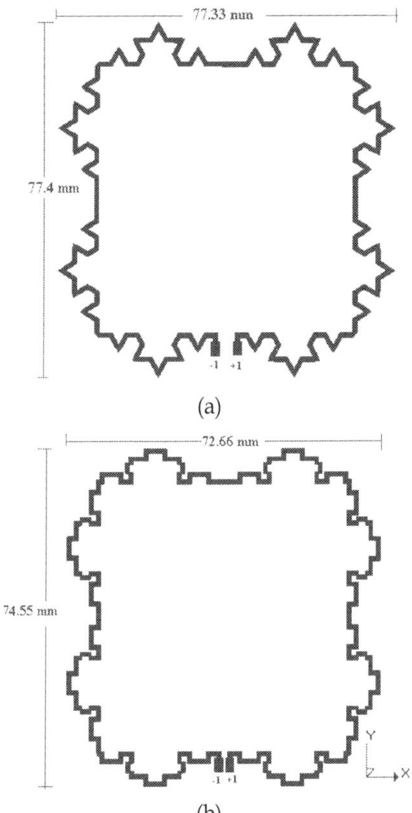

(a)

(b)

Fig. 20. The designed fractal loops: (a) Standard Koch fractal loop, (b) The new proposed fractal loop

A loop antenna responds mostly to the time varying magnetic flux density $\vec{B}$ of the incident EM wave. The induced voltage across the 2- terminal's loop is proportional to time change of the magnetic flux Φ through the loop, which in turns proportional to the area S enclosed by the antenna. In simple form it can be expressed as (Andrenko, 2005):

$$V \propto \frac{\partial \Phi}{\partial t} \propto \omega \left| \vec{B} \right| S \tag{12}$$

The induced voltage can be increased by increasing the area (S) enclosed by the loop, and thus the read range of the tag will be increased. The proposed fractal curve has a greater area under curve than the standard Koch curve in second iteration. Starting with an initiator of length (l), the second iterations area is ($0.0766\ l^2\ cm^2$) for the proposed curve and ($0.0688\ l^2\ cm^2$) for the standard Koch curve. According to equation (12) one can except to obtain a significant level of gain from proposed fractal loop higher than that from Koch fractal loop.

Fig.21 shows the return loss (RL) of the designed loop antennas of 50Ω balanced feed port, and Table 1 summarizes the simulated results of the designed loop antennas.

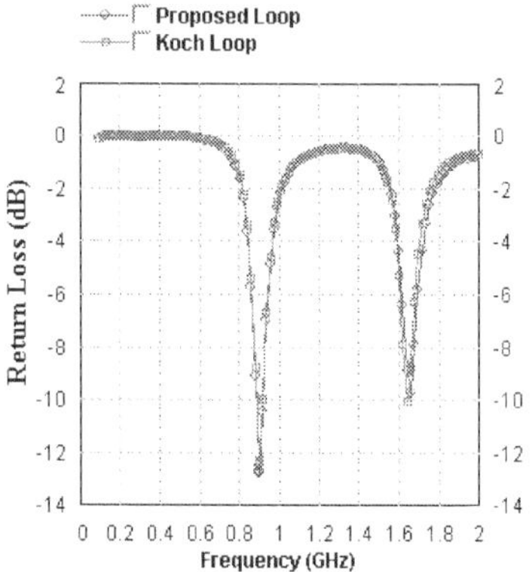

Fig. 21. Return loss of the two loop antennas.

Antenna type	Return Loss (dB)	BW (MHz)	Impedance (Ω)	eff. (%)	Gain (dBi)	Read Range (m)
Standard Koch Loop	-12.35	31.4	80.73-j7.3	78.5	1.74	6.287
Proposed Loop	-12.75	36	78.2-j8.9	81.8	1.97	6.477

Table 5. Simulated results of the designed loop antennas.

From Table 1 it can be seen that the proposed fractal loop has better radiation characteristics than the standard Koch fractal loop. As a result, higher read range is obtained. The proposed fractal loop also is smaller in size than the standard Koch fractal loop. The measured radiation pattern is in good agreement with the simulated one for the proposed fractal loop antenna as shown in Fig. 22.

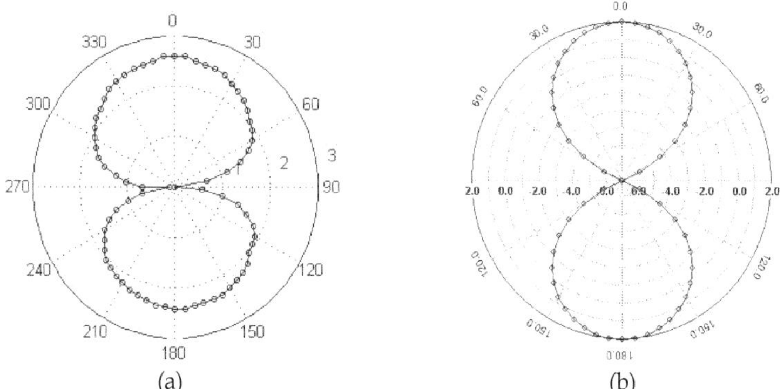

(a)   (b)

Fig. 22. The radiation pattern of the proposed fractal loop antenna (Salama et al., 2008). (a) measured. (b) simulated

The fabricated proposed fractal loop antenna is shown in Fig.23 below:

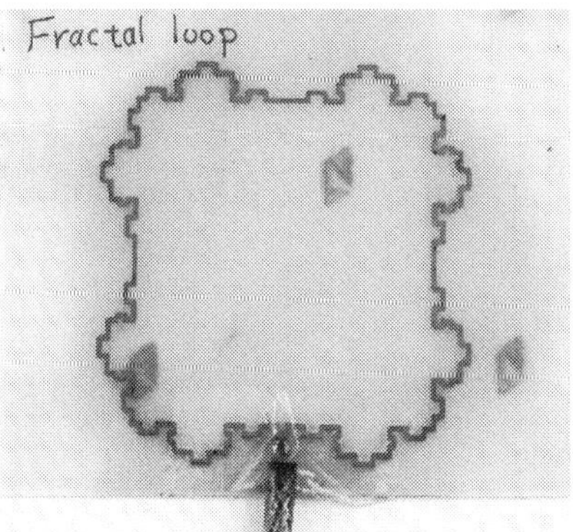

Fig. 23. The fabricated proposed fractal loop antenna.

3.4 Inverted-F antennas

A lot of different tag designs have been developed, but one of the biggest challenges nowadays is tagging objects that consist of metal or other conductive material. Printed dipole antennas may be used in RFID tags, but their performance is highly platform dependent. Conversely, microstrip patch antennas are more tolerant to the effects of the platform, but are very large in size. Metal strongly affects the performance of antennas for example by lowering the antenna's radiation efficiency. Metalized objects, such as aluminum cans, are common in most applications that utilize passive RFID systems. Therefore, tag antennas must be designed to enable passive tags to be read near and on metallic objects without performance degradation. Inverted-F antennas is a good solution which is a modification of a quarter-wave monopole antenna. The height of the antenna can be reduced by positioning the radiating element so as parallel with the ground plane while maintaining the resonant length. In Fig.24 the basic structures of IFA are presented. The microchip is attached to the feed point between the ground plane and the radiating element (Ukkonen et al., 2004a).

Inverted-F antennas can take various shapes and designs according to application and specifications. These types like: planar inverted-F antenna as shown in Fig.25 (Hirvonen et al., 2004) and wire-type inverted-F antenna as shown in Fig.26 (Ukkonen et al., 2004b).

4. Environmental effects and performance limitations

RFID tag performance can be affected by many factors. In particular, the electrical properties of objects near or in contact with the tag antenna will be changed. A tag is usually attached directly to the object to be identified. Many common materials, including metals and liquids, have strong effects on the performance of UHF tag antenna (Dobkin & Weigand, 2005). The effect of different materials has been studied where the RFID tags can be placed in free space (air), on cardboard, directly to metal, on plastic container filled with water and on wood ....etc. Figs. 26, 27 and 28 show the effect of some materials on the return loss of some practical antennas which are mentioned before in this chapter.

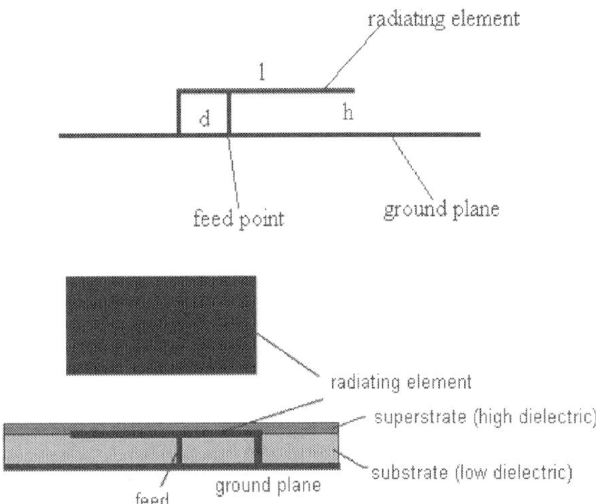

Fig. 24. The Basic structure of IFA.

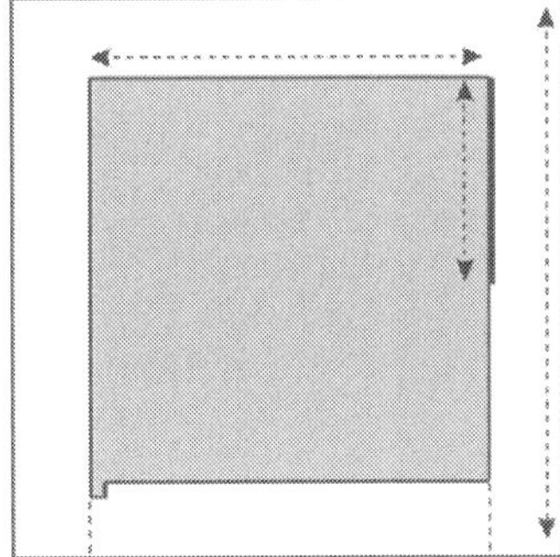

Fig. 24. Planar Inverted-F Antenna.

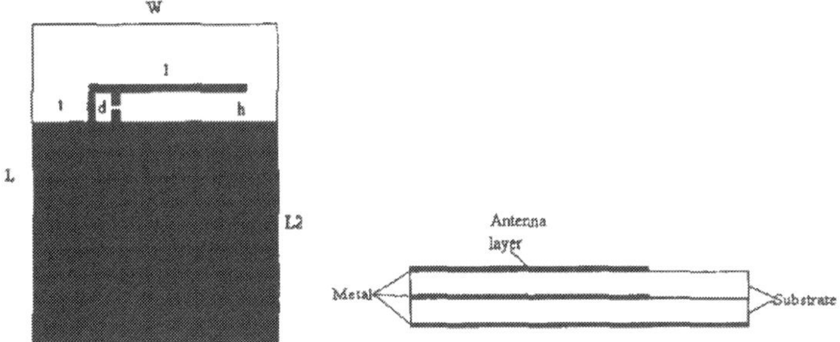

Fig. 25. Wire-type Inverted-F Antenna.

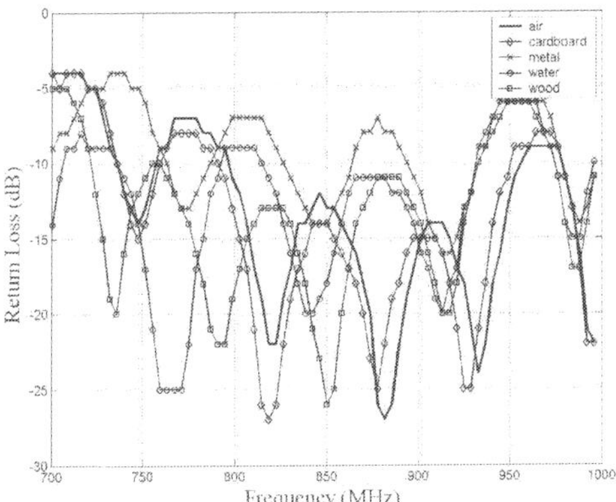

Fig. 26. Effect of different materials on Return loss of the Text Antenna.

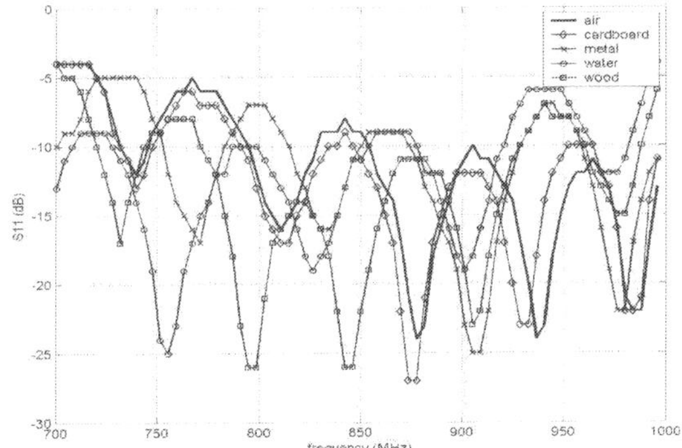

Fig. 27. Effect of different materials on Return loss of the Proposed Fractal Dipole antenna (P3).

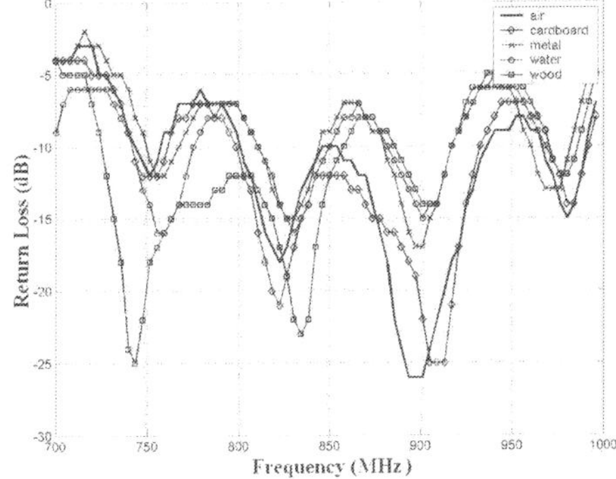

Fig. 28. Effect of different materials on Return loss of the Proposed Fractal Loop Antenna.

The results showed that the performance of the fractal loop antenna is practically accepted even if the antenna is attached to different materials and has better return loss with attaching materials when compared with other types.

5. References

Andrenko A. S., (2005). Conformal Fractal Loop Antennas for RFID Tag Applications, *Proceedings of the IEEE International Conference on Applied Electromagnetics and Communications, ICECom.*, Pages:1-6, Oct. 2008.

Baliarda, C. P.; Romeu J., & Cardama A., M. (2000), The Koch monopole: A small fractal antenna. *IEEE Trans. on Antennas and Propagation,* Vol.48, (2000) page numbers (1773-1781).

Curty J. P., Declerdq M., Dehollain C. & Joehl N., (2007). *Design and Optimization Of Passive UHF RFID Systems,* Springer, ISBN: 0-387-35274-0, New Jersey.

Dobkin D. M., Weigand S. M., (2005). Environmental Effects on RFID Tag Antennas, *Proceedings of IEEE Microwave Symposium Digest,* June 2005.

Hirvonen M., Pursula P., Jaakkola K., Laukkanen K., (2003). Planar Inverted-F Antenna For Radio Frequency Identification. *Electronics Letters,*Vol. 40, No. 14, (July 2004),

Hunt V., Puglia A. & Puglia M., (2007). *RFID A Guide To Radio Frequency Identification,* John Wiley & Sons Inc., ISBN: 978-0-470-10764-5 ,New Jersey.

Nikitin P. V., Rao K. V. S.,(2006). Performance Limitations of Passive UHF RFID Systems, *Proceedings of IEEE Antennas and propagation Symposium,* pp. 1011-1014, July 2006.

Rao K. V. S. , Nikitin P. V. & Lam S. F., (2005). Antenna Design For UHF RFID tags: A review and a practical application, *IEEE Trans. On Antennas and Propagation.,* Vol.53, No.12, Dec. 2005, pp. 3870-3876.

Salama A. M. A., Quboa K., (2008a). Text Antenna for Passive UHF RFID Tags, *Proceedings of The 5th Congress of Scientific Research Outlook in the Arab World (Scientific Innovation and sustained Development),* Morocco, Oct. 2008, Fez.

Salama A. M. A., Quboa K., (2008b). Fractal Dipoles As Meander Lines Antennas For Passive UHF RFID Tags, *Proceedings of The IEEE Fifth International Multi-Conference on Systems, Signals and Devices (IEEE SSD'08),* Page: 128, Jordan, July 2008.

Salama A. M. A., Quboa K., (2008c). A New Fractal Loop Antenna for Passive UHF RFID Tags Applications, *Proceedings of the 3rd IEEE International Conference on Information & Communication Technologies: from Theory to Applications (ICTTA'08),* Page: 477, Syria, April 2008, Damascus.

Ukkonen L., Sydanheimo L. & Kivikoski M., (2004a). A Novel Tag Design Using Inverted-F Antenna for Radio Frequency Identification of Metallic Objects, *Proceedings of IEEE/Sarnoff Symp. On Advances in Wired and Wireless Communication,* pp. 91-94, April 2004.

Ukkonen L., Engels D., Sydanheimo L. & Kivikoski M., (2004b). Planar Wire-type Inverted-F RFID Tag Antenna Mountable On Metallic Objects, *Proceedings of IEEE Int. Symp. on AP-S, Vol. 1, pp. 101–104, June 2004.*

Werner D. H. & Ganguly S. (2003). An Overview of Fractal Antenna Engineering Research. *IEEE Antennas and Propagation Magazine,* Vol.45, No.1, (Feb. 2003), page numbers (38-56).

Zhan J. Q., (2006). *RFID Tag Antennas Designed By Fractals and Manufactured By Printing Technologies, Institute of Communication Engineering, Tatung University,* June 2006.

7

Near Field On Chip RFID Antenna Design

Alberto Vargas and Lukas Vojtech
Czech Technical University in Prague
Czech Republic

1. Introduction

This chapter deals with the designing strategy and process integration for a small On-Chip-Antenna (OCA) with a small Radio Frequency Identification (RFID) tag on a chip-area 0.64 x 0.64 mm at 2.45 GHz for communication in near field. On the other hand, communication between Reader device and set of OCA-Tag is based on inductive coupling.

Embedded antenna is an important step down into the route of miniaturisation. A special micro-galvanic process that can take place on a normal CMOS wafer makes it possible. The coil could be placed directly onto the isolator of the silicon chip in the form of a planar (single layer) spiral arrangement and contacted to the circuit below by means conventional openings in the passivation layer (e.g., Usami, 2004). Dimension of the conductor flows in the range of 5-10 μm with a layer thickness of 15-30 μm (Finkenzeller, 2003).

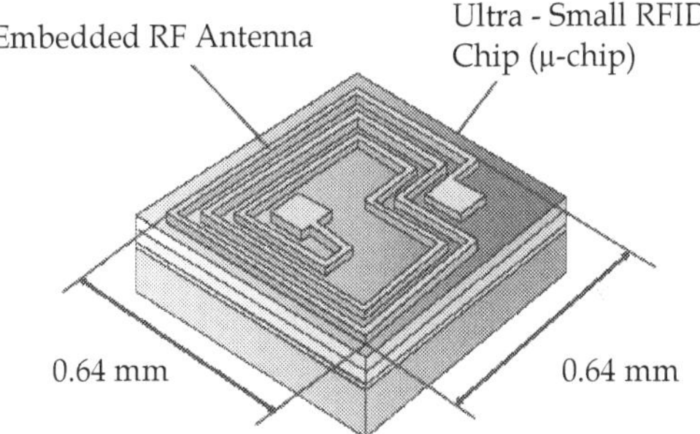

Fig. 1. Embedded antenna structure of ultra-small RFID chip.

RF signal can be radiated effectively if the linear dimension of the antenna is comparable with the wavelength of the operating frequency; however, the wavelength at 2.45 GHz is 12.24 centimetres. Due the size minimization of the transponder (reducing its area), a loop antenna in the shape of a coil that is resonating have to be used. Since the operating read range of these tags is relatively small compared to a wavelength, they operate in the near-field radiation region. The tag made up of OCA is preferred to save a major portion of the assembly cost in fabrication and it also enhances the system reliability.

Design strategy and communication principles are introduced by the description of its equivalent circuit parameters. The next step is matching network design and simulation based on the complex impedance of real RFID chip EM4222 and silicon chip materials constants. These preliminary steps are presented, the OCA structure design, simulation and final dimension optimisation duties are performed.

The resulting OCA transponder with an integrated antenna implemented on the chip thus requires only approx. 0.64 mm2 of silicone. Its small dimensions allow for expansion of applications used for marking various miniature objects as e.g. minicontainers for chemical or biological samples. Economical and application reasons force us to modify the existing object marking technologies. One of the development trends is the miniaturization of identifiers, and especially the reduction of their price. Low price and miniaturization allow for production of new RFID technology applications.

2. Tag antenna design

The efficiency of the antenna is limited by its allowable area, which is simply determined by the underlying tag chip area. The antenna for backscattering model is a far-field one (dipole, slot or path antenna) and its typical dimension should be $\lambda/2$ or $\lambda/4$. Understanding a symbol λ as wavelength. According to these, the size of the antenna should be around several centimetres. Since the chip dimension is about micrometers, it is too small to use the backscattering model although it is widely used in 2.45 GHz. In comparison, with an inductive-coupling model, the dimension of the inductor coils can be very small as in mm range. Therefore, OCA antenna becomes the best suitable technology to embed into Chip.

As magnetic-coupling model, OCA should be a coil on the Tag, a portion of the transformer, constructed together with another coil of the Reader collectively, based on Inductive Coupling technology as shown in figure 3. The tag either receives energy from the Reader or communicates the signal to Reader through those two coils. High coupling efficiency, or communication efficiency between the Reader and Tag, can be obtained by optimizing the design of the antenna coils. In addition to these, OCA's coil needs to be designed with a large inductive reactance, in order to obtain a high electric potential induced under a given intensity of magnetic field, generated by the Reader's antenna.

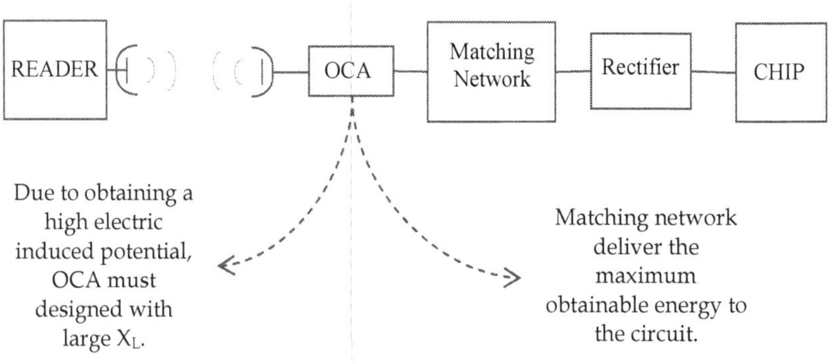

Fig. 2. Model scheme of the circuit to design

Going in depth into the entire design of the receiver part or the Tag, as shown in figure 2, antenna, matching network and doubler rectifier in series are implemented. The antenna must be matched with the input impedance of the subsequent circuit, Z_{RC}, in order to deliver the maximum obtainable energy to the Chip. Furthermore, this Matching Network has to be only constructed from capacitors since inductors will share the energy in OCA and thus reduce the efficiency of the coil. On the other hand, selected Chip contains Doubler Schottky rectifier (Prat, 1999) and internal capacitance, which are responsible for converting AC signal into DC. Indeed, not only rectifier capacitance is usually assumed integrated, but also parasitic capacitances of the Tag's Chip material.

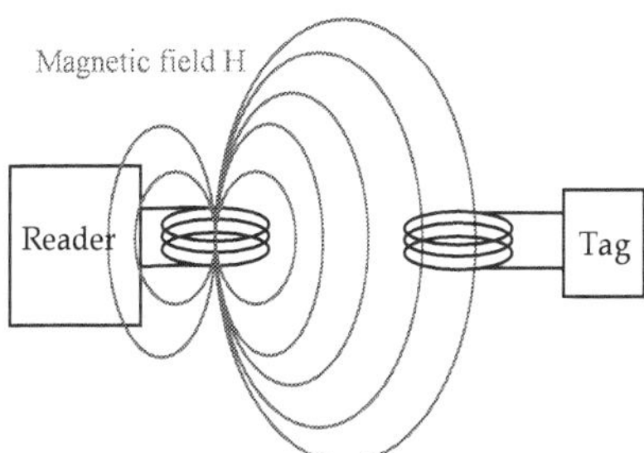

Fig. 3. Scheme of the Inductive Coupling Technology.

In terms of analysis, load process of the Chip is often modelled by a resistance at its highest current usage (Atmel, 2002) at power-up reset or during an EEPROM write.
Keeping the resistance of the coil to the minimum permits to enhance the quality factor, Q, of the antenna and increase the available energy to the Tag. In order to obtain the high electric potential induced, OCA must be designed with a large inductive reactance. Furthermore, OCA must be fabricated of a required Q-factor performance to satisfy the matching, and a rate around 3 is assumed (Guo et al, 2006).

2.1 Parameters

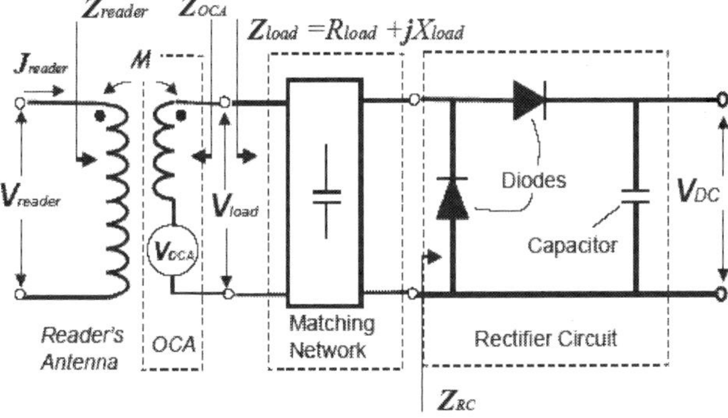

Fig. 4. Schematic diagram of the Reader and the Tag.

Figure 4 shows the equivalent circuit where the Reader and the Tag antennas are magnetically coupled via mutual inductance, M. The parallel resonant circuit of the antenna's Tag includes the inductance of the loop, L_{OCA}, and as well, its own impedance Z_{OCA} (1).

$$Z_{OCA} = R_{OCA} + j\, X_{OCA} \tag{1}$$

Voltage induced in the antenna, V_{OCA}, involves both the Tag and the Reader coils in a linear expresion, as it can be check in formula 2. J_{reader} is the current throughout the Reader's antenna.

$$/V_{OCA}/ = \omega\, K\, \sqrt{(L_{reader}\, L_{antena})}\, J_{reader} \tag{2}$$

Where ω is the circular frequency ($\omega = 2\Pi\, f$) and K is the coupling coefficient between OCA and reader's antenna. Power delivered to the Chip (3), understanding loading stage (figure 4) as the set of Matching network, Double rectifier, the Chip and its input voltage is assumed as follows:

$$P_{load} = \frac{1}{2}\left|V_{OCA}\right|^2 \frac{R_{load}}{\left(R_{load} + R_{OCA}\right)^2 + \left(X_{load} + X_{OCA}\right)^2} \tag{3}$$

$$V_{load} = \frac{R_{load} + jX_{load}}{\left(R_{OCA} + R_{load}\right) + j\left(X_{OCA} + X_{load}\right)} \cdot V_{OCA} \tag{4}$$

In terms of design process, when the Chip is properly matched with the antenna coil (5), the maximum voltage (7) and power (6) is delivered to the load.

$$Z_{OCA} = Z_{load}^{*} \tag{5}$$

$$P_{load\ max} = \frac{1}{8}\frac{\left|V_{OCA}\right|^2}{R_{OCA}} = \frac{1}{8}\frac{\left|V_{OCA}\right|^2}{R_{load}} \tag{6}$$

$$\left|V_{load\ max}\right| = \frac{1}{2R_{OCA}}\sqrt{R_{OCA}^{2} + X_{OCA}^{2}} \cdot \left|V_{OCA}\right| \tag{7}$$

According to the Power maximum load and the theory of the Q-factor in the resonant LC circuit, detailed below in section 4.1, both factors can be expressed in function of each other. Thus, the voltage supplied to the load depends on the voltage of the OCA. Achieving a higher Q-factor is faced, and 3 is assumed optimal value as reported in the initial part of this section.

$$P_{load\ max} = \frac{1}{8}\frac{\left|V_{OCA}\right|^2}{R_{OCA}} \cdot Q_{OCA} \tag{8}$$

$$\left|V_{load}\right| = \frac{1}{2}\sqrt{1 + Q_{OCA}^{2}} \cdot \left|V_{OCA}\right| \tag{9}$$

Finally, signal amplitude on the Reader side, ΔV_{reader}, which is sent back from the Tag to the Reader is proportional to the current throughout the Reader's antenna and the OCA's Q-Value:

$$\Delta V_{reader} = \frac{1}{2}\frac{\omega^2 M^2 Q_{OCA}}{X_{OCA}} \cdot J_{reader} \tag{10}$$

3. Circuit modelling

The suggested design scenario starts with co-designing of the reader's antenna (L_{reader}) and OCA (L_{OCA}) based on the available fabrication technology to meet the required specifications. Besides, Q-factor for the power conversion is specified on the based of three-dimensional electromagnetic simulation would determine the Tag's efficiency. Matching Network between OCA and Chip takes an important stage and it is considered through checking the input impedance of the circuit. Overall, goals must meet according to the specifications provided by manufacturer of the Chip.

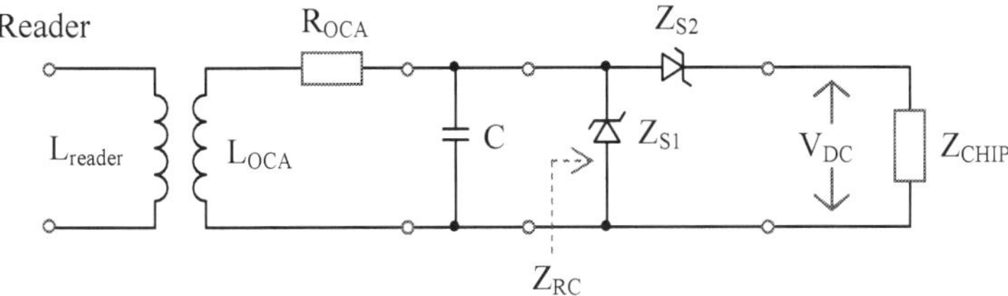

Fig. 5. Equivalent Schematic made by lumped elements

3.1 Design goals

Maximum power at the entrance of the Chip from the reader's antenna takes an important requirement to face within design process of the antenna. It is a full duplex system; this means that Tag must be designed to receive and send data, allowing signal communication between Reader and Chip. The RFID antenna receives signal from the reader's antenna and the signal is powering the Chip when OCA delivers voltage enough to wake up the Integrated Chip. In transmitting mode, Chip is serving as a source and also responsible for sending out its stored data through the RFID antenna. The most important and critical design-procedure is transmitting part, to such an extend if the best results in this mode are achieved, there are also the best outcomes in receiving mode.

The most important tag performance characteristic is read range. One limitation on the range is the maximum distance from which the tag receives just enough power to turn on from the reader. Another limitation is the maximum distance from which the reader can detect this signal from the tag. The read range is the smaller of two distances (typically, the first one since RFID reader sensitivity is usually high). Because reader sensitivity is typically high in comparison with the tag, the read range is defined by the tag response threshold. Read range is also sensitive to the tag orientation, the material of the tag is placed on, and to the propagation environment. Theoretical read range depends on the power reflection coefficient and can be calculated using the Friis free-space formula as:

$$r_{max} = \frac{\lambda}{4\pi} \sqrt{\frac{P_t G_t G_r \left(1 - |s|^2\right)}{P_{th}}} \tag{11}$$

where λ is the wavelength, Pt is the power transmitted by the RFID reader, Gt is the gain of the transmitting antenna (PtGt is EIRP, equivalent isotropic radiated power), Gr is the gain of the receiving tag antenna, Pth is the minimum threshold power necessary to power up the chip. Typically Pt, Gt, Gr and Pth are slow varying, and $|s|2$ is dominant in frequency dependence and primarily determines the tag resonance. Received power in decibel, formula 12, where L_{system} is the system losses that need to be taken into account during the measurement. This includes the cable and connector losses, temperature differences that cause internal losses in the instruments (i.e. antenna + transceiver of reader and tag).

Indeed, our main goal is to deliver the maximum power from the chip to the OCA during transmitting mode in terms of feeding our Tag; once internal impedance of the tag (Z_{RC}, figure 4 and 5) is known and the type of the antenna is chosen (coil), tuning step will support us to achieve the final goal in terms of matching dimension between OCA and Tag.

$$P_r = P_t - L_{system} + G_t + G_r + (1 - |s|^2) - 20\log_{10}\left(\frac{4\pi}{\lambda}\right) - 20\log_{10}(r_{max})$$ (12)

4. Matching network

Initially, as discussed before, Antenna's impedance is fixed to Tag's conjugated one at the frequency working (5).Both in figure 6 and 7 are useful to locate the input impedance of the double rectifier plus Chip (IC); Smith Diagram has been used through Mixed Series and Parallel Connections software, v. 2.8. Parameters such as equivalent impedance, admittance, its series and parallel scheme and its location in the Z-Smith Chart can be observed.

$$Z_{RC} = 80 - j232$$ (13)

OCA's resistance must be kept as small as possible in order to get a higher Q-factor inductor coil antenna. Since capacitors are exclusively employed in the matching network and trying

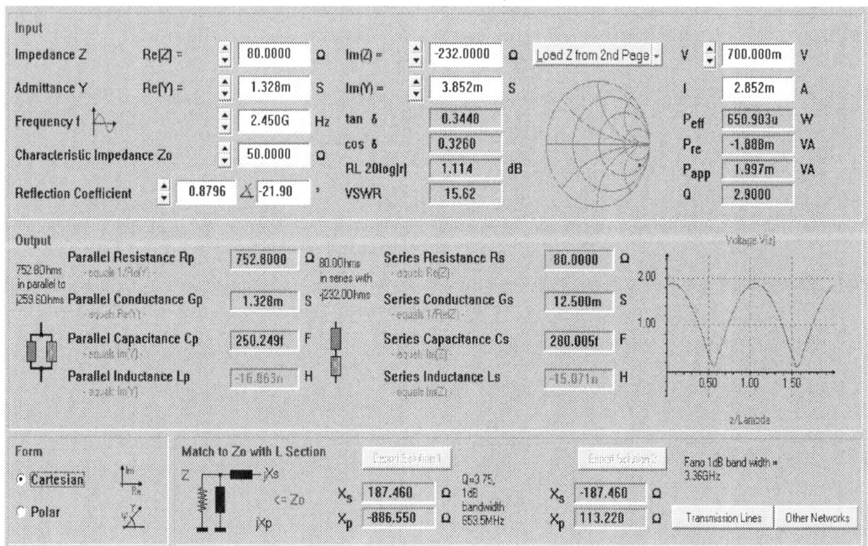

Fig. 6. Equivalent impedance, admittance, its series and parallel scheme and its location in the Z-Smith Chart at the frequency working 2.45 GHz. Characteristic impedance 50 Ω.

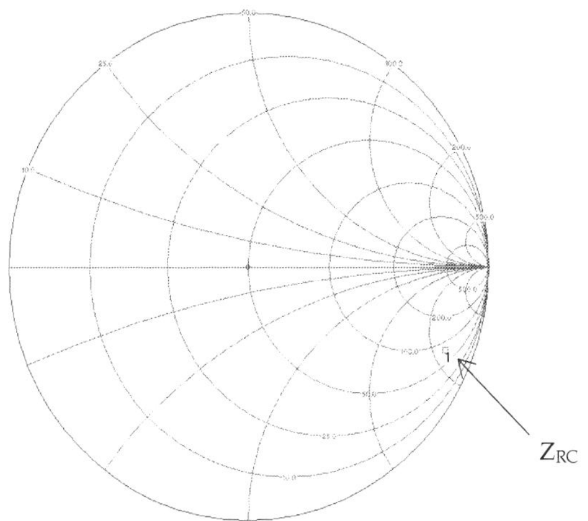

Fig. 7. Z-Smith Chart and the location of the input impedance of the RFID Tag used, Z_{RC}.

to get a good Q-factor of the OCA, suitable values for equivalent impedance will be computed and analyzed. Besides, capacitors influence only in imaginary part of the impedance within reactance part, but not in resistance. According to all of these OCA resistance can be designed equally to the Tag resistance, formula 14.

$$R_{OCA} = R_{RC} = 80 \ \Omega \tag{14}$$

On the other hand, relationship between reactance and resistance given below (15) together with Q-factor around 3, provide a consistent and a reliable design of OCA's impedance.

$$Q = \frac{X_{OCA}}{R_{OCA}} \approx 3 \ \rightarrow \ X_{OCA} \approx 3R_{OCA} \tag{15}$$

As a result, a consistent reactance for the tag antenna is $X_{OCA} \approx 240\Omega$ (L_{OCA}=15.6nH) is concluded. The impedance of OCA could be Z_{OCA}=80+j240 Ω, and formula 16 shows the normalized value according to the characteristic impedance, which is located at the up-hemisphere of the Smith Chart, figure 8:

$$\overline{Z_{OCA}} = \frac{Z_{OCA}}{Z_0} \approx 1,6 + j4,6 \tag{16}$$

As it can be seen, OCA impedance has an inductive feature, which is coherent with the type of antenna, a coil. It is actually an inductive set by both the reader and the Tag coils. In this case, the matching area is the region inside of the Middle-Line and the frontier of the resistance and conductance in the down-hemisphere.

An ideal case would be if OCA antenna Impedance is exactly as the Chip input one; then no Matching network would be needed because both stages would be already matched, figure 9.

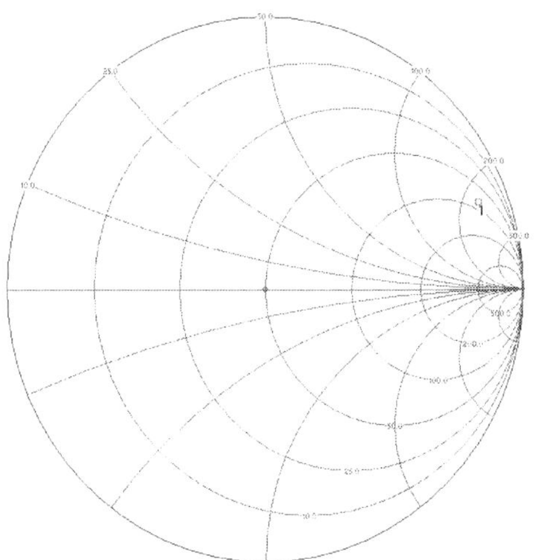

Fig. 8. DP-Nr. 1 (80.0 + j240.0)Ω, Q = 3.0 at 2.450 GHz.

Data Number Point	Impedance Ω	Q-Factor
1	80-j240	2.9
2	80+j240	2.9

Table 1. Data Point Numbers with its corresponding impedance and Q-factor at 2,45 GHz.

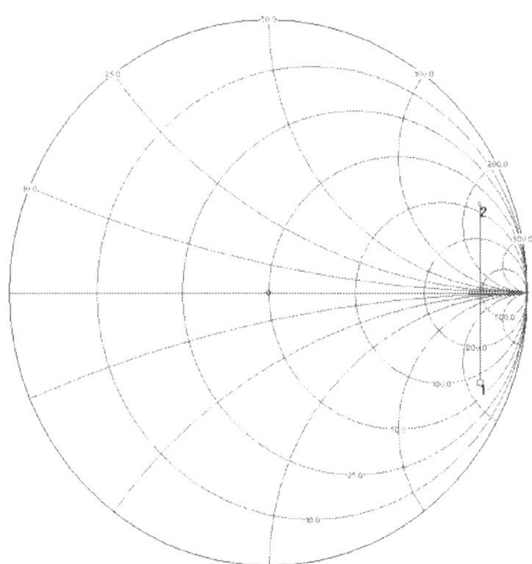

Fig. 9. Z-Smith Chart of the circuit impedance and its corresponding matching conjugated point.

Table 1 shows that the Q-factor for the circuit's impedance - 2,9 - is quite close to the value we want to reach. Anyway, we have designed a suitable matching network and for that reason the capacitors inside this network are so small. Actually, using this matching we can reach Q-factor equal to 3.

Matching Network is used to achieve formula 5 though the use of Smith Chart. According to the theory of RF adaption, a serial capacitor (C_1=6.6pF) plus a parallel capacitor (C_2=2.8pF) are chosen after analyzing some tests based on Smith Chart and considering desired goals.

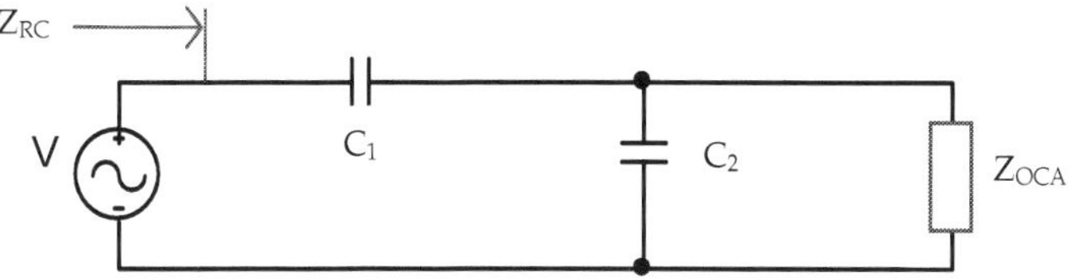

Fig. 10. Final Scheme of the Matching Network.

Impedance	Ω	Q-Factor
Z_{RC}	80-j232	2.9
Z_{OCA}	80+j240	3.0

Table 2. impedance and Q-factor at 2,45 GHz

In fact, the set of the Integrated Chip, matching network and transformer can be analyzed as a RLC resonant circuit. Then, in chapter 4.1, a study of the whole circuit and its response is going to be performed.

4.1 Resonant circuit RLC parallel process.

The equivalent Circuit model of the Reader plus the entire Tag can be modelled as a parallel resonant circuit LC (Yan et al, 2006). According to coupling volume theory the resonance is required to make good use of the power transferred to the label antenna.

$$f_{resonant} = \frac{1}{2\pi\sqrt{L \cdot C}} \tag{17}$$

Optimal power transferred to the label antenna is achieved through coil inductance at the carrier frequency resonance ($f_{resonant}$=f_0=2.45 GHz). The capacitor's chip is about 28pF, which fulfills our goals, also being acceptable compared to other researches (e.g., Sabri et al, 2006). On the other hand, designed capacitance is charged up to 1.5 V level in less than 50 s (Gregori et al, 2004) in order to deliver an output voltage higher than 1.2V to the Chip.

After analyzing whole equivalent impedance, set by lumped elements, on the right side of the Receiver (figure 11), Transformer's equivalent circuit will lead us to study and determine the OCA and reader antennas (figure 12).

$$n = \sqrt{\frac{L_{reader}}{L_{OCA}}} \tag{18}$$

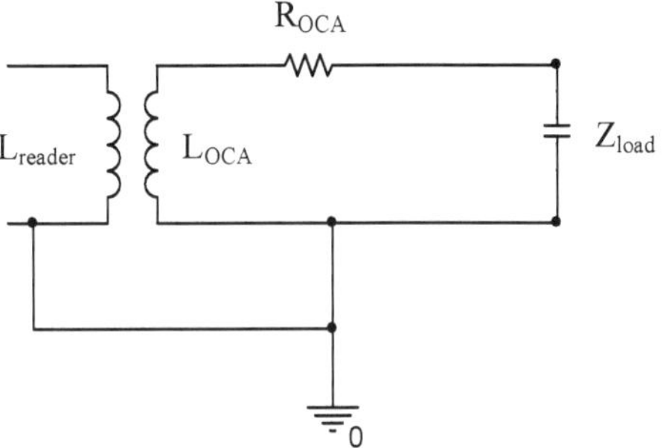

Fig. 11. The Reader, OCA and Tag equivalent scheme.

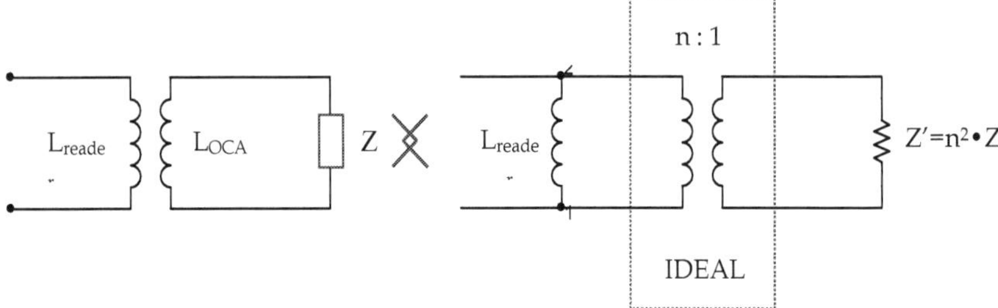

Fig. 12. The equivalent circuit of the transformer.

According to the electrical equivalent scheme of the transformer, both coils inductances can be substituted by the reader antenna introducing a new parameter n, which is the proportional to the reader antenna and inversely proportional to the Tag.

Assuming that Z is composed by R_{OCA} and X_{load} in series, as figure 7 shows. Thus, $Z = R_{OCA} + R_{load} \,||\, C_{load} = 160\,\Omega \,||\, 270\,pF$.

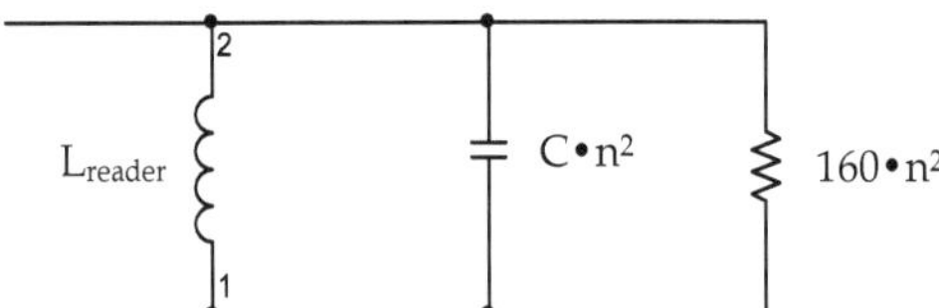

Fig. 13. Equivalent RLC circuit

Resonant RLC circuit provides (figure 13), through formula 18 a relationship between both coils and parameter n, a solution for the reader's coil at resonant frequency of the circuit, f_0.

$$L_{reader} = \sqrt{\frac{L_{OCA}}{C} \cdot \frac{1}{\left(2\pi \cdot f_0\right)^2}} \tag{19}$$

Actually, the final expression for the Reader's inductance depends on the value of the OCA's inductance and the capacitors located at the matching network. Thus, Tag's antenna has been finally designed by $L_{reader}=15{,}6\,nH$ and $n=1$.

4.2 Lumped element model

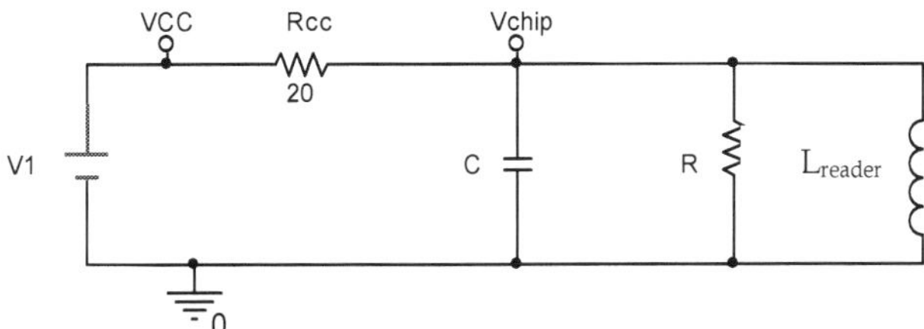

Fig. 14. Final equivalent RLC scheme.

As a result, figure 14 shows the equivalent circuit of the entire Tag, which describe the same behaviour, working at the ressonant frequency and delivering enough voltage to feed the Chip (wake-up voltage claimed by producer into datasheet is 1.4V). Therefore, the input signal, after rectifier stage, supplied to the subsequent circuit will be stable for establishing and mainteinance a communication between reader and Tag within a near field.

5. Tag antenna geometry

The type of chosen antenna is called a magnetic dipole antenna. Where the radius of the loop and the current (I) which has a Φ orientation and the radiation vector is:

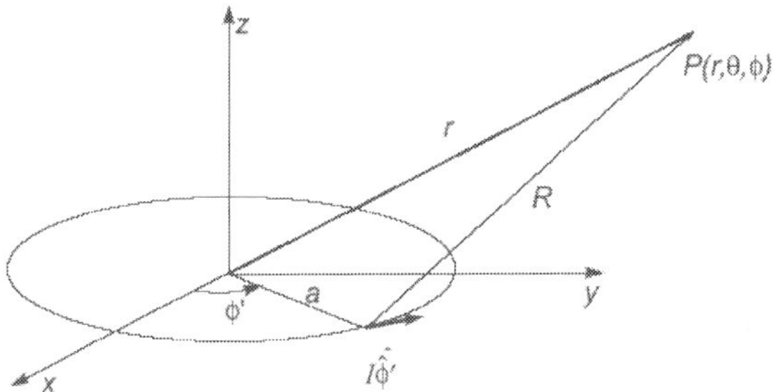

Fig. 15. total Power and Magnetic field radiated by a loop.

$$\vec{N} = \oint_{<2\pi>} I\hat{\phi}' e^{jk\hat{r}} a \, d\phi' \tag{20}$$

The expressions for the radiation vector in a loop in polar coordinates can be expressed (r, θ, Φ):

$$N_r = a I \, sen\theta \int_0^{2\pi} e^{jka\, sen\theta\, cos(\phi-\phi')} sen(\phi-\phi') d\phi' \tag{21}$$

$$N_\theta = a I \cos\theta \int_0^{2\pi} e^{jka\, sen\theta\, cos(\phi-\phi')} sen(\phi-\phi') d\phi \tag{22}$$

$$N_\varphi = a I \int_0^{2\pi} e^{jka\, sen\theta\, cos(\phi-\phi')} cos(\phi-\phi') d\phi' \tag{23}$$

These last expressions is illustration of the Bessel functions. Computing these formulas according to the geometry of antenna, we get this new formulas:

$$N_r = 0 \tag{24}$$

$$N_\theta = 0 \tag{25}$$

$$N_\phi = j2\pi a I J_1(ka\,sen\phi) \xrightarrow{\ when\ a<<\lambda\ } N_\phi = jk\pi a^2 I sen\theta \tag{26}$$

The correct proof for a circular loop with the uniform current. The total radiated power is:

$$P_r = \iint K\,d\Omega = 20\,\pi^2\,(k\,a)^4 \tag{27}$$

$$and \quad K = \frac{\eta}{4\lambda^2}\left|N_\phi\right|^2 = k^2\,\eta\,\pi\,a^2\,I\,\frac{e^{-jkr}}{4\pi r}\,sen\theta \tag{28}$$

Where η is the impedance wave and k is the wave number.
From the centre of the loop, its near magnetic field radiation falls off with r-3 and increases linearly with the number of turns N. Reminding that μ0 is the permeability of the free space (μ0=4π ·10-7).

$$B_z = \frac{\mu_0\,I\,N\,a^2}{2\left(a^2 + r^2\right)^{3/2}} \xrightarrow{\ r^2>>a^2\ } B_z = \frac{\mu_0\,I\,N\,a^2}{2\,r^3} \tag{29}$$

5.1 Structure

OCA schematic structure consists of a copper coil layer (Cu/USG single damascene process loop) with thickness of 1 μm to come up with both optimal Q-factor for the power conversion and match the antenna to the subsequent Chip. OCA was also covered by a 0,5 μm silicon nitride as passivation. Then, a thick dielectric undoped SiO_2 (USG) layer ~ 19,3 μm. The coil and Al-shielding layer are interspaced by a SiO_2 dielectric substrate containing deep vias, with a lower thickness of the metal. In order to resume the mutual EM interference between OCA and the tag's circuit an AL-shielding layer is used. It also enhances the Q-factor, reducing its rate. Finally, the silicon substrate is given by closed boundary (ground) to represent the backing plate.

5.2 Design

The optimal goal in terms of size, minimizing OCA in order to suit with the Integrated Chip dimension which is 0,64mm x 0.64 mm. Modelling is initiated by entering the substrate details of the layers such as shown in figure 16. Details depend upon whether the substrate layer is ossless, lossy, or a conducting layer, it is important to achieve these parameters in the l correct manner. For lossless substrates such as silicon dioxide (SiO_2) the relative permittivity (ε_r=4.1) and thickness is entered. For substrates with complex permittivity (and hence lossy) such as silicon (Si), the real part of the permittivity is entered together with a conductivity value in S/m. For metals, parameters are conductivity and thickness (Wilson, 2002). The variations of Q-factor pointed to the thickness of the substrate -19,3 μm

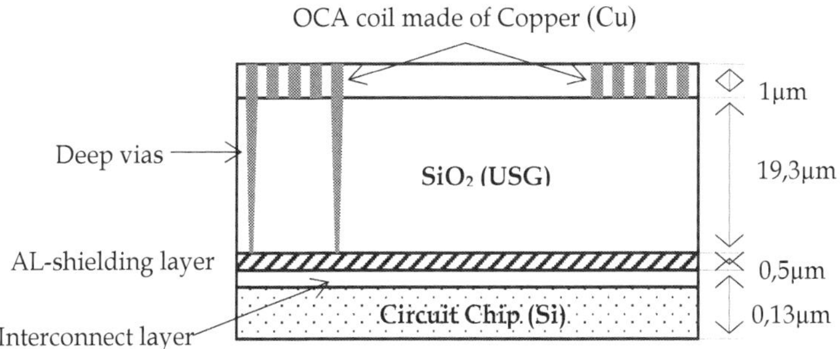

Fig. 16. The schematic cross-section of tag chip with OCA.

thickness of SiO$_2$- of OCA's Coil (Guo et al, 2004). Nevertheless, Metal thickness (Cu) is set to 1 μm in order to design the steady antenna, when longer copper increases the real part of the impedance.

OCA design is performed through software IE3D; all parameters given previously with adviced geometry of the coil. OCA impedance is characterized through reactance and resistance computed from the measured S_{11} parameter at the working frequency (f_0 = 2.45 GHz), and dimension optimisation is performed in order to reach global goals; both size and load impedance.

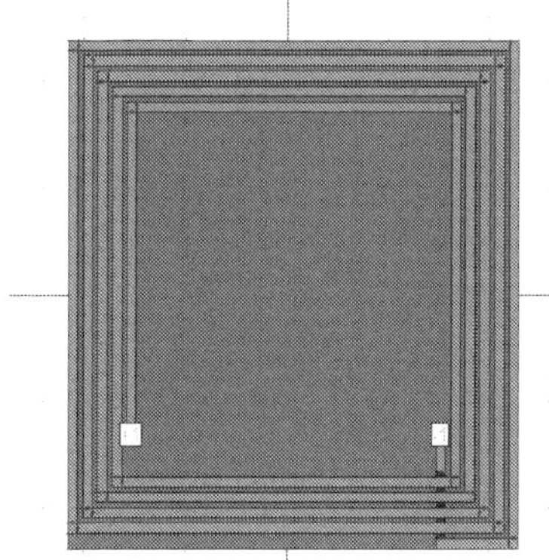

Fig. 17. Initial design.

Going in depth with the main goal, in terms of size, minimizing OCA in order to suit with the Tag dimension which is 0,64mm x 0.64 mm. Looking at the pattern, by changing the space of trace width and line spacing, input impedance of the coil becomes modified too. Actually, when playing with the width on the antenna, it means the breadth of the SiO$_2$ substrate layer, it is possible to modify the real part of the input impedance of the antenna. Final goal keeps on the same state; the challenge is to get 80 and 232 Ω as a real and imaginary part respectively. However, it is possible to play with the width (initially set to 12 μm) of the metals used - copper and aluminium layer - and change also the input impedance. Do not forget about the length of the inductors metallic layer, in this case copper. If it is getting longer, then the real part is increasing. Space line is set to 6 μm. It is important to point it out that the ports are located in the exact place as Pad1 and Pad2, in order to get the most optimal-real antenna.

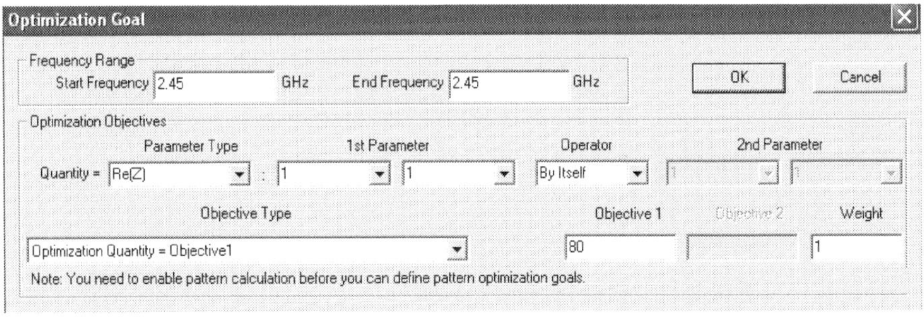

Fig. 18. Optimization goal for the real part of S_{11} at 2.45 GHz. Objective: $Re[S_{11}] = 80\Omega$.

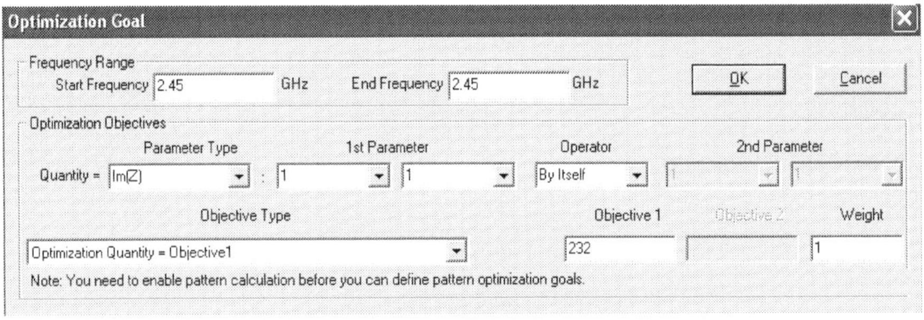

Fig. 19. Optimization goal for the imaginary part of S_{11} at 2.45 GHz. Objective: $Im[S_{11}] = 232\Omega$.

Indeed, figure 21 shows the final Antenna's model after optimization process, a rectangular coil; 0.7 mm x 0.75 mm, which is quite close to the optimal one introduced initially. Looking at its properties and behaviour, Figure 22 shows the equivalent input impedance, computed through the S_{11} parameter, which fully accomplish the desired properties.

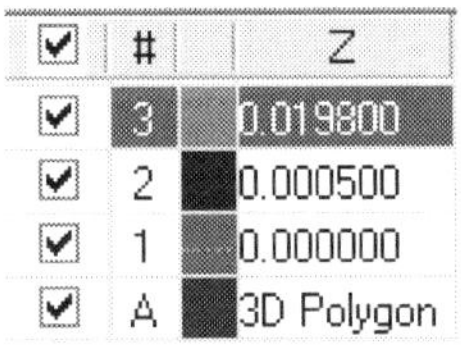

Fig. 20. Layers' depths

6. Conclusion

The process of fabricating the antenna on the top of the RFID chip eliminates the need for a separated and costly expensive process for antenna printing and assemblage, compulsory for a separated "off-chip" antenna which is much more times larger than the chip itself. This

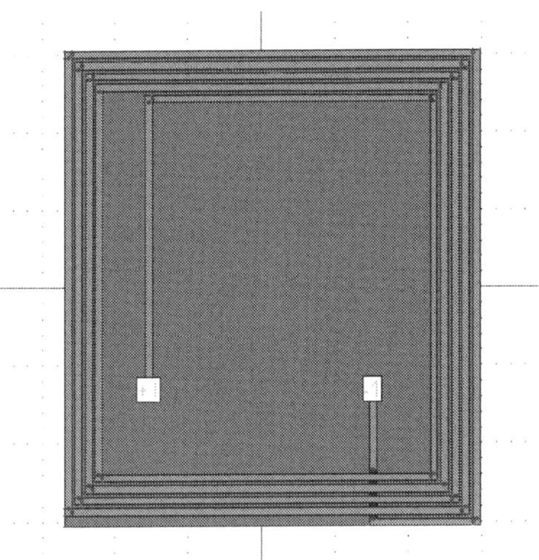

Fig. 21. Optimized design.

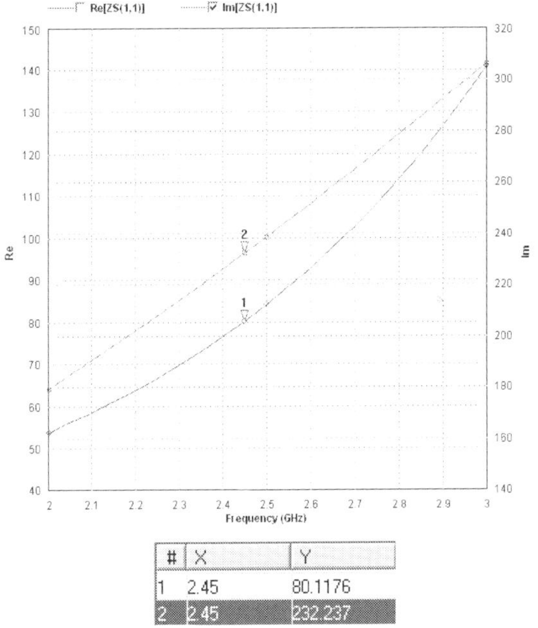

#	X	Y
1	2.45	80.1176
2	2.45	232.237

Fig. 22. Final input impedance of the antenna designed.

technology requires a layer of a suitable dielectric to be deposited on the chip surface and isolates the antenna from the circuits below. Overall, conventional RFID tags are typically of a few cm² in size and more expensive as well. In comparison, this newly developed chip

(EM4222) is considerably miniaturized, less than 1 mm^2 and at a lower cost, which is packed with powerful functions too. Furthermore, designed On-Chip Antenna (OCA) is based on inductive coupling technology resonating at the working frequency selected and embedded into the Chip. The antenna is performed by the coil and it is modelled by lumped elements and it is implemeted through 3D EM simulation tool; the software used to perform this design in IE3D.

Design at glance, both the Reader and the Tag coils are drawn together in order to deliver the maximum signal permitted to wake up the Chip. The matching network stage must be designed in order to deliver maximum power from the Tag coil to the Chip. However, the final Antenna designed (0.7 mm x 0.75 mm) is a bigger than the Chip (0,64mm x 0.64 mm), but considered optimal since it is refered to a milimeter scale and its features accomplish main goals in the terms of adaptation. Our results, according to settings and chosen materials, bring into agreement with the tittle of this chapter, a small OCA antenna operating within the near field.

7. Acknowledgement

This work has been conducted in the Department of Telecommunication Engineering at the Czech Technical University in Prague and supported by the Ministry of Education, Youth, and Sport of the Czech Republic under research program MSM6840770038.

8. References

Atmel Corporation. (2002). Atmel release document. *Tag Tuning/RFID, application Note*. 2055A–RFID–07/02. San Jose. USA.

Lauss Finkenzeller. (2003). *RFID Handbook, fundamentals and Applications in Contactless Smart Cards and Identification*. Editorial Wiley, pp 20.

Stefano Gregori, Yunlei Li, Huijuan Li, Jin Liu, Franco Maloberti,. (2004). ISLPED'04, August 9-11 2.45 GHz Power and Data Transmission for a Low-Power Autonomous Sensors Platform. California. USA.

L. H. Guo, W. G. Yeoh, Y. B. Choi, X. S. Chen, H. Y. Li, M. Tang, G. Q. Lo, N. Balasubramanian, and D. L. Kwong. (2004). *Design and Manufacturing of Small Area On-Chip-Antenna (OCA) for RFID Tags*. Institute of Microelectronics, 11 Science Park Road, Singapore.

L. H. Guo, A. P. Popov, H. Y. Li, Y. H. Wang, V. Bliznetsov, G. Q. Lo, N. Balasubramanian, and D.-L. Kwong. (2006). *A Small OCA on a 1×0.5-mm2 2.45-GHz RFID Tag – Design and Integration Based on a CMOS-Compatible Manufacturing Technology*. IEEE electron device letters, vol. 27, no. 2, february 2006.

Lluis Prat Viñas. (1999). *Circuits and electronics devices. Electronic's theory*. UPC Publishing. pp 169. Barcelona. Spain.

Sabri Serkan Basat, Dr. Manos M. Tentzeris, Dr. John Papapolymerou, Dr. Joy Laskar. (2006). *Design and characterization of RFID modules inmultilayer configurations*. Georgia Institute of Technology, December. Georgia.

M. Usami. (2004). *An ultra small RFID chip: μ-chip*. Proc. RFIC, 2004, pp. 241–244.. Japan.

H. Yan, J.G. Marcias Montero, A. Akhnoukh, L.C.N. de Vreede, J.N. Burghartz. (2006). *An Integration for RF Power Harvesting*. DIMES institute of the Delf. University of Technology. Netherlands.

M P Wilson. (2002). *Modelling of integrated VCO resonators using Momentum*. Tality,UK.

8

RFID TAGs Coil's Dimensional Parameters Optimization As Excitable Linear Bifurcation System

Ofer Aluf

Department of Physics, Ben-Gurion University of the Negev, Be'er-Sheva, Israel

1. Introduction

In this article, Very Crucial subject discussed in RFID TAG's Coil design. RFID Equivalent circuits of a Label can be represent as Parallel circuit of Capacitance (Cpl), Resistance (Rpl), and Inductance (Lpc). The Label measurement principal is as follow: Label positioned in defined distance to measurement coil, Low current or voltage source, Measuring of $|Z|$ and Teta of measurement coil, Resonance frequency fro at Teta = 0, Calculation of unloaded quality factor Q0 out of measured bandwidth B0. The Coil design procedure is based on three important steps. The first is Preparations: Definition of limits and estimations, Calculation of parameters. The second is Matrix Run 1: Definition of matrix, Equations and Calculations, Sample production, evaluation of samples, and Re calculation of parameters. The third is Matrix Run 2: Definition coil parameters, samples production, evaluation of samples, and decision on best parameters. Any RFID Coil design include definition of Limits and estimations, Maximum dimensions of coil (Maximum overall length, width), definition of the minimum gap between tracks and track thickness. It is very important to emphasis that basic Label IC (NXP I CODE for example), equivalent circuit is Capacitor (Cic) and Resistor (Ric) in parallel. The additional coil traces give the complete RFID equivalent circuit (Capacitor, Resistor, and inductor in parallel). The RFID equivalent circuit can be represent as a differential equation which depending on variable parameter. The investigation of RFID's differential equation based on bifurcation theory, the study of possible changes in the structure of the orbits of a differential equation depending on variable parameters. The article first illustrate certain observations and analyze local bifurcations of an appropriate arbitrary scalar differential equation. Since the implicit function theorem is the main ingredient used in these generalizations, include a precise statement of this theorem. Additional analyze the bifurcations of a RFID's differential equation on the circle. Bifurcation behavior of specific differential equations can be encapsulated in certain pictures called bifurcation diagrams. All of that for optimization of RFID TAG's dimensional parameters optimization – to get the best performance.

2. RFID TAG equivalent circuit

RFID TAG can be represent as a parallel Equivalent Circuit of Capacitor and Resistor in parallel. For example see below NXP/PHILIPS ICODE IC Parallel equivalent circuit.

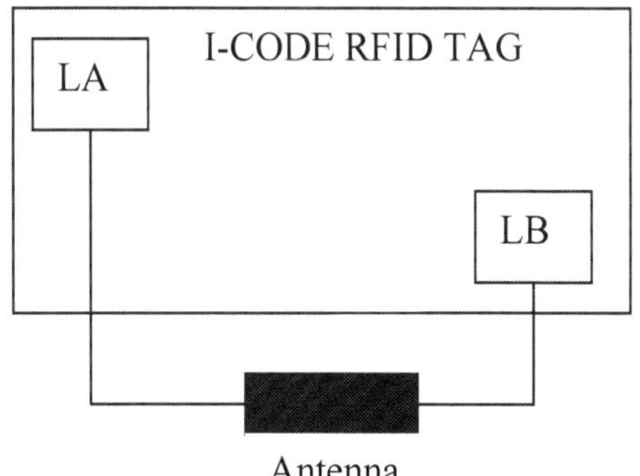

Fig. 1.

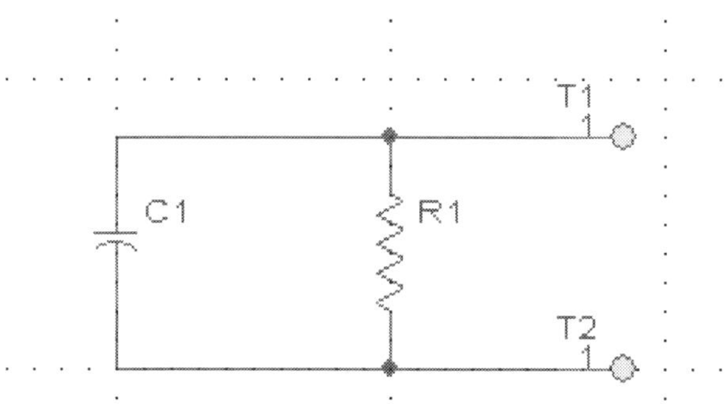

Fig. 2.

The RFID TAG Antenna can be represents as Parallel inductor to the basic RFID Equivalent Circuit. The simplified complete equivalent circuit of the label is as below:

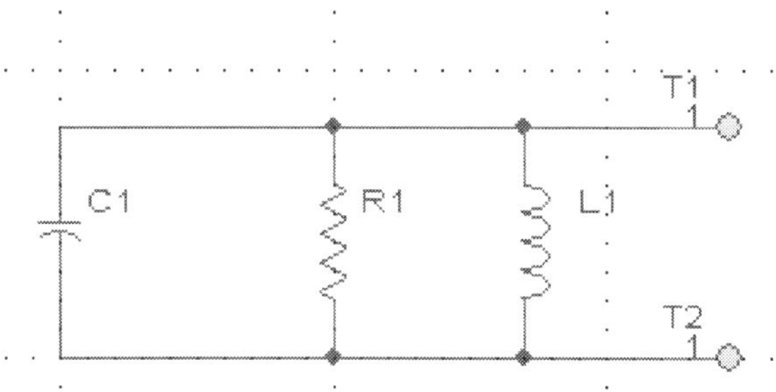

Fig. 3.

$$C1 = Cic + Ccon + Cc, \; R1 = (Ric * Rpc)/(Ric + Rpc) .$$

$$Vl1 = L * \frac{dIl}{dt}, \; Ic1 = C * \frac{dVc}{dt}$$

$$Il1 = \frac{1}{L1} * \int_0^{t1} Vl1 * dt$$

$$\sum_{i=1}^{i=3} Ii = 0, \frac{V}{R1} + C1 * \frac{dV}{dt} + \frac{1}{L1} * \int_{t=0}^{t=t1} V * dt = 0$$

$$\frac{1}{R1} * \frac{dV}{dt} + C1 * \frac{d^2V}{dt^2} + \frac{1}{L1} * V = 0$$

we get differential equation of RFID TAG sys which describe the evolution of the sys in continues time. V = V(t).

Now I define the following Variable setting definitions:

$$V2 = \frac{dV1}{dt} = \frac{dV}{dt}, V1 = V$$

And get the dynamic equation system:

$$\frac{dV1}{dt} = V2$$

$$\frac{dV2}{dt} = -\frac{1}{C1 * R1} * V2 - \frac{1}{C1 * L1} * V1$$

The system shape is as Non linear system equations:

$$\frac{dV1}{dt} = f1(V1, V2..., Vn), \quad \frac{dV2}{dt} = f2(V1, V2..., Vn)$$

The V1 and V2 variables are the phase space dimension two.

Now Lets Move to three variables system – which the time (t) is the third variable, V3 = t.

$$\frac{dV1}{dt} = V2, \quad \frac{dV2}{dt} = -\frac{1}{C1 * L1} * V1 - \frac{1}{C1 * R1} * V2+, \quad \frac{dV3}{dt} = 1$$

Now we get the RFID's coil dimensional parameters:

$$d = 2 * (t + w) / \Pi, Aavg = a0 - Nc * (g + w), Bavg = b0 - Nc * (g + w)$$

a0, b0 – Overal dimensions of the coil.

Aavg, Bavg – Average dimensions of the coil.

t – Track thickness, w – Track width, g – Gap between tracks.

Nc – Number of turns, d – Equivalent diameter of the track.

Average coil area; -Ac=Aavg * Bavg. Integrating all those parameters give the equations for inductance calculation:

$$X1 = Aavg * \ln\left(\frac{2 * Aavg * Bavg}{d * (Aavg + \sqrt{Aavg^2 + Bavg^2})} \right)$$

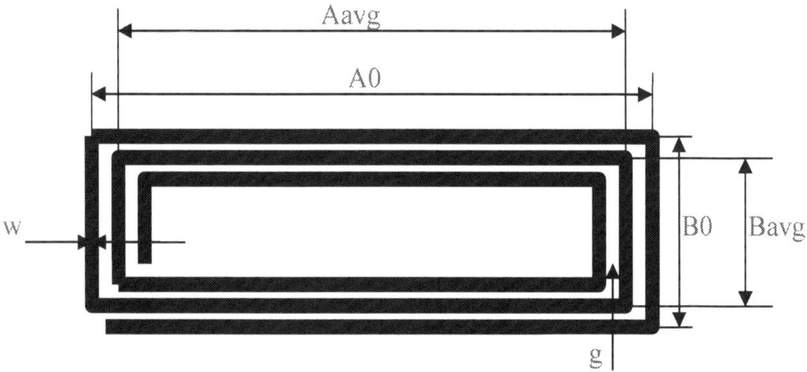

Fig. 4.

$$X2 = Bavg * \ln\left(\frac{2 * Aavg * Bavg}{d * (Bavg + \sqrt{Aavg^2 + Bavg^2})}\right)$$

$$X3 = 2 *\left[Aavg + Bavg - \sqrt{\left[Aavg^2 + Bavg^2\right]}\right]$$

$X4 = (Aavg + Bavg) / 4$, The RFID's coil calculation inductance expression is

$$Lcalc = \left[\frac{\mu0}{\pi} * [X1 + X2 - X3 + X4] * Nc^p\right]$$

, L1 = Lcalc

Definition of limits, Estimations: Track thickness t, Al and Cu coils (t > 30um). The printed coils as high as possible. Estimation of turn exponent p is needed for inductance calculation.

Coil manufacturing technology	P
Wired	1.8 – 1.9
Etched	1.75 – 1.85
Printed	1.7 – 1.8

Table 1.

Now I integrate the Lcalc value inside the differential equations which characterize the RFID system with the Coil inductance.

$$\frac{dV1}{dt} = 0 * V1 + 1 * V2 + 0 * V3$$

$$\frac{dV2}{dt} = -\frac{1}{C1 *\left[\frac{\mu0}{\pi} * [X1 + X2 - X3 + X4] * Nc^p\right]} * V1 - \frac{1}{C1 * R1} * V2 + 0 * V3$$

From above K1 condition we get RFID overall relationship.

$$[X1 + X2 - X3 + X4] = \frac{\pi}{C1 * \mu0 * Nc^{p}}$$

then

$$\left[Aavg * \ln\left(\frac{2 * Aavg * Bavg}{d * (Aavg + \sqrt{Aavg^2 + Bavg^2})} \right) + Bavg * \ln\left(\frac{2 * Aavg * Bavg}{d * (Bavg + \sqrt{Aavg^2 + Bavg^2})} \right) -2 * \left[Aavg + Bavg - \sqrt{\left[Aavg^2 + Bavg^2 \right]} \right] + (Aavg + Bavg) / 4 = \frac{\pi}{C1 * \mu0 * Nc^{p}} \right]$$

In case the A matrix is $\begin{bmatrix} 0 & 1 \\ 0 & 0 \end{bmatrix}$

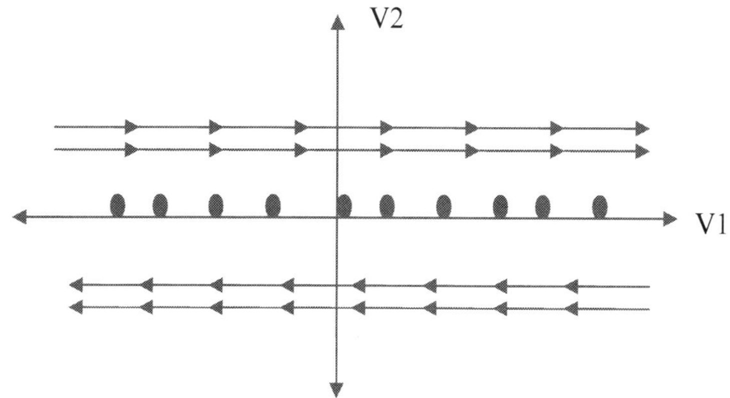

Fig. 6.

to fulfill those A matrix conditions , the following K1, K2 condition must exist:

$$K2 \to 0, \forall\{-\frac{1}{C1 * R1}\} \to 0, \forall C1, R1 \to \infty$$

$$K1 = 0, \forall\{-\frac{1}{C1 * \left[\frac{\mu0}{\pi} * [X1 + X2 - X3 + X4] * Nc^{p} \right]}\} = 0$$

$$C1 * \left[\frac{\mu0}{\pi} * [X1 + X2 - X3 + X4] * Nc^{p} \right] \to \infty$$

4. RFID TAG behavior based on eigensolutions and eigenvalues characterization

Sketch a typical phase portrait for the case

$$\lambda 2 < \lambda 1 < 0$$

$$\frac{1}{2}*K2 - \frac{1}{2}*\sqrt{K2^2 + 4*K1} < \frac{1}{2}*K2 + \frac{1}{2}*\sqrt{K2^2 + 4*K1} < 0$$

$$\lambda 1 < 0 ...... \rightarrow ...... -K2 > \sqrt{K2^2 + 4*K1}$$

$$-\{-\frac{1}{C1*R1}\} > \sqrt{\{-\frac{1}{C1*R1}\}^2 + 4*\{-\frac{1}{C1*\left[\frac{\mu 0}{\pi}*[X1+X2-X3+X4]*Nc^p\right]}\}}$$

$$\frac{1}{C1*R1} > \sqrt{\{\frac{1}{C1*R1}\}^2 - \{\frac{4}{C1*\left[\frac{\mu 0}{\pi}*[X1+X2-X3+X4]*Nc^p\right]}\}}$$

$$\lambda 2 < \lambda 1$$

$$\frac{1}{2}*K2 - \frac{1}{2}*\sqrt{K2^2 + 4*K1} < \frac{1}{2}*K2 + \frac{1}{2}*\sqrt{K2^2 + 4*K1}$$

$$0 < \sqrt{K2^2 + 4*K1} .... \rightarrow ..... K2^2 + 4*K1 > 0 ..... \rightarrow ..... K2^2 > -4*K1$$

$$\{-\frac{1}{C1*R1}\}^2 > -4*\{-\frac{1}{C1*\left[\frac{\mu 0}{\pi}*[X1+X2-X3+X4]*Nc^p\right]}\}$$

$$\{\frac{1}{C1*R1}\}^2 > \frac{4}{C1*\left[\frac{\mu 0}{\pi}*[X1+X2-X3+X4]*Nc^p\right]}$$

Then both eigensolutions decay exponentially. The fixed point is a stable node, except eigenvectors are not mutually perpendicular, in general. Trajectories typically approach the origin tangent to the slow eigendirection, defined as the direction spanned by the eigenvector with the smaller $|\lambda|$. In backward time $t \rightarrow \infty$ the trajectories become parallel to the fast eigendirection.

if we reverse all the arrows in the above figure, we obtain a typical phase portrait for an unstable node. Now I investigate the case when eigenvalues are complex number. If the eigenvalues are complex, the fixed point is either a center or a spiral. The origin is surrounded by a family of closed orbits. Note that centers are neutrally stable, since nearby trajectories are neither attracted to nor repelled from the fixed point. A spiral would accur if the RFID sys were lightly damped. Then the trajectory would just fail to close, because the RFID sys loses a bit of energy on each cycle. To justify these statements, recall that the eigenvalues are

$$\lambda_{1,2} = \frac{1}{2}*K2 \pm \frac{1}{2}*\sqrt{K2^2 + 4*K1}$$

$$K2^2 + 4*K1 < 0$$

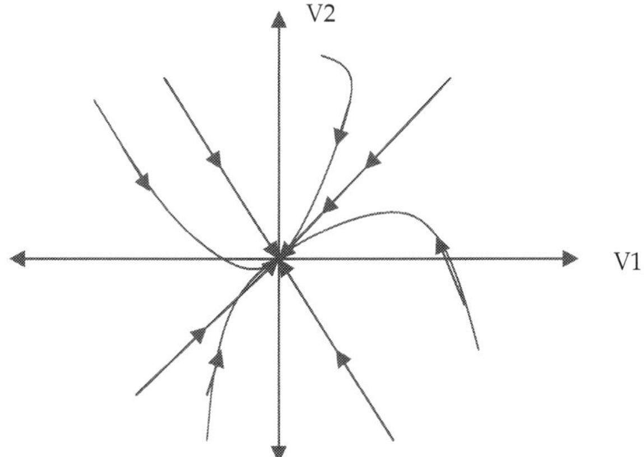

Fig. 7.

To simplify the notation, lets write the eigenvalues as

$$\lambda_{1,2} = \alpha \pm i\omega, \alpha = \frac{1}{2} * K2, \omega = -\frac{1}{2} * \sqrt{K2^2 + 4 * K1}$$

$$\omega \neq 0..\exists..V(t) = C1 * e^{\left[\frac{1}{2}*K2 + \frac{1}{2}*\sqrt{K2^2 + 4*K1}\right]*t} * S1 + C2 * e^{\left[\frac{1}{2}*K2 - \frac{1}{2}*\sqrt{K2^2 + 4*K1}\right]*t} * S2$$

$$C's, S's...complex,...since,...\lambda's...complex$$

$$V(t) = C1 * e^{[\alpha + i\omega]*t} * S1 + C2 * e^{[\alpha - i\omega]*t} * S2$$

$$Euler's..formula..\rightarrow...e^{[i\omega]*t} = \cos[\omega * t] + i * \sin[\omega * t]$$

Hence V(t) is a combination of terms involving

$$e^{\alpha*t} * \cos[\omega * t], e^{\alpha*t} * \sin[\omega * t]$$

Such terms represent exponentially decaying oscillations if $\alpha = \mathrm{Re}(\lambda) < 0$
And growing if $\alpha > 0$. The corresponding fixed points are stable and unstable spirals, respectively. If the eigenvalues are pure imaginary $\alpha = 0$, then all the solutions are periodic with period $T = \frac{2 * \pi}{\omega}$. The oscillators have fixed amplitude and the fixed point is center. For both centers and spirals, its easy to determine whether the rotation is clockwise or counterclockwise.

$$\alpha = \frac{1}{2} * K2 = \{-\frac{1}{2 * C1 * R1}\}$$

$$Decaying..oscillators..\forall..\alpha < 0 \rightarrow \{-\frac{1}{2 * C1 * R1}\} < 0 \rightarrow \frac{1}{2 * C1 * R1} > 0$$

$$Growing..oscillators..\forall..\alpha > 0 \rightarrow \{-\frac{1}{2 * C1 * R1}\} > 0 \rightarrow \frac{1}{2 * C1 * R1} < 0$$

C1, R1 > 0 always then only the first behavior, decaying oscillator can exist in our RFID system.

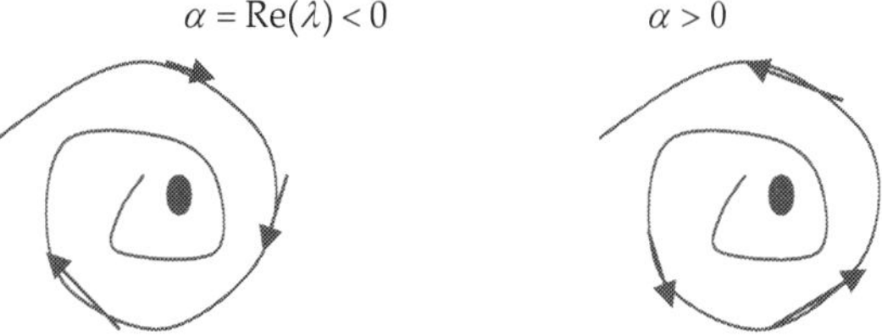

Fig. 8.

In all analysis until now, we have been assuming that the eigenvalues are distinct. What happens if the eigenvalues are equal ? Suppose eigenvalues are equal $\lambda 1 = \lambda 2 = \lambda$ then there are two possibilities: either there are two independent eigenvectors corresponding to λ , or there's only one. If there are two independent eigenvectors, then they span the plane and so every vector is an eigenvector with this same eigenvalue λ. To see this, lets write an arbitary vector X0 as a linear combination of the two eigenvectors: X0=C1*S1+C2*S2. Then

$$A * X0 = A * (C1 * S1 + C2 * S2) = C1 * \lambda * S1 + C2 * \lambda * S2 = \lambda * X0$$

X0 is also an eigenvector with eigenvalue λ. Since the multiplication by A simply stretches every vector by a factor λ, the matrix must be a multiple of the identity :

$A = \begin{bmatrix} \lambda & 0 \\ 0 & \lambda \end{bmatrix}$ then if $\lambda \neq 0$, all trajectories are straight lines through the origin

$X(t) = e^{\lambda * t} * X0$ and the fixed point is a star node. On the other hand, if $\lambda = 0$ the whole plane is filled with fixed points.

Lets now sketch the above options with RFID Overal parameters restriction.
$\lambda 1 = \lambda 2 = \lambda \neq 0$ then

$$\frac{1}{2} * K2 + \frac{1}{2} * \sqrt{K2^2 + 4 * K1} = \frac{1}{2} * K2 - \frac{1}{2} * \sqrt{K2^2 + 4 * K1}$$

$$\sqrt{K2^2 + 4 * K1} = 0 \rightarrow K2^2 + 4 * K1 = 0 \rightarrow K2^2 = -4 * K1$$

$$\frac{\mu 0}{\pi} * [X1 + X2 - X3 + X4] * Nc^p = C1 * 4 * R1^2$$

Now lets summarize the classification of fixed points in RFID system based on all investigation I did. It is easy to show the type and stability of all the different fixed points on a single diagram.

$$\tau^2 - 4 * \Delta = K2^2 + 4 * K1 = 0, \rightarrow K2 = 2 * \sqrt{-K1}$$

$$\tau = trace(A) = K2, \Delta = \det(A) = -K1$$

$$\lambda^2 - K2 * \lambda - K1 = 0, \lambda_{1,2} = \frac{1}{2} * \left[\tau \pm \sqrt{\tau^2 - 4 * \Delta} \right]$$

$$\tau = \lambda 1 + \lambda 2 = K2, \Delta = \lambda 1 * \lambda 2 = -K1$$

$$Charecteristic..equation...(\lambda - \lambda 1) * (\lambda - \lambda 2) = \lambda^2 - \tau * \lambda + \Delta = 0$$

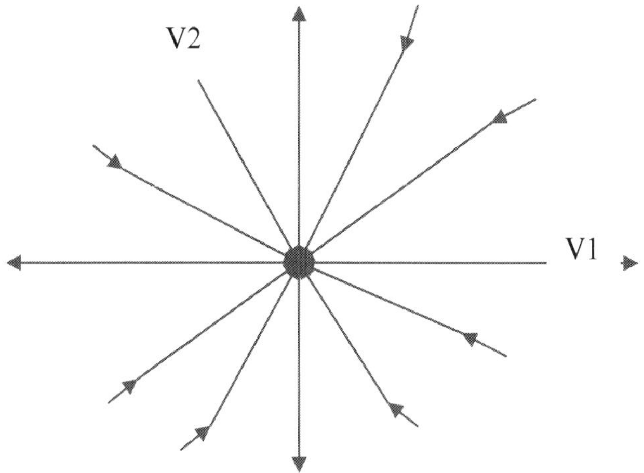

Fig. 9.

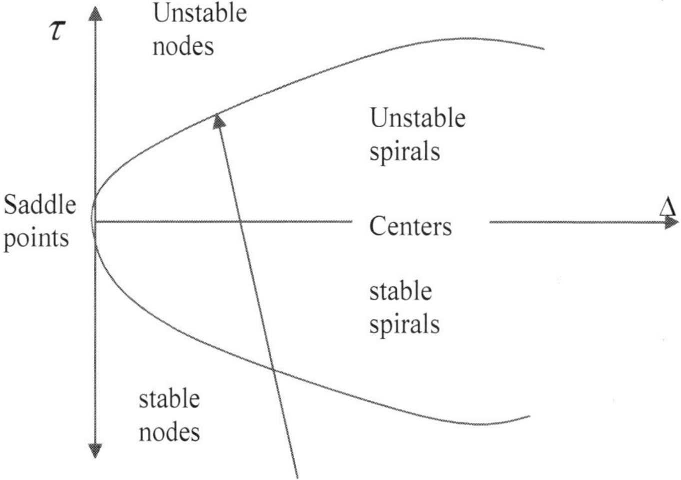

Fig. 10.

$\Delta < 0....(K1 > 0)$	$\Delta > 0....(K1 < 0)$	$\Delta = 0....(K1 = 0)$
The eigenvalues are real and have opposite sign hence the fixed point is a saddle point,	The eigenvalues are either real with the same sign (nodes), or complex conjugate (spiral & centers).	At least one of the eigenvalues is zero. Then the origin is not an isolated fixed point. There is either a whole line of a fixed points, or a plane of fixed point

Table 2.

Nodes satisfy $\tau^2 - 4*\Delta > 0$ and spirals satisfy $\tau^2 - 4*\Delta < 0$. The parabola $\tau^2 - 4*\Delta = 0$ is the borderline between nodes and spirals. Star nodes and degenerate nodes live on this parabola. The stability of the nodes and spirals is determined by τ value. When $\tau < 0$, both eigenvalues have negative real parts, so the fixed point is stable. Unstable spirals and nodes have $\tau > 0$. Neutrally stable centers live on the borderline $\tau = 0$, where eigenvalues are purely imaginary.

5. RFID system phase plan

For our RFID system the general form of a vector field on the phase plan is

$$\dot{V1} = f1(V1, V2)$$

$$\dot{V2} = f2(V1, V2)$$

Where V1 is the voltage on RFID system (V) and V2 is the derivative of that voltage respect with time (dV/dt). The f1 and f2 are given functions and this system can be written more compactly in vector notation as:

$$\frac{dV}{dt} = f(V)$$

Where V=(V1,V2) and f(V)=(f1(V), f2(V)). Here V represents a point in the phase plane (RFID voltage at time t), and dV/dt is the voltage change in time at that point. By flowing the vector field, a phase point traces out a solution V(t), corresponding to a trajectory winding through the phase plan.

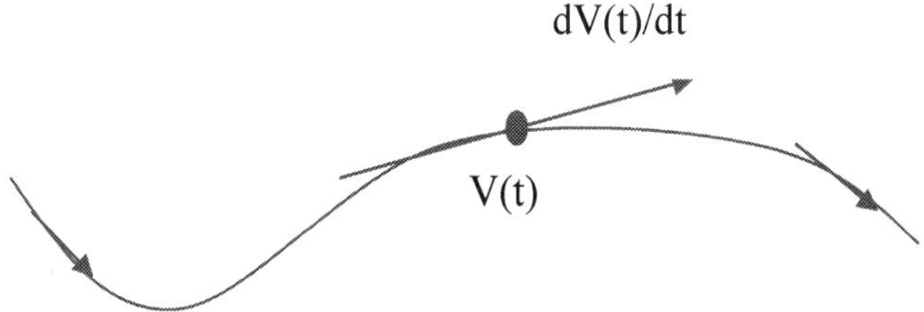

Fig. 11.

Furthermore, the entire phase plane is filled with trajectories, since each point can play the role of an initial condition. The most salient features of phase portrait are:

1. Fixed points satisfy f(V*)=0, and correspond to steady states or equilibria of the system.
2. The closed orbits, these correspond to periodic solutions, i.e. solutions for which V(t+T) = V(t) for all t values, and T>0.
3. The stability or instability of the fixed points and closed orbits. Some fixed points are unstable, because nearby trajectories tend to move away from them, where the closed orbit is stable.

Now lets sketch the numerical computation of RFID phase portraits. Using Runge - Kutta method, which in vector form is the following:

$$V_{n+1} = V_n + \frac{1}{6} * \left(\varsigma_1 + 2 * \varsigma_2 + 2 * \varsigma_3 + \varsigma_4 \right)$$

$$\varsigma_1 = f(V_n) * \Delta t$$

$$\varsigma_2 = f(V_n + \frac{1}{2} * \varsigma_1) * \Delta t$$

$$\varsigma_3 = f(V_n + \frac{1}{2} * \varsigma_2) * \Delta t$$

$$\varsigma_4 = f(V_n + \varsigma_3) * \Delta t$$

$$\Delta t = 0.1 \ldots (stepsize)$$

The step 0.1 usually provides sufficient accuracy for our RFID system. When plotting the phase portrait, there is a grid of representative vectors in the vector field. The arrowheads and different lengths of the vectors tend to clutter such pictures. A plot of the direction field is clearer, short line segments are used to indicate the local direction of flow. Existence and Uniqueness of RFID system: We consider in our system the initial state dV/dt=f(V) and V(0)=V0 which is the initial RFID voltage before investigation the system dynamic change with time. Suppose that f is continuous and that all partial derivatives

$$\frac{\partial f_i}{\partial V_j}, i,j = 1,2,\ldots,n$$

Are continuous for V in some open connected set $D \subset \mathbb{R}^n$. Then for $V_0 \in D$, the initial value problem has a solution V(t) on some time interval about t=0 , and the solution is unique. Existence and uniqueness of solutions are guaranteed if f is continuously differentiable. We will assume that all our vector fields are smooth enough to ensure the existence and uniqueness of solutions, starting from any point in the RFID phase space. The existence and uniqueness theorem has an important corollary, different trajectories never intersect. If two trajectories did intersect, then there would be two solutions starting from the same point (the crossing point) and would violate the uniqueness part of the theorem. Trajectories cant move in two directions at once in our RFID system. Because trajectories cant intersect, phase portraits always have a well groomed look to them. Otherwise they might degenerate into a snarl of criss crossed curves and the existence and uniqueness theorem prevents this from happening. In our RFID two dimensional phase spaces, these results have especially strong topological consequences. For example, suppose there is a closed orbit behavior
in the RFID phase plan. Then any trajectory starting inside the closed orbit is trapped in there forever. If there are fixed points inside that closed orbit, then of course the trajectory might eventually approach one of them. For vector fields on the plane, the Poincare-Bendixson theorem states that if a trajectory is confined to a closed, bounded region and there are no fixed points in the region, then trajectory must eventually approach a closed orbit. Now Lets suppose that V1* and V2* is fixed point (V1*,V2*) then we get

$$f_1(V1^*, V2^*) = 0$$

$$f_2(V1^*, V2^*) = 0$$

$$Lets........$$

$$\xi_1 = V1 - V1*$$

$$\xi_2 = V2 - V2*$$

ξ_1, ξ_2 - denote the components of a small disturbance from the fixed point. To see whether the disturbance grows or decays, we need to derive differential equations for ξ_1, ξ_2. Lets do the ξ_1 equation first.

$$\dot{\xi}_1 = \dot{V}_1 = f1(V1* + \xi 1, V2* + \xi 2) =$$

$$f1(V1*, V2*) + \xi_1 * \frac{\partial f1}{\partial V1} + \xi_2 * \frac{\partial f1}{\partial V2} + O(\xi_1^2, \xi_2^2, \xi_1 * \xi_2)$$

$$\forall .... f1(V1*, V2*) = 0...Fixed..point..condition$$

$$f1 = V2, f2 = V1 * K1 + V2 * K2$$

$$\frac{\partial f1}{\partial V1} = 0, \frac{\partial f1}{\partial V2} = 1$$

Similarly we can write:

$$\dot{\xi}_2 = \dot{V}_2 = f2(V1* + \xi 1, V2* + \xi 2) =$$

$$f2(V1*, V2*) + \xi_1 * \frac{\partial f2}{\partial V1} + \xi_2 * \frac{\partial f2}{\partial V2} + O(\xi_1^2, \xi_2^2, \xi_1 * \xi_2)$$

$$\forall .... f2(V1*, V2*) = 0...Fixed..point..condition$$

$$f1 = V2, f2 = V1 * K1 + V2 * K2$$

$$\frac{\partial f2}{\partial V1} = K1, \frac{\partial f2}{\partial V2} = K2$$

Hence the disturbance ξ_1, ξ_2 evolve according to

$$\xi_1, \xi_2$$

$$\begin{bmatrix} \dot{\xi}_1 \\ \dot{\xi}_2 \end{bmatrix} = \begin{bmatrix} \frac{\partial f1}{\partial V1} & \frac{\partial f1}{\partial V2} \\ \frac{\partial f2}{\partial V1} & \frac{\partial f2}{\partial V2} \end{bmatrix} * \begin{bmatrix} \xi_1 \\ \xi_2 \end{bmatrix} + Quadratic..terms$$

$$Matrix$$

$$A = \begin{bmatrix} \frac{\partial f1}{\partial V1} & \frac{\partial f1}{\partial V2} \\ \frac{\partial f2}{\partial V1} & \frac{\partial f2}{\partial V2} \end{bmatrix}_{(V1*, V2*)}$$

$$Linearized..system$$

$$\begin{bmatrix} \dot{\xi}_1 \\ \dot{\xi}_2 \end{bmatrix} = \begin{bmatrix} \dfrac{\partial f1}{\partial V1} & \dfrac{\partial f1}{\partial V2} \\ \dfrac{\partial f2}{\partial V1} & \dfrac{\partial f2}{\partial V2} \end{bmatrix} * \begin{bmatrix} \xi_1 \\ \xi_2 \end{bmatrix}$$

As we move from one dimensional to two dimensional systems, still fixed points can be created or destroyed or destabilized as parameters are varied – in our system RFID global TAG parameters. We can describe the ways in which oscillations can be turned on or off. The exact meaning of bifurcation is: if the phase portrait changes its topological structure as a parameter is varied, we say that a bifurcation has occurred. Examples include changes in the number or stability of fixed points, close orbits, or saddle connections as a parameter is varied.

6. RFID TAG with losses as a dynamic system

RFID TAG system is not an ideal and pure solution. There are some Losses which need to be under consideration. The RFID TAG losses can be represent first by the equivalent circuit. The main components of RFID TAG simple equivalent circuit are Capacitor in Parallel to Resistor and additional Parallel inductance (Antenna Unit). The RFID equivalent circuit Under Losses consideration is as describe below:

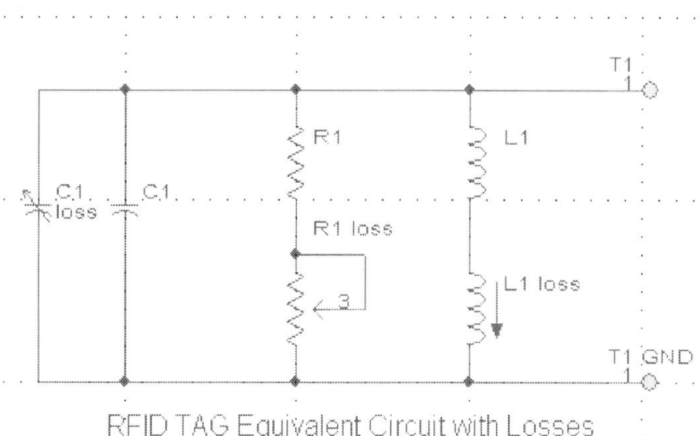

Fig. 12.

C1loss, R1loss and L1loss need to be tuned until we get the desire and optimum dynamic behavior of RFID system. Now, Lets investigate the RFID TAG system under those losses. The C1, R1, L1 (Lcalc) move value displacement due to those losses: C1 >> C1+C1loss, R1 >> R1+R1loss, L1 >> L1+L1loss. We consider in all analysis that L1 is Lcalc and depend in many parameters.

$$Lcalc = Lcalc(X1, X2, X3, X4, N_c^p) = \left[\frac{\mu0}{\pi} * [X1 + X2 - X3 + X4] * N_c^p \right]$$

$$X1 \rightarrow X1 + X1loss, X2 \rightarrow X2 + X2loss$$

$$X3 \rightarrow X3 + X3loss, X4 \rightarrow X4 + X4loss$$

$$then.....Lcalc \rightarrow Lcalc + Lcalcloss$$

$$Lets..go..back..to..each..RFID..Coil..Parameter..and..his..loss..value$$

$$d \rightarrow d + dloss,....Aavg \rightarrow Aavg + Aavgloss,...Bavg \rightarrow Bavg + Bavgloss$$

$$a0 \rightarrow ao + a0loss,...bo \rightarrow bo + boloss,....t \rightarrow t + tloss, w \rightarrow w + wloss$$

$$g \rightarrow g + gloss$$

Now Lets sketch the X1...X4 graphs depend on Aavg and Bavg:
X1=X1(Aavg, Bavg), X2=X2(Aavg, Bavg), X3=X3(Aavg, Bavg),
X4=X4(Aavg, Bavg).

$$X1 = Aavg * \ln\left(\frac{2 * Aavg * Bavg}{d * (Aavg + \sqrt{Aavg^2 + Bavg^2})} \right) = X1(Aavg, Bavg) \text{ , 3D sketch}$$

Fig. 13.

$$X2 = Bavg * \ln\left(\frac{2 * Aavg * Bavg}{d * (Bavg + \sqrt{Aavg^2 + Bavg^2})} \right) = X2(Aavg, Bavg) \text{ , 3D sketch}$$

Fig. 14.

$$X3 = 2 * \left[Aavg + Bavg - \sqrt{\left[Aavg^2 + Bavg^2 \right]} \right] = X3(Aavg, Bavg) \text{ , 3D sketch}$$

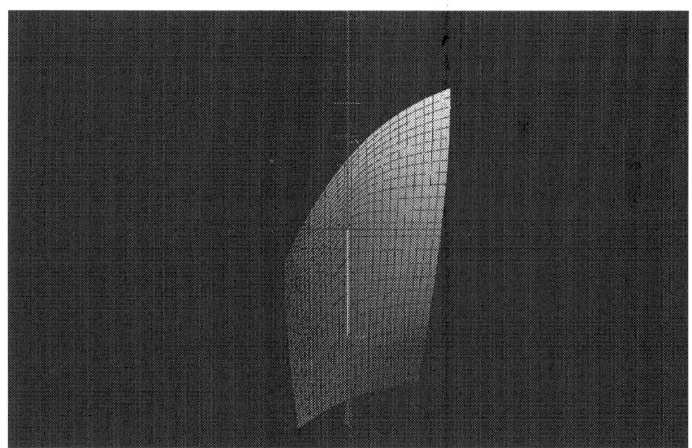

Fig. 15.

$$X4 = (Aavg + Bavg) / 4 = X4(Aavg, Bavg) \text{ , 3D sketch}$$

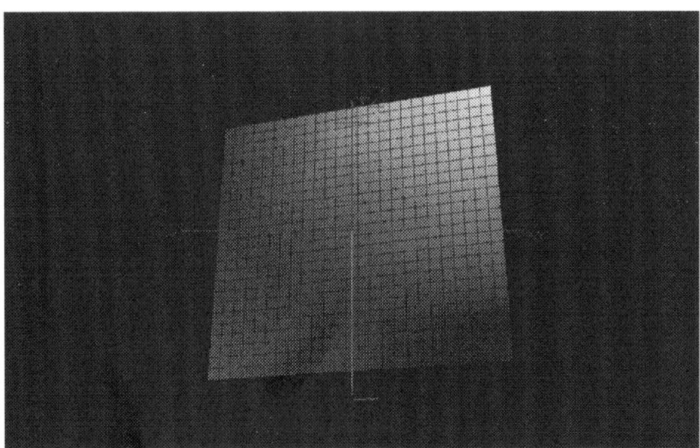

Fig. 16.

All X1, ... X4 draw in one 3D coordinate system

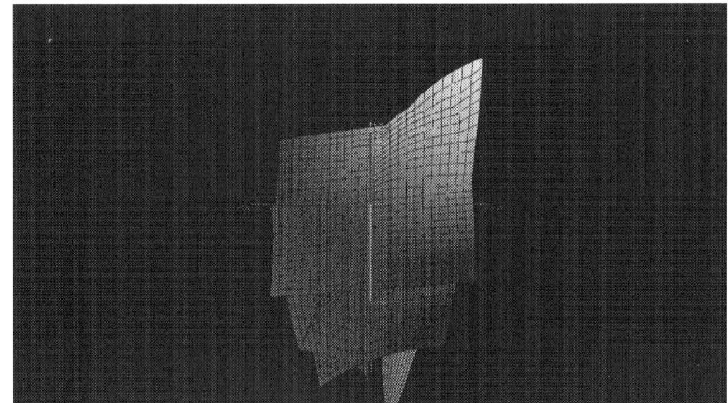

Fig. 17.

Now lets sketch 3D diagram of Lcalc = Lcalc (Aavg, Bavg)

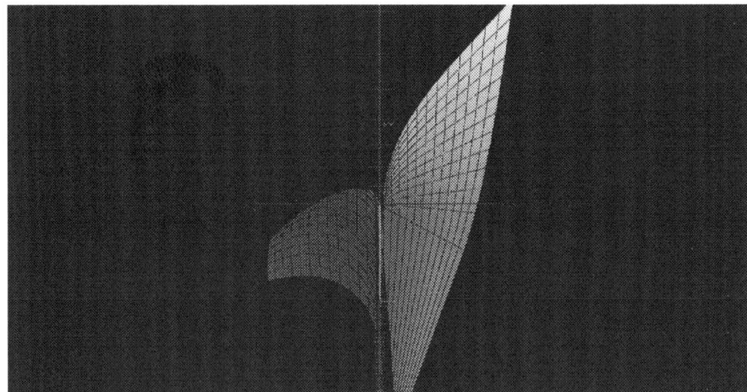

Fig. 18.

$$K1 = K1(a0, b0, w, g, d, N_c, t, p, C1, R1) = \{-\frac{1}{C1 * \left[\frac{\mu 0}{\pi} * [X1 + X2 - X3 + X4] * Nc^p\right]}\}$$

K1 = K1(Aavg, Bavg) 3D Sketch graph:

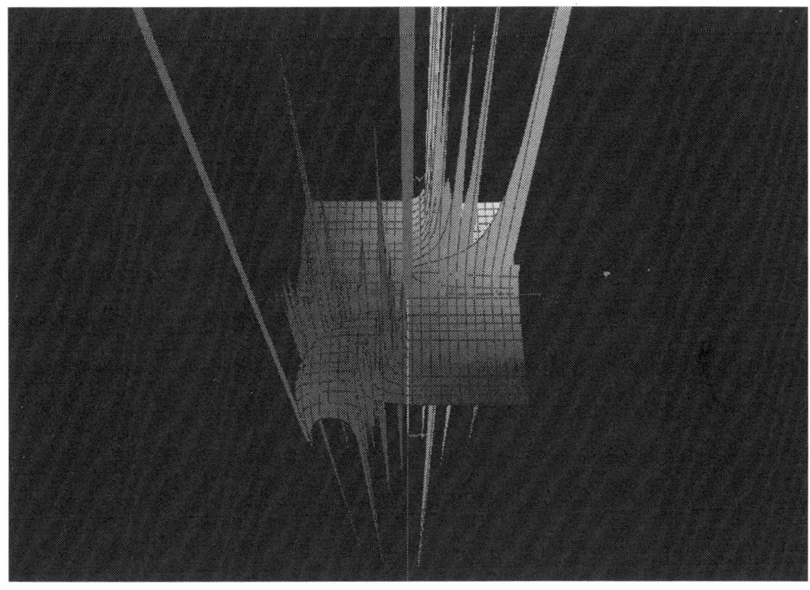

Fig. 19.

K1 is a critical function in all RFID Bifurcation system. Calculation of Aavgloss, Bavgloss and dlossgives:

$$Aavg \rightarrow Aavg + Aavgloss =$$
$$a0 + a0loss - (Nc + Ncloss) * (g + gloss + w + wloss) =$$
$$a0 + a0loss - Nc * g - Nc * gloss - Nc * w - Nc * wloss$$
$$-Ncloss * g - Ncloss * gloss - Ncloss * w - Ncloss * wloss =$$
$$a0 - Nc * (g + w) + aloss - Nc * (gloss + wloss)$$
$$-Ncloss * (g + gloss + w + wloss) = Aavg + aloss - Nc * (gloss + wloss)$$
$$-Ncloss * (g + gloss + w + wloss)..........then$$
$$Aavgloss = a0loss - (Nc + Ncloss) * (gloss + wloss) - Ncloss * (g + w)$$
$$and..in..the..same..way..get..Bavgloss..value..Bavg \rightarrow Bavg + Bavgloss$$
$$Bavgloss = b0loss - (Nc + Ncloss) * (gloss + wloss) - Ncloss * (g + w)$$
$$d \rightarrow d + dloss = 2 * (t + tloss + w + wloss) / \pi = d + \frac{2 * (tloss + wloss)}{\pi}$$
$$dloss = \frac{2 * (tloss + wloss)}{\pi}$$

Lets now describe the X1, .., X4 , Lcalc internal function parameter under Losses.

$$X1 \rightarrow X1 + X1loss =$$

$$\left[Aavg + Aavgloss\right] * \ln\left[\frac{2*(Aavg + Aavgloss)(Bavg + Bavgloss)}{(d + dloss)*\left[Aavg + Aavgloss + \sqrt{(Aavg+Aavgloss)^2 + (Bavg+Bavgloss)^2}\right]}\right] =$$

$$X1 + \ln\left[\frac{\left\{\dfrac{2*(Aavg+Aavgloss)(Bavg+Bavgloss)}{(d+dloss)*\left[Aavg+Aavgloss+\sqrt{(Aavg+Aavgloss)^2+(Bavg+Bavgloss)^2}\right]}\right\}^{[Aavg+Aavgloss]}}{*\left[\dfrac{d*\left[Aavg+\sqrt{Aavg^2+Bavg^2}\right]}{2*Aavg*Bavg}\right]^{Aavg}}\right]$$

$$X1loss = \ln\left[\frac{\left\{\dfrac{2*(Aavg+Aavgloss)(Bavg+Bavgloss)}{(d+dloss)*\left[Aavg+Aavgloss+\sqrt{(Aavg+Aavgloss)^2+(Bavg+Bavgloss)^2}\right]}\right\}^{[Aavg+Aavgloss]}}{*\left[\dfrac{d*\left[Aavg+\sqrt{Aavg^2+Bavg^2}\right]}{2*Aavg*Bavg}\right]^{Aavg}}\right]$$

$$X2 \rightarrow X2 + X2loss =$$

$$\left[Bavg + Bavgloss\right] * \ln\left[\frac{2*(Aavg + Aavgloss)(Bavg + Bavgloss)}{(d + dloss)*\left[Bavg + Bavgloss + \sqrt{(Aavg+Aavgloss)^2 + (Bavg+Bavgloss)^2}\right]}\right] =$$

$$X2 + \ln\left[\frac{\left\{\dfrac{2*(Aavg+Aavgloss)(Bavg+Bavgloss)}{(d+dloss)*\left[Bavg+Bavgloss+\sqrt{(Aavg+Aavgloss)^2+(Bavg+Bavgloss)^2}\right]}\right\}^{[Bavg+Bavgloss]}}{*\left[\dfrac{d*\left[Bavg+\sqrt{Aavg^2+Bavg^2}\right]}{2*Aavg*Bavg}\right]^{Bavg}}\right]$$

$$X2loss = \ln\left[\frac{\left\{\dfrac{2*(Aavg+Aavgloss)(Bavg+Bavgloss)}{(d+dloss)*\left[Bavg+Bavgloss+\sqrt{(Aavg+Aavgloss)^2+(Bavg+Bavgloss)^2}\right]}\right\}^{[Bavg+Bavgloss]}}{*\left[\dfrac{d*\left[Bavg+\sqrt{Aavg^2+Bavg^2}\right]}{2*Aavg*Bavg}\right]^{Bavg}}\right]$$

$$X3 \to X3 + X3loss =$$

$$2*[Aavg + Aavgloss + Bavg + Bavgloss - \sqrt{(Aavg+Aavgloss)^2 + (Bavg+Bavgloss)^2}\,] =$$

$$2*\left[\frac{Aavg + Bavg + (Aavgloss + Bavgloss) - \sqrt{(Aavg+Aavgloss)^2 + (Bavg+Bavgloss)^2} +}{\sqrt{Aavg^2 + Bavg^2} - \sqrt{Aavg^2 + Bavg^2}}\right] =$$

$$2*\left[Aavg + Bavg - \sqrt{Aavg^2 + Bavg^2}\right] + 2*\left[\frac{(Aavgloss + Bavgloss) - \sqrt{(Aavg+Aavgloss)^2 + (Bavg+Bavgloss)^2} +}{\sqrt{Aavg^2 + Bavg^2}}\right] =$$

$$X3 + 2*\left[\frac{(Aavgloss + Bavgloss) - \sqrt{(Aavg+Aavgloss)^2 + (Bavg+Bavgloss)^2} +}{\sqrt{Aavg^2 + Bavg^2}}\right]$$

$$X3loss = 2*\left[\frac{(Aavgloss + Bavgloss) - \sqrt{(Aavg+Aavgloss)^2 + (Bavg+Bavgloss)^2} +}{\sqrt{Aavg^2 + Bavg^2}}\right]$$

$$X4 \to X4 + X4loss = \left[Aavg + Aavgloss + Bavg + Bavglos\right] / 4 = \frac{Aavg + Bavg}{4} + \frac{Aavgloss + Bavgloss}{4}$$

$$X4loss = \frac{Aavgloss + Bavgloss}{4}$$

7. Summery

RFID TAG system can be represent as Parallel Resistor, Capacitor, and Inductance circuit. Linear bifurcation system explain RFID TAG system behavior for any initial condition V(t) and dV(t)/dt. RFID's Coil is a very critical element in RFID TAG
functionality. Optimization can be achieved by Coil's parameters inspection and System bifurcation controlled by them. Spiral, Circles, and other RFID phase system behaviors can be optimize for better RFID TAG performance and actual functionality. RFID TAG losses also controlled for best performance and maximum efficiency.

8. References

[1] Yuri A. Kuznetsov, Elelments of Applied Bifurcation Theory. Applied Mathematical Sciences.
[2] Jack K. Hale. Dynamics and Bifurcations. Texts in Applied Mathematics, Vol. 3
[3] Steven H. Strogatz, Nonlinear Dynamics and Chaos. Westview press
[4] John Guckenheimer, Nonlinear Oscillations, Dynamical Systems, and Bifurcations of Vector Fields. Applied Mathematical Sciences Vol 42.
[5] Stephen Wiggins, Introduction to Applied Nonlinear Dynamical Systems and Chaos. Text in Applied Mathematics (Hardcover).
[6] Syed A.Ahson and Mohammad Ilyas, RFID Handbook: Applications, Technology, Security, and Privacy. CRC; 1 edition (March 18, 2008).
[7] Dr klaus Finkenzeller, RFID Handbook: Fundamentals and Applications in Contactless Smart Cards and Identification 2nd edition. Wlley; 2 edition (May 23, 2003).

[8]Klaus Finken zeller and Rachel Waddington, RFID Handbook: Radio-Frequency Identification Fundamentals and Applications. John wiley & Sons (January 2000).

9

Active RFID TAGs System Analysis of Energy Consumption As Excitable Linear Bifurcation System

Ofer Aluf

Department of Physics, Ben-Gurion University of the Negev, Be'er-Sheva, Israel

1. Introduction

In this article, Very Critical and useful subject is discussed: Active RFID TAGs system energy analysis as excitable linear bifurcation system. Active RFID TAGs have a built in power supply, such as a battery, as well as electronics that perform specialized tasks. By contrast, passive RFID TAGs do not have a power supply and must rely on the power emitted by a RFID Reader to transmit data. Thus, if a reader is not present, the passive TAGs cant communicate an data. Active TAGs can communicate in the absence of a reader. Active RFID TAGs system energy consumption can be function of many variables : q(m), u(m), z(m), t(m), tms (m), when m is the number of TAG IDs which are uniformly distributed in the interval [0,1). It is very important to emphasis that basic Active RFID TAG, equivalent circuit is Capacitor (Cic), Resistor (Ric), L (RFID's Coil inductance as a function of overall Coil's parameters) all in parallel and Voltage generator Vs(t) with serial parasitic resistance. The Voltage generator and serial parasitic resistance are in parallel to all other Active RFID TAG's elements (Cic, Ric, and L (Coil inductance)). The Active RFID TAG equivalent circuit can be represent as a differential equation which depending on variable parameters. The investigation of Active RFID's differential equation based on bifurcation theory, the study of possible changes in the structure of the orbits of a differential equation depending on variable parameters. The article first illustrate certain observations and analyze local bifurcations of an appropriate arbitrary scalar differential equation. Finally investigate Active RFID TAGs system energy for the best performance using excitable bifurcation diagram.

2. Energy aware anti collision protocol for active RFID TAGs system

Active RFID TAGs have a built in power supply, such as a battery. The major advantages of an active RFID TAGs are: It can be read at distances of one hundred feet or more, greatly improving the utility of the device. It may have other sensors that can use electricity for power. The disadvantages of an active RFID TAGs are: The TAG cannot function without battery power, which limits the lifetime of the TAG. The TAG is typically more expensive. The TAG is physically larger, which may limit applications. The long term maintenance costs for an active RFID tag can be greater than those of a passive TAGs if the batteries are replaced. Battery outages in an active TAGs can result in expensive misreads. Active RFID TAGs may have all or some of the following features: Longest communication range of any TAG. The capability to perform independent monitoring and control.

The capability of initiating communications. The capabilities of performing diagnostics. The highest data bandwidth. The active RFID TAGs may even be equipped with autonomous networking ; the TAGs autonomously determine the best communication path. Mainly active RFID TAGs have a built in power supply, such as battery, as well as electronics that perform specialized tasks. By By contrast, passive RFID TAGs do not have a power supply and must rely on the power emitted by a RFID Reader to transmit data. There is an arbitration while reading TAGs (TAGs anti collision problem). First identify and then read data stored in RFID TAGs.

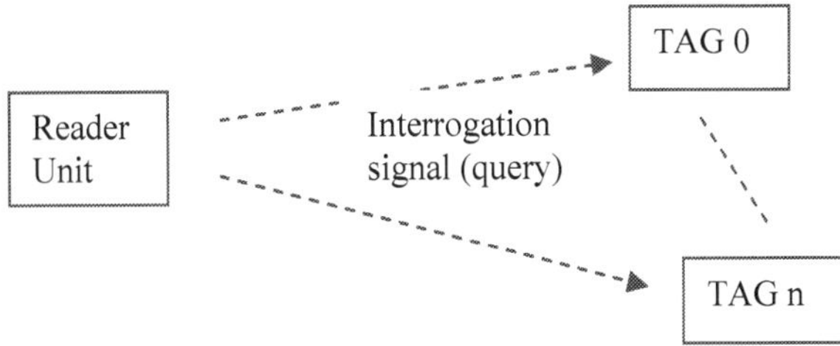

Fig. 1.

It is very important to read TAG IDs of all. The Anti collision protocol based on two methods: ALOHA and its variants and Binary tree search. ALOHA protocol reducing collisions by separating TAG responds by time (probabilistic and simple). TAG ID may not be read for a very long time. The Binary tree search protocol is deterministic in nature. Read all TAGs by successively querying nodes at a different levels of the tree with TAG IDs distributed on the tree based on there prefix. Guarantee that all TAGs IDs will be read within a certain time frame. The binary tree search procedure, however, uses up a lot of reader queries and TAG responses by relying on colliding responses of TAGs to determine which sub tree to query next. Higher energy consumption at readers and TAGs (If they are active TAGs). TAGs cant be assumed to be able to communicate with each other directly. TAGs may not be able of storing states of the arbitration process in their memory. There are three anti collision protocols: Alls include and combine ideas of a binary tree search protocol with frame slotted ALOHA, deterministic schemes, and energy aware. The first anti collision protocol is Multi Slotted (MS) scheme, multiple slots per query to reduce the chances of collision among the TAG responses. The second anti collision protocol is Multi Slotted with Selective sleep (MSS) scheme, using sleep commands to put resolved TAGs to sleep during the arbitration process. Both MS and MSS have a probabilistic flavor, TAGs choose a reply slot in a query frame randomly. The third anti collision protocol is Multi Slotted with Assigned slots (MAS), assigning TAGs in each sub tree of the search tree to a specific slot of the query frame. It's a deterministic protocol, including the replay behavior of TAGs. All three protocols can adjusting the frame size used per query. Maximize energy savings at the reader by reducing collisions among TAG responses. The frame size is also chosen based on a specified average time constraint within which all TAGs IDs must be read. The binary search protocols are Binary Tree (BT) and Query Tree (QT). Both work by splitting TAG IDs using queries from the reader until all TAGs are read.

Binary Tree (BT) relies on TAGs remembering results of previous inquiries by the readers. TAGs susceptible to their power supply. Query Tree (QT) protocol, is a deterministic TAG anti collision protocol, which is memory less with TAGs requiring no additional memory except that required to store their ID.

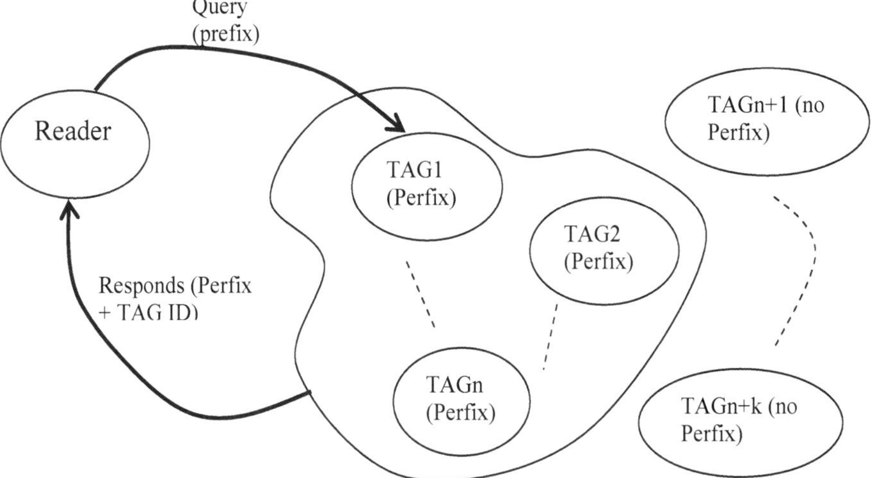

Fig. 2.

The approach to energy aware anti collision protocols for RFID systems is to combine the deterministic nature of binary search algorithms along with the simplicity of frame slotted ALOHA to reduce the number of TAG response collisions. The QT protocol relies on colliding responses to queries that are sent to internal modes of a tree to determine the location of TAG ID. Allow TAGs to transmit responses within a slotted time frame and thus, try to avoid collisions with responses from other TAGs. The energy consumption at the reader is a function of the number of queries it sends, and number of slots spent in the receive mode. Energy consumption at an active TAG is function of the number of queries received by the TAG and the number of responses it sends back. Neglect the energy spent in modes other than transmit and receive for simplicity. Assumption: Time slot in which a reader query or message is sent is equal to the duration as that of a TAG response. The

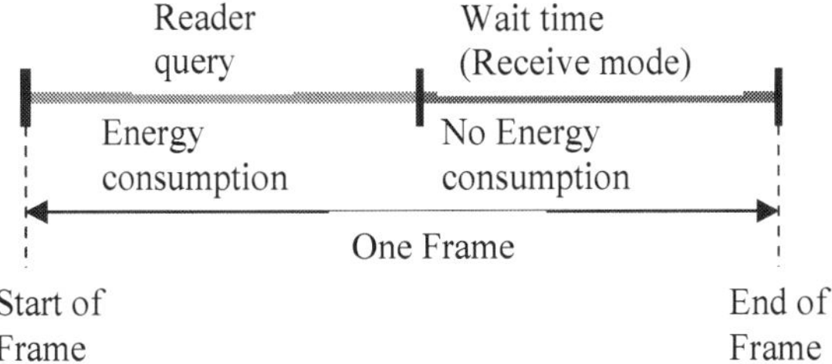

Fig. 3.

energy model of the reader is based upon a half duplex operation. Reader transmits energy and its query for a specific period and then waits in receive mode with no more energy transmission until end of frame. The flow chart for reader query and TAGs response mechanism is as below:

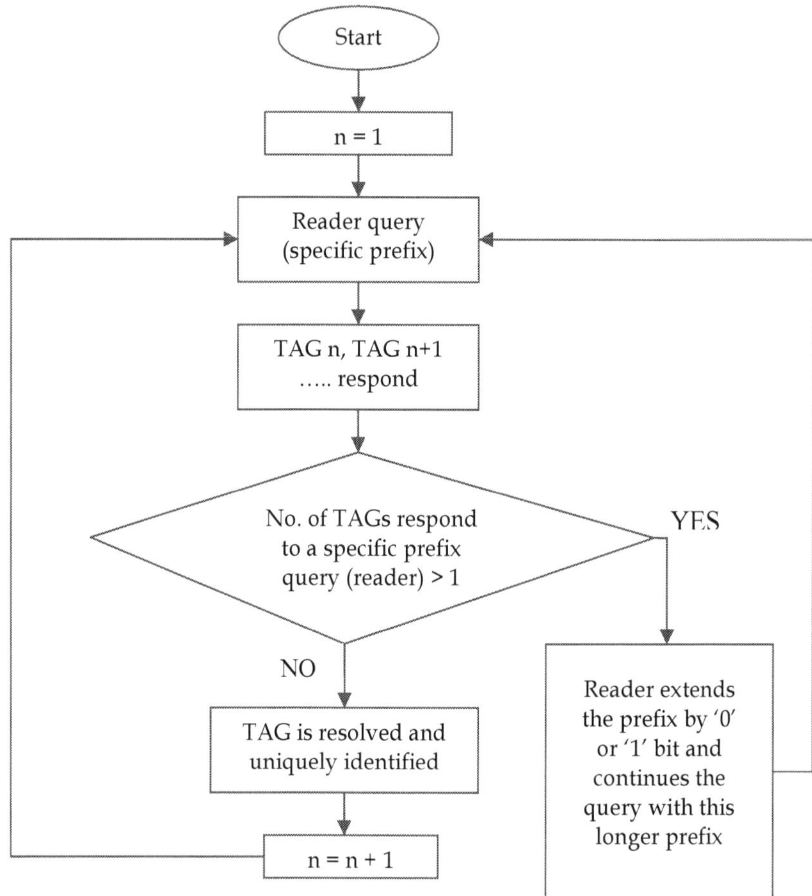

Fig. 4.

Pulse based half duplex operation is termed as sequential (SEQ) operation.

Power required by the reader to transmit	Power required by the reader to receive
PRtx	PRrx

Table 1.

And

Power required by an active TAG to transmit	Power required by an active TAG to receive
PTtx	PTrx

Table 2.

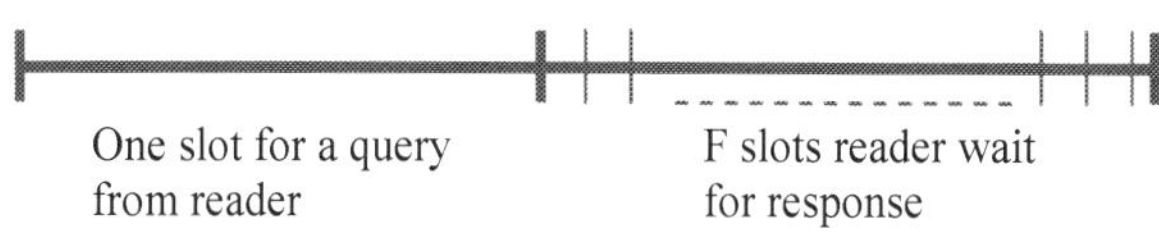

Fig. 5.

Reader energy consumption: q(m)*(PRtx + PRrx*F) when q(m) is the number of queries for read m TAGs. The energy consumption of all active TAGs: q(m)*PTrx + u(m)*PTtx when q(m) is the number of reader queires, u(m) is the number of TAG responses. For MSS scheme (include sleep command) the reader energy consumption is

$$q(m) * (PRtx + PRrx * F) + z(m) * PRtx.$$

The total energy consumption for all active TAGs is

$$q(m)*PTrx + u(m)*PTtx + z(m) * PTrx,$$

when z(m) is the number of sleep commands issued by the reader. The average analysis of energy consumption:

$$\bar{q}(m) --- average..number..of..reader..queires.$$

$$\bar{u}(m) - average\ number\ of\ TAG\ responses.$$

$$\bar{z}(m) - average\ number\ of\ sleep\ commands$$
$$issued\ by\ the\ reader\ (only\ for\ MSS\ Scheme)$$

$$\bar{t}(m) - average\ number\ of\ time\ slots\ required$$
$$to\ read\ all\ TAGs.$$

$$\bar{t}_{MS}(m) - average\ number\ of\ time\ slots\ required\ to\ read\ m\ TAGs$$

m TAG IDs are uniformly distributed in the interval [0.1].
I get the expression for One active RFID TAG total energy consumption:

$$TAG\ Power = \frac{1}{m} * \left[q(m) \cdot P_{Trx} + U(m) \cdot P_{Ttx} + Z(m) \cdot P_{Trx} \right]$$

3. Active RFID TAG equivalent circuit

Active RFID TAG can be represent as a parallel Equivalent Circuit of Capacitor and Resistor in parallel with Supply voltage source (internal resistance).

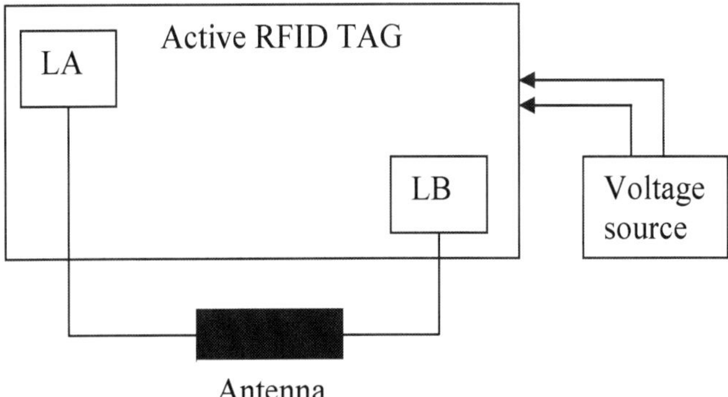

Fig. 6.

The Active RFID TAG Antenna can be represents as Parallel inductor to the basic Active RFID Equivalent Circuit. The simplified complete equivalent circuit of the label is as below:

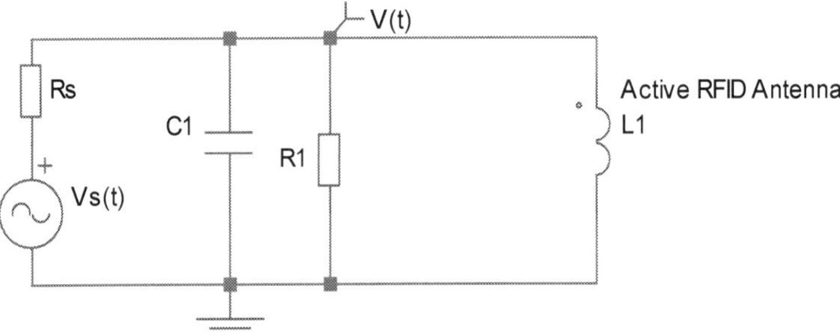

Active RFID's Equivalent circuit

Fig. 7.

$$V_{L1} = L1 \cdot \frac{dI}{dt} \Rightarrow I_{L1} = \frac{1}{L1} \cdot \int_0^{t1} V_{L1} \cdot dt, \quad I_{C1} = C1 \cdot \frac{dVc1}{dt}, \quad \sum_{j=1}^{4} Ij = 0$$

$$V = Vc1 = V_{L1} = V_{R1}$$

$$\frac{V}{R1} + C1 \cdot \frac{dV}{dt} + \frac{1}{L1} \cdot \int_0^{t1} V \cdot dt + \frac{V - Vs(t)}{Rs} = 0$$

$$\{\frac{1}{R1} \cdot \frac{dV}{dt} + C1 \cdot \frac{d^2V}{dt^2} + \frac{1}{L1} \cdot V + \frac{dV}{dt} \cdot \frac{1}{Rs}\} \xrightarrow{\frac{dVs(t)}{dt} \to \varepsilon(0<\varepsilon<<1)} 0$$

$$\varepsilon \gg 1 \Rightarrow \ddot{V} \cdot C1 + (\frac{1}{R1} + \frac{1}{Rs}) \cdot \dot{V} + \frac{1}{L1} \cdot V = \frac{1}{Rs} \cdot \dot{V}_s(t)$$

$$\frac{1}{R1} \cdot \frac{dV}{dt} + C1 \cdot \frac{d^2V}{dt^2} + \frac{1}{L1} \cdot V + [\frac{dV}{dt} - \frac{dVs(t)}{dt}] \cdot \frac{1}{Rs} = 0$$

$$V2 = \frac{dV1}{dt} = \frac{dV}{dt}, \quad V1=V$$

$$\frac{dV1}{dt} = V2, \quad \frac{dV2}{dt} = -[\frac{1}{C1 \cdot R1} + \frac{1}{Rs \cdot C1}] \cdot V2 - \frac{1}{C1 \cdot L1} \cdot V1 + \frac{1}{Rs \cdot C1} \cdot \frac{dVs(t)}{dt}$$

$$\begin{pmatrix} \frac{dV1}{dt} \\ \frac{dV2}{dt} \end{pmatrix} = \begin{pmatrix} 0 & 1 \\ -\frac{1}{C1 \cdot L1} & -[\frac{1}{C1 \cdot R1} + \frac{1}{Rs \cdot C1}] \end{pmatrix} \cdot \begin{pmatrix} V1 \\ V2 \end{pmatrix} + \begin{pmatrix} 0 \\ \frac{1}{Rs \cdot C1} \cdot \frac{dVs(t)}{dt} \end{pmatrix}$$

$$Lcalc = \left[\frac{\mu 0}{\pi} * [X1 + X2 - X3 + X4] * Nc^p \right]$$

, L1 = Lcalc

$$X1 = Aavg * \ln \left(\frac{2 * Aavg * Bavg}{d * (Aavg + \sqrt{Aavg^2 + Bavg^2})} \right)$$

$$X2 = Bavg * \ln \left(\frac{2 * Aavg * Bavg}{d * (Bavg + \sqrt{Aavg^2 + Bavg^2})} \right)$$

$$X3 = 2 * \left[Aavg + Bavg - \sqrt{Aavg^2 + Bavg^2} \right]$$

$X4 = (Aavg + Bavg) / 4$, The RFID's coil calculation inductance expression is

Definition of limits, Estimations: Track thickness t, Al and Cu coils (t > 30um). The printed coils as high as possible. Estimation of turn exponent p is needed for inductance calculation.

Coil manufacturing technology	P
Wired	1.8 – 1.9
Etched	1.75 – 1.85
Printed	1.7 – 1.8

Table 3.

Active RFID can be considered as Van der Pol's system. Van der Pol's equation provides an example of an oscillator with nonlinear damping, energy being dissipated at large amplitudes and generated at low amplitudes. Such systems typically posses limit cycles, sustained oscillations a round a state at which energy generation and dissipation balance. The basic Van der Pol's equation can be written in the form:

$$\ddot{X} + \alpha \cdot \phi(x) \cdot \dot{X} + X = \beta \cdot \rho(t)$$

$$\varepsilon \gg 1 \implies \ddot{V} \cdot C1 + (\frac{1}{R1} + \frac{1}{Rs}) \cdot \dot{V} + \frac{1}{L1} \cdot V = \frac{1}{Rs} \cdot \dot{V}s(t)$$

$$\varepsilon \gg 1 \;\Rightarrow\; \ddot{V} + \frac{1}{C1}\bullet(\frac{1}{R1}+\frac{1}{Rs})\bullet\dot{V} + \frac{1}{L1\bullet C1}\bullet V = \frac{1}{Rs\bullet C1}\bullet\dot{V}s\,(t)$$

$$X \to V, \;\; \alpha\bullet\phi(x) \to \frac{1}{C1}\bullet(\frac{1}{R1}+\frac{1}{Rs})$$

$$\frac{1}{L1\bullet C1} \to 1, \;\; \frac{1}{Rs\bullet C1}\bullet\dot{V}s\,(t) \to \beta\bullet\rho(t)$$

Lets define:

$$f_s(t) = \dot{V}s\,(t) \;\; \text{then} \; \triangleleft$$

$$\varepsilon \gg 1 \;\Rightarrow\; \ddot{V}\bullet C1 + (\frac{1}{R1}+\frac{1}{Rs})\bullet\dot{V} + \frac{1}{L1}\bullet V = \frac{1}{Rs}\bullet f_s(t)$$

then "f " is a "T" periodic function of the independent variable t, and $\;\; \lambda = \dfrac{1}{Rs}$

The term $\;\lambda\bullet f_s(t) = \dfrac{1}{Rs}\bullet\dot{V}s\,(t)\;$ is called the forcing function.

$\lambda \to 0 \Rightarrow \dfrac{1}{Rs} \to 0 \Rightarrow Rs \to \infty \;$ there is no forcing and the system act as Van Der Pol Oscillator.

It is necessary to examine the trajectories (V1,V2,t) of the non-autonomous Active RFID system in $\mathbb{R}^2 x\mathbb{R}$ rather than the orbits in $\mathbb{R}^2$. Equivalently, we may consider the orbits of the Active RFID TAGs three dimensional autonomous system.

$$\frac{dV1}{dt} = V2$$

$$\frac{dV2}{dt} = -[\frac{1}{C1\bullet R1}+\frac{1}{Rs\bullet C1}]\bullet V2 - \frac{1}{C1\bullet L1}\bullet V1 + \frac{1}{Rs\bullet C1}\bullet f_s(t) \;\; \forall \; f_s(t) = \dot{V}s\,(t)$$

$$\frac{dV3}{dt} = 1 \;\; \forall \; (V3(t){=}t)$$

First examine the case of $\;\lambda = 0 \Rightarrow Rs\bullet C1 \to \infty$, C1=const, then Rs $\to \infty$

The limit cycle, the isolated periodic orbit, of the unforced oscillator of Van Der Pol becomes a cylinder; that is, topologically it is homeomorphism to $S^1 x\mathbb{R}$. The cylinder is an invariant manifold in the sense that any solution starting on the cylinder remains on it for all positive time. This invariant cylinder attracts all nearby solutions. For $\lambda = 0$, $\lambda \to 0$, $Rs \to \infty$ the Active RFID TAG invariant cylinder is filled with a family of periodic solutions. The cylinder under the projection $\mathbb{R}^2 x\mathbb{R} \to \mathbb{R}^2$ simply becomes the limit cycle. Actually Active RFID TAGs act as periodic forcing with small amplitude, that $|\lambda|$ small. In this case, there is still a cylinder in $\mathbb{R}^2 x\mathbb{R}$ close to the invariant cylinder of the unforced oscillator. This new cylinder is an invariant manifold of solutions of the forced equation and attracts all nearby solutions. The flow on the invariant cylinder of the forced equation can be quite different from the one of the unforced oscillator. In Active RFID TAG concern to Van Der Pol's equation we get the equation:

$$\ddot{X} + \alpha \bullet \phi(x) \bullet \dot{X} + X = \lambda \bullet f_s(t)$$

$$\varepsilon \gg 1 \;\Rightarrow\; \ddot{V} + (\frac{1}{R1} + \frac{1}{Rs}) \bullet \frac{1}{C1} \bullet \dot{V} + \frac{1}{L1 \bullet C1} \bullet V = \frac{1}{Rs \bullet C1} \bullet f_s(t)$$

$$\varepsilon \gg 1 \;\Rightarrow\; \ddot{V} + (\frac{1}{R1} + \frac{1}{Rs}) \bullet \frac{1}{C1} \bullet \dot{V} + \frac{1}{L1 \bullet C1} \bullet V = \frac{1}{Rs \bullet C1} \bullet \dot{V}_s(t)$$

$$\text{then } \phi(x) = 1, \; \alpha = ((\frac{1}{R1} + \frac{1}{Rs}) \bullet \frac{1}{C1}), \; \frac{1}{L1 \bullet C1} \to 1 \; (L1 \bullet C1 \approx 1)$$

$\phi(x) = 1 > 0 \; \forall \; |t| > 1\text{sec}, \; f_s(t)$ is T periodic and α, β are non negative parameters. $\alpha = (\frac{1}{R1} + \frac{1}{Rs}) \bullet C1, \; \beta = \frac{1}{Rs \bullet C1}$

Unforced investigation: $\lambda = 0 \Rightarrow \dfrac{1}{Rs} \to 0 \Rightarrow Rs \to \infty$ then we return to Passive RFID TAG since the battery has a very high serial resistance – disconnected status.

4. Active RFID TAG as a dynamic energy analysis

Active RFID equivalent circuit total TAG power is a summation of all element's power.

$$P_{total} = \sum_{i=1}^{N} p_i = \text{TAG Power} \;,\; \sum_{i=1}^{N} p_i = \frac{1}{m} * \left[q(m) \bullet P_{Trx} + U(m) \bullet P_{Ttx} + Z(m) \bullet P_{Trx} \right]$$

$$\sum_{i=1}^{N} p_i = P_{Rs} + P_{C1} + P_{R1} + P_{L1,} \;,\; energy \Rightarrow W(t_0, t) \triangleq \int_{t0}^{t} p(t') dt' = \int_{t0}^{t} v(t') \cdot i(t') dt'$$

$$P(t)_{total} = \frac{dW(t_0, t)}{dt} = \frac{d}{dt} [\sum_{i=1}^{N} w_i], \; energy \Rightarrow w_{inductor} = \frac{1}{2} \cdot L \cdot I^2$$

$$energy \Rightarrow w_{capacitor} = \frac{Q^2}{2 \cdot C} \;,\; P_{resistor} = I^2 \cdot R \;,\; P_{R1} = I_{R1}^2 \cdot R1, \; P_{Rs} = I_{Rs}^2 \cdot Rs$$

$$energy \Rightarrow w_{L1} = \frac{1}{2} \cdot L_1 \cdot I_{L1}^2 \Rightarrow P_{L1} = \frac{d}{dt} w_{L1} = L \cdot I_{L1} \cdot \dot{I}_{L1}$$

$$energy \Rightarrow w_{C1} = \frac{Q_{C1}^2}{2 \cdot C1} \Rightarrow P_{C1} = \frac{d}{dt} w_{C1} = \frac{Q_{C1} \cdot \dot{Q}_{C1}}{C1}$$

$$energy \Rightarrow w_{C1} = \frac{C_1 \cdot V_{C1}^2}{2} \Rightarrow P_{C1} = \frac{d}{dt} w_{C1} = C_1 \cdot V_{C1} \cdot \dot{V}_{C1}$$

$$I_{L1} = \frac{1}{L1} \cdot \int_{0}^{t} V_{L1} \cdot dt \Rightarrow \dot{I}_{L1} = \frac{V_{L1}}{L1}$$

$$\sum_{i=1}^{N} P_i = I_{R1}^2 \cdot R1 + I_{Rs}^2 \cdot Rs + L \cdot I_{L1} \cdot \dot{I}_{L1} + \frac{Q_{C1} \cdot \dot{Q}_{C1}}{C1}$$

$$\sum_{i=1}^{N} P_i = \frac{V^2}{R1} + \frac{[V - Vs(t)]^2}{Rs} + L \cdot I_{L1} \cdot \dot{I}_{L1} + C_1 \cdot V_{C1} \cdot \dot{V}_{C1}$$

$$\sum_{i=1}^{N} P_i = V^2 \cdot [\frac{1}{R1} + \frac{1}{Rs}] - \frac{2 \cdot V \cdot Vs(t)}{Rs} + \frac{[Vs(t)]^2}{Rs} + \frac{V}{L1} \cdot \int_0^t Vdt + C_1 \cdot V \cdot \dot{V}$$

$$\frac{1}{m} * \left[q(m) \bullet P_{Trx} + U(m) \bullet P_{Ttx} + Z(m) \bullet P_{Trx} \right] =$$

$$V^2 \cdot [\frac{1}{R1} + \frac{1}{Rs}] - \frac{2 \cdot V \cdot Vs(t)}{Rs} + \frac{[Vs(t)]^2}{Rs} + \frac{V}{L1} \cdot \int_0^t Vdt + C_1 \cdot V \cdot \dot{V}$$

5. Active RFID TAG fixed points and linearization

$$\frac{dV1}{dt} = V2$$

$$\frac{dV2}{dt} = -[\frac{1}{C1 \bullet R1} + \frac{1}{Rs \bullet C1}] \bullet V2 - \frac{1}{C1 \bullet L1} \bullet V1 + \frac{1}{Rs \bullet C1} \bullet \dot{V}_S(t)$$

Now we consider linear system: $\frac{dV1}{dt} = f(V1, V2)$, $\frac{dV2}{dt} = g(V1, V2)$

And suppose that (V_1^*, V_2^*) is a fixed point: $f(V_1^*, V_2^*) = 0$, $g(V_1^*, V_2^*) = 0$

Let $U1 = V1 - V_1^*$, $U2 = V2 - V_2^*$ Denote the components of a small disturbance from the fixed point. To see whether the disturbance grows or decays, we need to derive differential equations for U1 and U2. Lets do the U1 equation first:

$$\frac{dU1}{dt} = \frac{dV1}{dt} \quad \text{since} \quad V_1^* \text{ is constant.}$$

$$\frac{dU1}{dt} = \frac{dV1}{dt} = f(U1 + V_1^*, U2 + V_2^*) = f(V_1^*, V_2^*) + U1 \cdot \frac{\partial f}{\partial V1} + U2 \cdot \frac{\partial f}{\partial V2} + O(U_1^2, U_2^2, U1 \cdot U2)$$

(Taylor series expansion)

To simplify the notation, we have written $\frac{\partial f}{\partial V1}$ and $\frac{\partial f}{\partial V2}$ these partial derivatives are to be

evaluated at the fixed point (V_1^*, V_2^*); thus they are numbers, not functions. Also the short

hand notation $O(U_1^2, U_2^2, U1 \cdot U2)$ denotes quadratic terms in U1 and U2. Since U1 and U2

are small, these quadratic terms are extremely small. Similarly we find

$\frac{dU2}{dt} = U1 \cdot \frac{\partial g}{\partial V1} + U2 \cdot \frac{\partial g}{\partial V2} + O(U_1^2, U_2^2, U1 \cdot U2)$, Hence the disturbance (U1, U2) evolves

according to :

$$\begin{pmatrix} \dfrac{dU1}{dt} \\ \dfrac{dU2}{dt} \end{pmatrix} = \begin{pmatrix} \dfrac{\partial f}{\partial V1} & \dfrac{\partial f}{\partial V2} \\ \dfrac{\partial g}{\partial V1} & \dfrac{\partial g}{\partial V2} \end{pmatrix} \cdot \begin{pmatrix} U1 \\ U2 \end{pmatrix} + \text{Quadratic terms}.$$

The Matrix $A = \begin{pmatrix} \dfrac{\partial f}{\partial V1} & \dfrac{\partial f}{\partial V2} \\ \dfrac{\partial g}{\partial V1} & \dfrac{\partial g}{\partial V2} \end{pmatrix}_{(V_1^*,V_2^*)}$ is called the Jacobian matrix at the fixed point

(V_1^*,V_2^*) and the Quadratic terms are tiny, its tempting to neglect them altogether.

If we do that, we obtain the linearized system. $\begin{pmatrix} \dfrac{dU1}{dt} \\ \dfrac{dU2}{dt} \end{pmatrix} = \begin{pmatrix} \dfrac{\partial f}{\partial V1} & \dfrac{\partial f}{\partial V2} \\ \dfrac{\partial g}{\partial V1} & \dfrac{\partial g}{\partial V2} \end{pmatrix} \cdot \begin{pmatrix} U1 \\ U2 \end{pmatrix}$ whose

dynamic can be analyzed by the general methods.

$$f(V1,V2) = V2$$

$$g(V1,V2) = -[\frac{1}{C1 \cdot R1} + \frac{1}{Rs \cdot C1}] \cdot V2 - \frac{1}{C1 \cdot L1} \cdot V1 + \frac{1}{Rs \cdot C1} \cdot \dot{V}_s(t)$$

$$\frac{\partial f}{\partial V1} = 0, \quad \frac{\partial f}{\partial V2} = 1, \quad \frac{\partial g}{\partial V1} = -\frac{1}{C1 \cdot L1}, \quad \frac{\partial g}{\partial V2} = -(\frac{1}{C1 \cdot R1} + \frac{1}{Rs \cdot C1})$$

$$\begin{pmatrix} \dfrac{dU1}{dt} \\ \dfrac{dU2}{dt} \end{pmatrix} = \begin{pmatrix} 0 & 1 \\ -\dfrac{1}{C1 \cdot L1} & -[\dfrac{1}{C1 \cdot R1} + \dfrac{1}{Rs \cdot C1}] \end{pmatrix} \cdot \begin{pmatrix} U1 \\ U2 \end{pmatrix}$$

6. Active RFID TAG stability analysis based on forced Van Der Pol's system

The basic Active RFID Forced Van Der Pol's equation

$$\varepsilon \gg 1 \;\Rightarrow\; \ddot{V} + (\frac{1}{R1} + \frac{1}{Rs}) \cdot \frac{1}{C1} \cdot \dot{V} + \frac{1}{L1 \cdot C1} \cdot V = \frac{1}{Rs \cdot C1} \cdot \dot{V}_s(t)$$

$$\text{then } \phi(x) = 1, \quad \alpha = ((\frac{1}{R1} + \frac{1}{Rs}) \cdot \frac{1}{C1}), \quad \frac{1}{L1 \cdot C1} \to 1 \; (L1 \cdot C1 \approx 1)$$

$$\beta = \frac{1}{Rs \cdot C1}$$

In our case $\phi(V) = 1$, $\phi(V) > 0$ for $|V| > 1$ and $\dot{V}_s(t)$ is T periodic and $(\frac{1}{R1} + \frac{1}{Rs}) \cdot \frac{1}{C1}$, $\frac{1}{Rs \cdot C1}$

are non negative parameters. It is convenient to rewrite the Active RFID forced Van Der

Pol's equation as an autonomous system. $\theta = t \;\Rightarrow\; \dfrac{d\theta}{dt} = 1$

$$\dot{V} = Y - (\frac{1}{R1} + \frac{1}{Rs}) \cdot \frac{1}{C1} \cdot \phi(V)$$

$$\dot{Y} = -V + \frac{1}{R1 \cdot C1} \cdot \dot{V}_s(\theta)$$

$$\dot{\theta} = 1 \quad (V, Y, \theta) \in \mathbb{R}^2 x S^1$$

$\phi(V) = 1$ remain strictly positive as $|V| \to \infty$ for unforced system $\frac{1}{R1 \cdot C1} \cdot \dot{V}_s(\theta) \to 0$

but $\frac{1}{R1 \cdot C1} \neq 0$ then $\dot{V}_s(\theta) = 0$ no energy is supply to the Active RFID TAG, become

Passive RFID TAG. First we suppose that $\alpha \ll 1$ $((\frac{1}{R1} + \frac{1}{Rs}) \cdot \frac{1}{C1} \ll 1)$ is a small parameter, so

the autonomous system is a perturbation of linear oscillator. $\dot{V} = Y, \dot{Y} = -V$ which has a phase plane filled with circular periodic orbits each of period $2 \cdot \pi$. Using regular perturbation or averaging methods, we can show that precisely one of these orbits is preserved under the perturbation. Selecting the invertible transformation:

$$\begin{pmatrix} \xi 1 \\ \xi 2 \end{pmatrix} = \begin{pmatrix} \cos(t) & -\sin(t) \\ -\sin(t) & -\cos(t) \end{pmatrix} \cdot \begin{pmatrix} V \\ Y \end{pmatrix}$$

which "freezes" the unperturbed system and the autonomous system become :

$$\dot{\xi 1} = -(\frac{1}{R1} + \frac{1}{Rs}) \cdot \frac{1}{C1} \cdot \cos t \cdot [(\xi 1 \cdot \cos(t) - \xi 2 \cdot \sin(t))^3 / 3 - (\xi 1 \cdot \cos(t) - \xi 2 \cdot \sin(t))]$$

$$\dot{\xi 2} = -(\frac{1}{R1} + \frac{1}{Rs}) \cdot \frac{1}{C1} \cdot \sin t \cdot [(\xi 1 \cdot \cos(t) - \xi 2 \cdot \sin(t))^3 / 3 - (\xi 1 \cdot \cos(t) - \xi 2 \cdot \sin(t))]$$

this transformation is orientation reversing approximation the function $\xi 1, \xi 2$ which vary slowly because $\dot{\xi} 1, \dot{\xi} 2$ are small. Integrating each function with respect to time (t) from 0 to T=$2 \cdot \pi$, holding $\xi 1, \xi 2$ fixed we obtain:

$$\dot{\xi 1} = (\frac{1}{R1} + \frac{1}{Rs}) \cdot \frac{1}{C1} \cdot \xi 1 \cdot [1 - (\xi 1^2 + \xi 2^2) / 4] / 2$$

$$\dot{\xi 2} = (\frac{1}{R1} + \frac{1}{Rs}) \cdot \frac{1}{C1} \cdot \xi 2 \cdot [1 - (\xi 1^2 + \xi 2^2) / 4] / 2$$

this system is correct at first order, but there is an error of $O([(\frac{1}{R1} + \frac{1}{Rs}) \cdot \frac{1}{C1}]^2)$. In polar coordinates, we therefore have

$$\dot{r} = \frac{r}{2} \cdot (\frac{1}{R1} + \frac{1}{Rs}) \cdot \frac{1}{C1} \cdot (1 - \frac{r^2}{4}) + O([(\frac{1}{R1} + \frac{1}{Rs}) \cdot \frac{1}{C1}]^2)$$

$$\dot{\varphi} = 0 + O([(\frac{1}{R1} + \frac{1}{Rs}) \cdot \frac{1}{C1}]^2)$$

Neglecting the $O([(\frac{1}{R1}+\frac{1}{Rs})\bullet\frac{1}{C1}]^2)$ terms this system has an attracting circle of fixed points at r = 2 reflecting the existence of a one parameter family of almost sinusoidal solutions: $V = r(t)\cdot\cos(t+\varphi(t))$ with   slowly   varying   amplitude

$$r(t) = 2 + O([(\frac{1}{R1}+\frac{1}{Rs})\bullet\frac{1}{C1}]^2)$$ and the phase $\varphi(t) = \varphi^0 + O([(\frac{1}{R1}+\frac{1}{Rs})\bullet\frac{1}{C1}]^2)$

Constant φ^0 being determined by initial conditions. When the value of $(\frac{1}{R1}+\frac{1}{Rs})\bullet\frac{1}{C1}$

Is not small the averaging procedure no longer works and other methods must be used.
The investigation can be done for Active RFID's system forced Van Der Pole. Lets consider

$\dot{V}_s(t)\neq 0$ we suppose $(\frac{1}{R1}+\frac{1}{Rs})\bullet\frac{1}{C1},\frac{1}{Rs\bullet C1}\ll 1$ and use the same transformation as we

use in the unforced system $\dot{V}_s(t)=0$. when we interest in the periodic forced response we

use the $\dfrac{2\cdot\pi}{\omega}$ periodic transformation.

$$\begin{pmatrix}\xi 1\\ \xi 2\end{pmatrix} = \begin{pmatrix}\cos(\omega t) & -\dfrac{1}{\omega}\cdot\sin(\omega t)\\ -\sin(\omega t) & -\dfrac{1}{\omega}\cdot\cos(\omega t)\end{pmatrix}\cdot\begin{pmatrix}V\\ Y\end{pmatrix}$$

$$\dot{\xi}1 = -(\frac{1}{R1}+\frac{1}{Rs})\bullet\frac{1}{C1}\cdot\phi(V)\cdot\cos(\omega\cdot t)-(\frac{\omega^2-1}{\omega})\cdot V\cdot\sin(\omega\cdot t)-\frac{1}{Rs\cdot C1\cdot\omega}\cdot\sin(\omega\cdot t\cdot\dot{V}_s(t))$$

$$\dot{\xi}2 = (\frac{1}{R1}+\frac{1}{Rs})\bullet\frac{1}{C1}\cdot\phi(V)\cdot\sin(\omega\cdot t)-(\frac{\omega^2-1}{\omega})\cdot V\cdot\cos(\omega\cdot t)-\frac{1}{Rs\cdot C1\cdot\omega}\cdot\cos(\omega\cdot t\cdot\dot{V}_s(t))$$

$\dfrac{1}{C1\bullet L1} \to 1$, $\phi(V)=1$ in our case.

$$\dot{\xi}1 = -(\frac{1}{R1}+\frac{1}{Rs})\bullet\frac{1}{C1}\cdot\cos(\omega\cdot t)-(\frac{\omega^2-1}{\omega})\cdot V\cdot\sin(\omega\cdot t)-\frac{1}{Rs\cdot C1\cdot\omega}\cdot\sin(\omega\cdot t\cdot\dot{V}_s(t))$$

$$\dot{\xi}2 = (\frac{1}{R1}+\frac{1}{Rs})\bullet\frac{1}{C1}\cdot\sin(\omega\cdot t)-(\frac{\omega^2-1}{\omega})\cdot V\cdot\cos(\omega\cdot t)-\frac{1}{Rs\cdot C1\cdot\omega}\cdot\cos(\omega\cdot t\cdot\dot{V}_s(t))$$

7. Summery

Active RFID TAG system can be represent as Voltage source (internal resistance) , Parallel Resistor, Capacitor, and Inductance circuit. Linear bifurcation system explain Active RFID TAG system behavior for any initial condition V(t) and dV(t)/dt. Active RFID's Coil is a very critical element in Active RFID TAG functionality. Optimization can be achieved by Coil's parameters inspection and System bifurcation controlled by them. Spiral, Circles, and other Active RFID phase system behaviors can be optimize for better Active RFID TAG performance and actual functionality. Active RFID TAG losses also controlled for best performance and maximum efficiency.

8. References

[1] Yuri A. Kuznetsov, Elelments of Applied Bifurcation Theory. Applied Mathematical Sciences.

[2] Jack K. Hale. Dynamics and Bifurcations. Texts in Applied Mathematics, Vol. 3

[3] Steven H. Strogatz, Nonlinear Dynamics and Chaos. Westview press

[4] John Guckenheimer, Nonlinear Oscillations, Dynamical Systems, and Bifurcations of Vector Fields. Applied Mathematical Sciences Vol 42.

[5] Stephen Wiggins, Introduction to Applied Nonlinear Dynamical Systems and Chaos. Text in Applied Mathematics (Hardcover).

[6] O. Aluf "RFID TAGs COIL's Dimensional Parameters Optimization As Excitable Linear Bifurcation Systems" (IEEE COMCAS2008 Conference, May 2008).

10

RFID Tag Antennas Mountable on Metallic Platforms

Byunggil Yu[1], Frances J. Harackiewicz[2] and Byungje Lee[1]

[1]Kwangwoon University
[2]Southern Illinois University Carbondale
[1]Korea
[2]USA

1. Introduction

Auto identification provides information without direct contacts and human intervention errors. Auto identification technology has become very popular in industries, such as the service industry, inventory control, distribution logistics, security systems, transportation and manufacturing process control. So far, the bar code technology leads the auto identification industry, but it has several limitations such as low storage capacity, required line-of-sight contact with the reader, and physical positioning of the scanned objects.

Recently, the radio frequency identification (RFID) has been an attractive alternative identification technology to the barcode. The numerous potential applications of the RFID system make ubiquitous identification possible at frequency bands of 125 KHz (LF), 13.56 MHz (HF), and 860-960 MHz (UHF). The RFID system generally consists of two basic components: the reader and the tag, which communicate with each other by electromagnetic waves. The reader can be a read or a read/write device that uses an antenna to send an electromagnetic wave to wake up the tags. The tag is the data carrying device located on the object being identified. In general, the performance of the tag seriously affects the performance of the whole RFID system. The tag consists of the tag antenna and the microchip. Since good connection and power transmission between the tag antenna and the microchip directly impact on the RFID system performance, the tag antenna has to be designed considering its operating environments or platforms.

As the use of RFID systems increases, manufacturers are pushing toward higher operating frequencies (UHF band) for long reading range, high reading speed, capable multiple accesses, anti-collision, and small antenna size compared to the LF or HF band RFID system. As the operating frequency of the RFID system becomes higher, the major part of the RFID system that mostly affects the ability to read the tag is the antenna.

There are several possible antenna types which can be used for RFID tags in this frequency band. The dipole types of antennas such as folded dipoles and meandered dipoles are used in many applications since they can be printed on a very thin film. However, when they are mounted on the metallic objects, the antenna performance is seriously decreased because of the reactance variation on the antenna impedance. Particularly, the UHF band RFID system is a passive system where a tag does not contain its own power source. Therefore, the reader antenna sends a radio signal into the air to activate the tag, then listens for a backscatter from the tag, and reads the data transmitted by the tag. Passive tag antenna must be designed to transmit maximum power to the microchip without possible losses. Therefore, near perfect impedance matching is required between the tag antenna and the microchip. Designing a passive tag antenna matched with the complex microchip impedance is the most challenging factor, since a microchip has very high Q (quality factor) due to its small resistance and large capacitive reactance. Also, the impedance of an RFID tag antenna varies when it is mounted on different objects. Especially, metallic objects strongly affect the antenna performance by lowering the tag's efficiency. Therefore, tag antennas have to be designed to enable tags to be read near and on metallic objects without severe performance degradation. In order to obtain stable antenna performance on various metallic platforms, minimizing the effect of the metallic supporting object is meaningful work. In this chapter, several types of antennas which are mountable on metallic platforms are introduced and analyzed.

2. Electromagnetic waves near metallic platforms

An RFID system communicates by electromagnetic waves. When designing the RFID tag antennas mountable on metallic platforms, it is very important to understand the behaviour of the electromagnetic fields near metallic surfaces since the antenna parameters (the input impedance, gain, radiation pattern, and radiation efficiency) can be seriously affected by metallic platforms. In this section, the behaviour of electromagnetic fields near metallic surfaces will be considered.

2.1 Boundary conditions for a general case

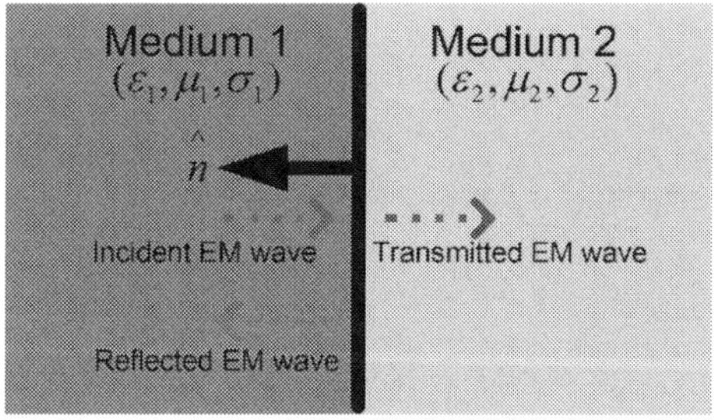

Fig. 2.1 Electromagnetic boundary between two media

Now we want to see how the electromagnetic fields behave at the boundary between a pair of dielectrics or between a dielectric and a conductor. Fig. 2.1 shows the electromagnetic boundary for a general interface between two media. The amplitude and phase of the incident and reflect waves are changed by the material (ε, μ, σ) properties. A UHF-band passive RFID system uses the modulated backscatter method, so the amplitude and phase of the reflected signal are very important. The electromagnetic boundary conditions for a general case can be expressed as follows:

$$\hat{n} \times (E_1 - E_2) = 0 \tag{2-1}$$

$$\hat{n} \cdot (D_1 - D_2) = 0 \tag{2-2}$$

$$\hat{n} \times (H_1 - H_2) = J_s \tag{2-3}$$

$$\hat{n} \cdot (B_1 - B_2) = 0 \tag{2-4}$$

where
$\hat{n}$ is the unit normal vector to the boundary directed from medium 2 to medium 1
E is the electric field intensity (V/m), D is the electric flux density (C/m²)
H is the magnetic field intensity (A/m), B is the magnetic flux density (W/m²)
ρ_s is the surface charge density (C/m) , J_s is the surface current density (A/m²)
By using above boundary conditions, we can also find the electromagnetic boundary conditions for the cases of PEC (Perfect Electric Conductor).

2.2 Boundary conditions at the PEC interface

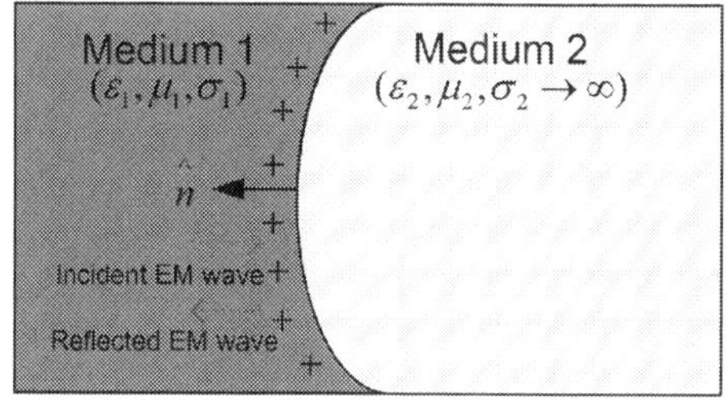

Fig. 2.2 Boundary conditions at the interface of PEC

If medium 2 is a PEC with infinite conductivity, all field components must be zero inside of the PEC. Then, we can express the boundary conditions at the interface as follows:

$$\hat{n} \times E_1 = 0 \qquad \text{or} \qquad E_{1t} = 0 \tag{2-5}$$

$$\hat{n} \cdot D_1 = \rho_s \qquad \text{or} \qquad D_{1n} = \rho_s \tag{2-6}$$

$$\hat{n} \times H_1 = J_s \qquad \text{or} \qquad H_{1t} = J_s \tag{2-7}$$

$$\hat{n} \cdot B_1 = 0 \qquad \text{or} \qquad B_{1n} = 0 \tag{2-8}$$

It is noticed that there are no tangential components of the electric field on a PEC boundary, and there are only normal components of the electric field for oscillation. On the other hand, there are no normal components of the magnetic field on a PEC boundary. There are only tangential components of the magnetic field. In addition, normal incident waves are totally reflected from the interface because the skin depth of the PEC is zero. Therefore, the amplitude of incident wave and reflected wave are the same, but their phases are 180⁰ different. In other words, while the total of the incident and reflected electric fields at the PEC boundary will be zero, the total magnetic field (tangential component) will be doubled at the PEC boundary surface.

3. Effects of metallic platforms on RFID tag antenna

Since RFID systems frequently apply near the metallic environment, the effect of metallic platforms should be considered in designing the tag antenna. As mentioned in the previous section, there are only the normal component of the electric field and tangential component of the magnetic field near the surface of the metallic platform. Therefore, any RFID tag antenna whose performance mostly depends on either the tangential component of the electric field or the normal component of the magnetic field may be faced with considerable performance degradation when it is attached to or close to a metallic platform. In addition, the tag antenna parameters such as the input impedance, resonant frequency, gain, radiation pattern, and the efficiency will be changed. The maximum power transmission can be realized only if the tag antenna impedance is equal to the conjugate of the microchip impedance. The impedance of the microchip is not the normal 50 ohm or 75 ohm, and it may be a random value, or vary with frequency and driving power. A microchip has also a high Q (quality factor) at its terminals, which makes it not easy to attain the conjugate match between the tag antenna and the microchip. In other words, a small variation in the impedance causes serious antenna performance degradation. A metal or liquid based platform also causes the shifting of resonant frequency and degradation of radiation efficiency. To solve these problems, some special types of tag antennas that will not be affected too much when attached to a metallic platform should be designed. In general, UHF-band RFID systems have used dipole-type tag antennas for non-metallic platform. However, if this type of tag antenna is mounted on the metallic platforms, then the reading range is significantly decreased. So, we need another tag structure for metallic platforms. One simple solution is to use an antenna which has its own ground plane to operate. Then, the microstrip antenna may be a good choice for identifying metallic objects.

3.1 Dipole type of RFID tag antenna

In practical applications of a passive UHF-band RFID system, the tag antenna should be designed with low profile, so that its vertical current is limited. The label-type tag antenna where the dipole is printed on a thin film has been used in many non-metallic platforms. When it is mounted near or on metallic platforms, its radiation will be damaged by an inductive current excited in opposite direction. Now we will consider the performance degradation of dipole type antenna near the metallic platform. Fig. 3.1 shows a meandered dipole tag antenna above the metallic platform. Fig. 3.2 shows the simulated antenna impedance by varying the distance (H) of a dipole antenna from a $2\lambda \times 2\lambda$ metallic platform at UHF band. This simulation is done by Ansoft HFSS Ver. 11. One can see that the impedance is varied due to a parasitic capacitance between the tag antenna and the metallic platform. Fig. 3.3 shows the radiation efficiency by varying frequency and the distance (H) of the antenna from a metallic platform. It is noticed that the radiation efficiency is decreased significantly when a tag is located close to the metallic platform. To maintain a certain level of radiation efficiency, the label-type tags where the dipole is printed on very thin film generally should be kept the proper distance from the metallic platform. However, this makes the size of a tag antenna larger and limits its applications.

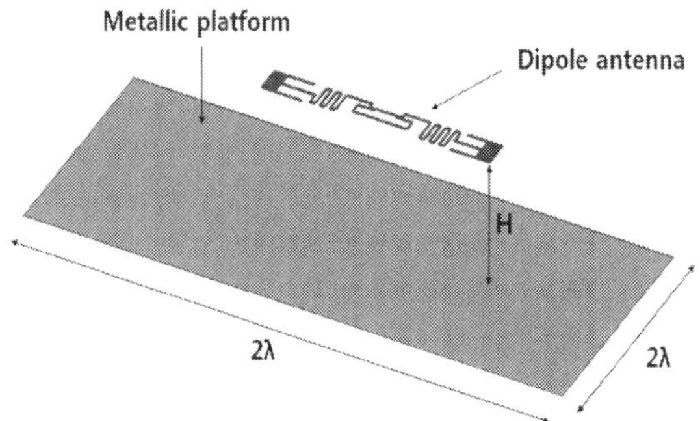

Fig. 3.1 Conventional dipole tag antenna above the metallic platform

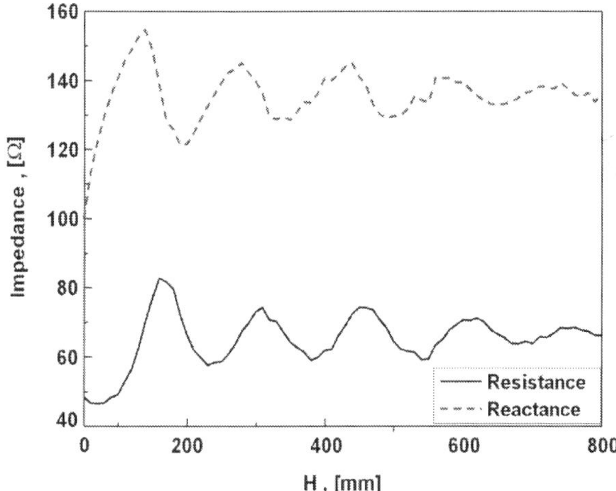

Fig. 3.2 Impedance variation as a function of the distance (H) between a dipole antenna and a metallic platform at UHF band

3.2 Microstrip patch antenna

Some studies have proposed using a microstrip patch tag antenna for metallic platforms. Even if these microstrip patch tag antenna can be applied easily to metallic platforms, there are several things to consider. Those are the size and shape of the metallic platform and attached position. In general, a microstrip patch antenna has stable performance when it has a ground plane size of more than 0.25 λ from the radiating patch. However, a microstrip patch antenna with such a ground size makes the antenna larger in dimension and more expensive.

Fig. 3.4 shows a conventional microstrip patch antenna designed by Ansoft HFSS with 50 Ω input impedance on a dielectric substrate (ε_r=1). It has a dimension (L x W x h) of 140 mm x 154 mm x 10 mm, respectively, and its center frequency is 900 MHz. Now mounting this patch antenna shown in Fig. 3.4 on the metallic platform as shown in Fig. 3.5, the antenna input impedance is observed by varying the size (A) of the metallic platform. Fig. 3.6 notices

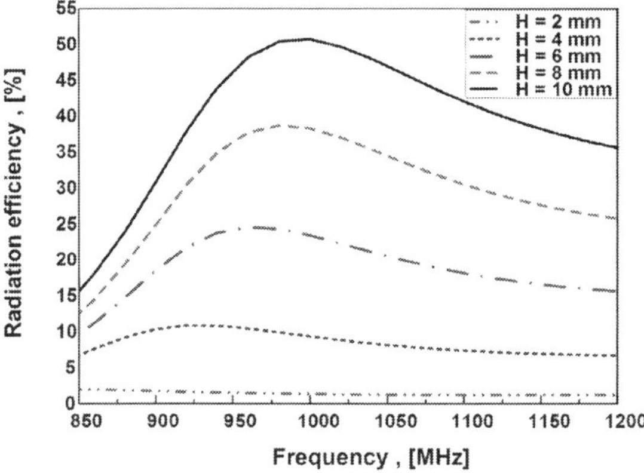

Fig. 3.3 Radiation efficiency as a function of the distance (H) between a dipole antenna and a metallic platform for different frequencies

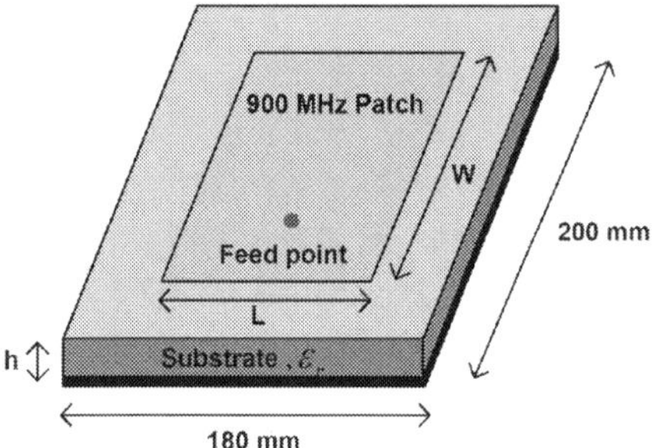

Fig. 3.4 Conventional microstrip patch antenna operating at 900 MHz

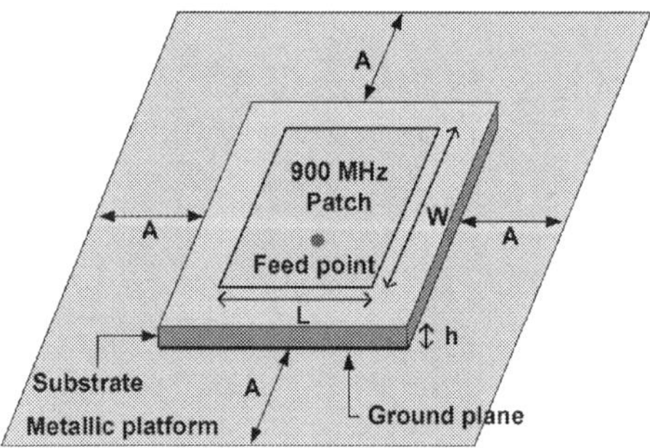

Fig. 3.5 Microstip patch antenna mounted on the metallic platform

that the input impedance and the resonant frequency change with different sizes of metallic platforms. The characteristic of the input impedance changes rapidly when the size (A) of the metallic platform becomes 0.2 λ. Designing a passive tag antenna matched with the complex microchip impedance is the most challengeable factor, since a microchip has very high Q(quality factor) because of its small resistance and large capacitive reactance. Therefore, tag antennas have to be designed to enable tags to be read near and on metallic platforms without severe performance degradation.

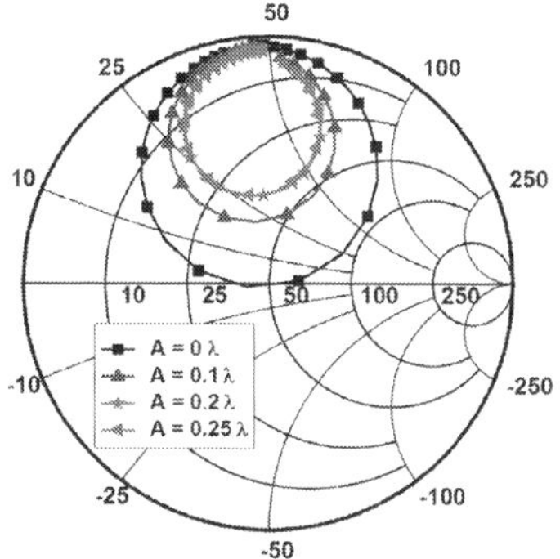

Fig. 3.6 Impedance characteristic with varying the size of the metallic platform

4. RFID tag antennas mountable on metallic platforms

In the previous section, effects of metallic platforms on RFID tag antennas are considered. Conventional tag antennas suffer degradation in performance when attached near or to metallic platforms. To solve the problem brought by the metallic objects, some special tag antennas should be designed. These antennas usually have a metallic ground. Some metallic platforms, which make the performance of the tag antenna worse, are modified to be as an extended part of the antenna to improve its performance. Therefore, in order to obtain stable antenna performance on various metallic platforms, minimizing the effect of the metallic supporting object is a very meaningful work. In this section, a number of RFID tag antennas suitable for mounting on metallic platforms will be discussed. Brief design concepts and some results will also be included for several tag antennas.

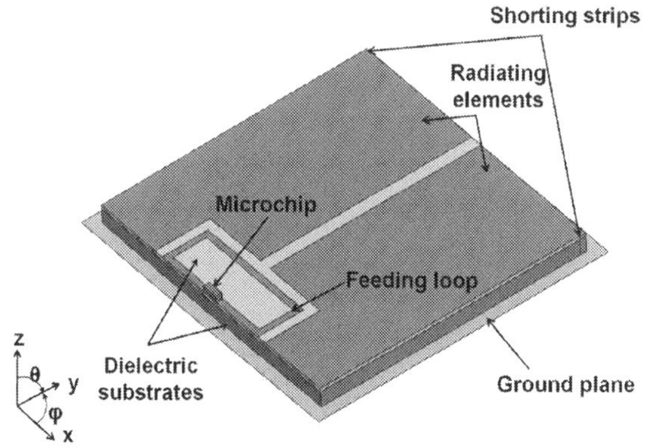

Fig. 4.1 Structure of the balanced-type microstrip patches for tag antennas

4.1 Balanced-type microstrip patches

The direction of the fringing field of a PIFA-type antenna is always from the radiating element to the ground plane, and vice versa. Although this type of an antenna has its own ground plane, its performance will be affected when attached to the metallic platform. To make up for this drawback, the balanced-type microstrip patch antenna (Yu et al., 2007) as shown in Fig. 4.1 was proposed. The proposed tag antenna consists of two symmetric shorted radiating elements and a feeding loop. Two symmetric radiating elements are etched on a substrate layer, and electrically shorted to the ground plane through the shorting strips. The feeding loop, which is connected to the microchip, is inductively coupled so that the currents on patches are out of phase with equal amplitude. The

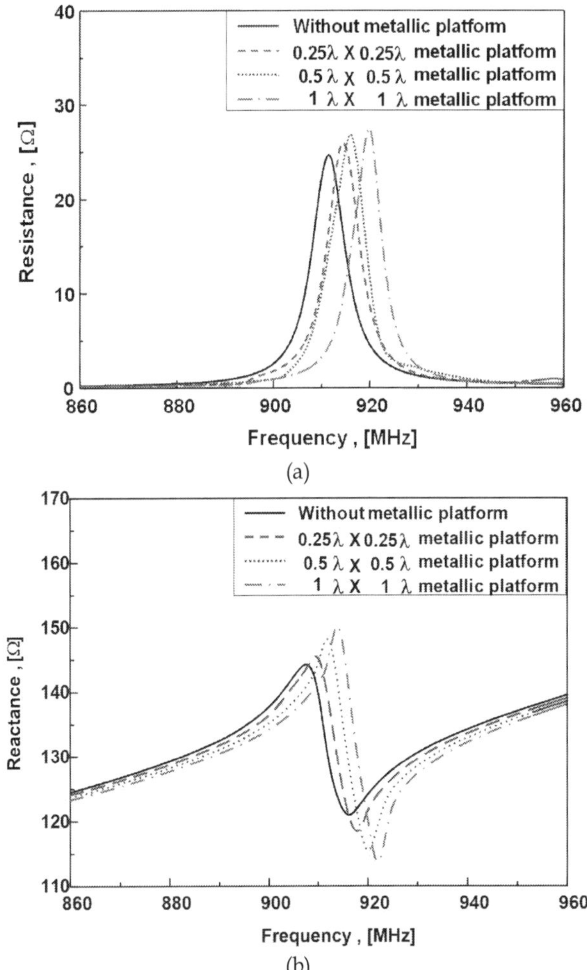

Fig. 4.2 Simulated impedance characteristics with different sizes of metallic platforms

conjugate match is achieved between antenna and microchip by adjusting the perimeter of the feeding loop and the gap between the radiating elements. Then, the proposed tag antenna gives a smaller variation of the antenna performance than that of conventional tag antennas when the tag is mounted on the various sizes of the metallic platforms.

Fig. 4.2 shows the simulated impedance characteristics of the tag antenna with different sizes of metallic platforms. One can see that the impedance variation is small without metallic platform and with various sizes of metallic platforms. Therefore, we can expect that this tag antenna gives smaller variation in the antenna performance than that of conventional tag antennas when the tag is mounted on the various sizes of the metallic platforms.

Although the currents on the radiating elements excited by the feeding loop are out of phase with equal amplitude, the direction of the surface current is very important so as to obtain the performance of a perfectly balanced antenna. Therefore, the symmetric shorting strips with respect to the y-axis are used to achieve more balanced current distributions as shown in Fig. 4.3. The main direction of the electric field is along with the x-axis since two symmetric patches are excited out of phase. This is the major difference from the radiation mechanism of the conventional PIFAs or IFAs, which cause the performance variation and reduction due to the electrical coupling between the radiator and ground plane. The proposed antenna has its main electrical coupling between two radiating elements rather than between the radiator and ground plane. This means the radiation of this antenna comes mainly from the two adjacent radiating elements. Therefore, considerable reduction of the effect of the metallic platform can be achieved. Fig. 4.4 shows the radiation efficiency for various sizes of metallic platforms. One can see that the reduction of radiation efficiency due to size variation of metallic platforms has not reached values that impede operation.

Fig. 4.5 shows the measured power bandwidth for different sizes of the metallic platforms. All the peaks have been normalized to 0 dB. The power bandwidth is defined as the half-power bandwidth of the antenna aperture, which is equivalent to +3 dB in required transmitted power P_{tx}. HPBW (Half Power Band Width) is 902 MHz ~ 928 MHz, and the variation of resonant frequency is less than 5.5 MHz. These variations are much smaller than those of the conventional tag antennas. The bandwidth within the 3 dB power variation shows that this antenna has a very good tolerance for different sizes of metallic platforms. Fig. 4.6 shows the radiation patterns. It is noticed that the direction of the antenna's main beam does not vary with the size of the metallic platform.

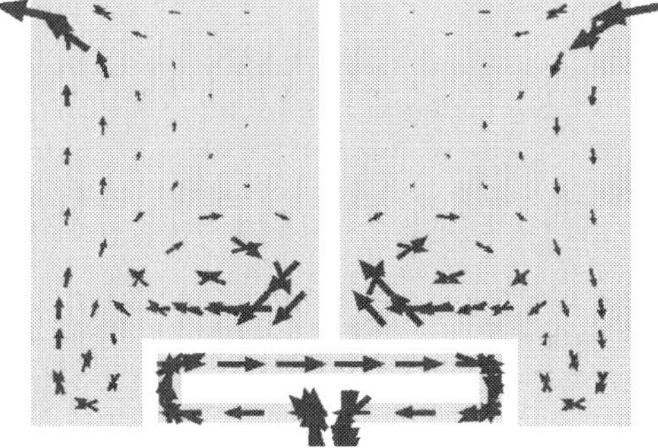

Fig. 4.3 Surface current distribution of balanced-type microstrip patches

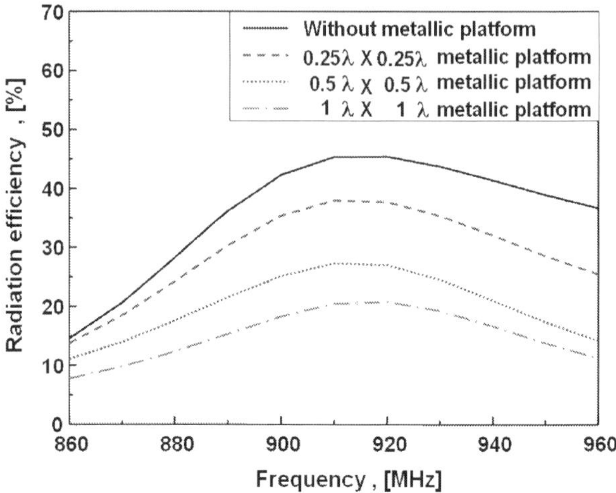

Fig. 4.4 Simulated radiation efficiency for different sizes of metallic platforms

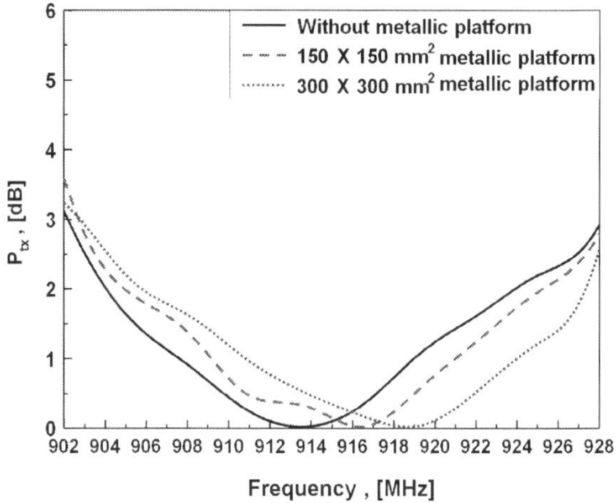

Fig. 4.5 Measured power bandwidth versus for different sizes of the metallic platforms

4.2 Compact microstrip patch

As mentioned, performance of a RFID tag antenna can becomes worse under the impact of a metallic environment. To overcome this problem, several PIFAs, IFAs, or microstrip patch antennas have been proposed. However, they still have the complexity of manufacturing because of the vertical feeding structure along with a microchip and use thick or multi-layered substrates. When it comes to designing RFID tag antenna for metallic platforms the dimension and complexity of the antenna are very important factors as they relate to the manufacturing cost. One way to reduce manufacturing costs is to keep the tag antenna design as simple as possible.

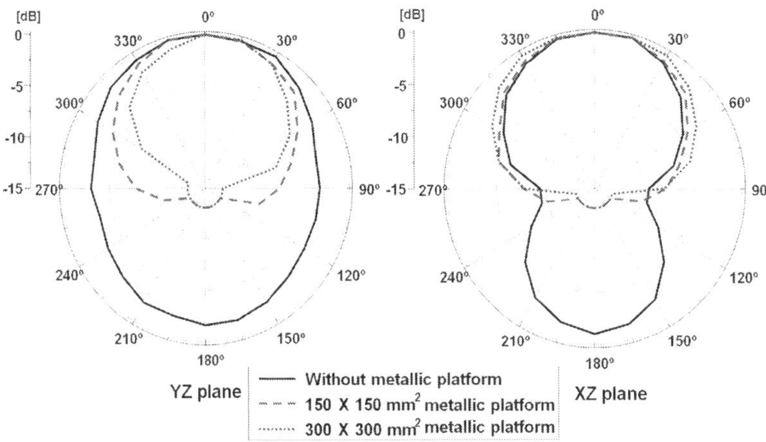

Fig. 4.6 Measured radiation patterns with different sizes of metallic platforms

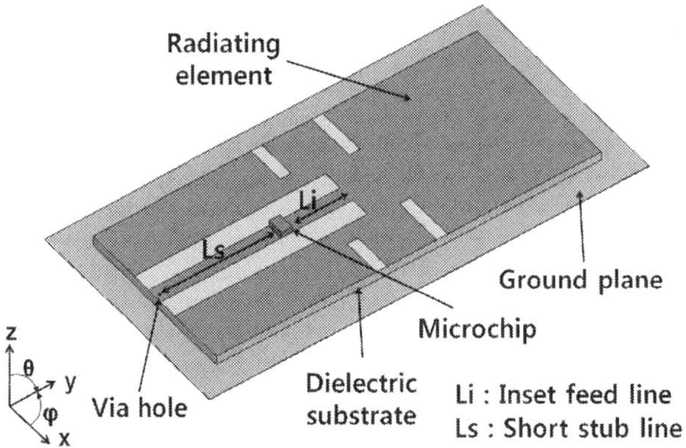

Fig. 4.7 Structure of the compact patch-type tag antenna

A new type of RFID tag antenna mountable on metallic objects in UHF band is proposed (Lee & Yu, 2008). This antenna can reduce the complexity of manufacturing and thickness of the antenna by using a microstrip patch type structure which has a single layer and the feed line on the same layer of the simple radiating patch. Moreover, this antenna makes the conjugate impedance match between the antenna and the microchip easy without additional matching networks. Fig. 4.7 shows the geometry of the compact patch-type tag antenna (Lee & Yu, 2008). The feed line is divided into the inset feed line (length of L_i) and the short stub line (length of L_s). The short stub line is electrically shorted to the ground plane by a via hole. The slits are symmetrically embedded on the radiating patch along the y-axis to reduce antenna size. The complex antenna impedance can be controlled by varying the length of the feed line (length of the inset feed line: L_i, length of short stub line: L_s). The conjugate match between the antenna and microchip can be achieved by adjusting the length of the inset feed line (L_i) and the length of the short stub line (L_s), which is much easier than previously reported techniques. Impedance matching can be achieved without major modification of the radiator and additional matching networks. It should be mentioned that changing L_i mainly affects the resistance while changing L_s mainly affect the reactance.

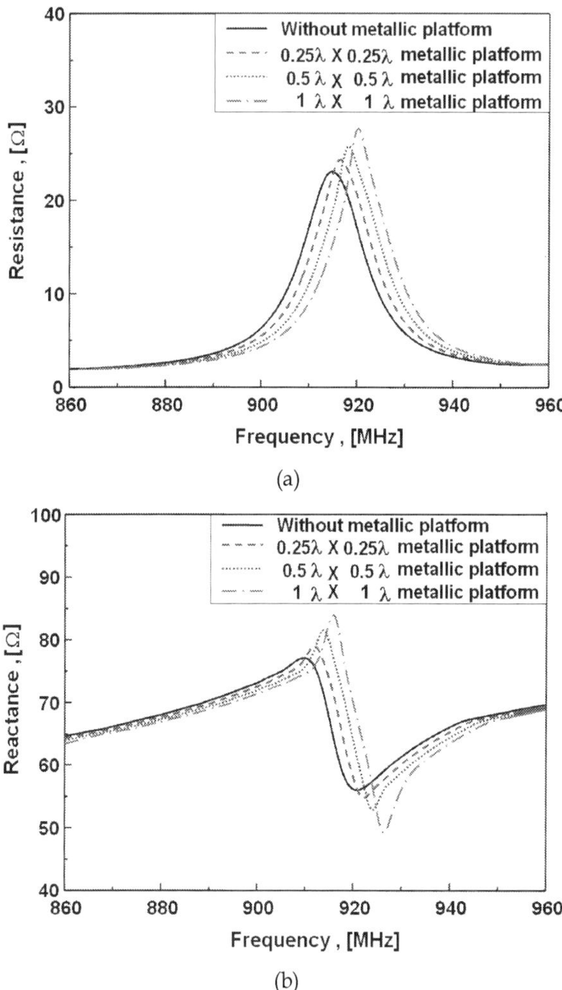

(a)

(b)

Fig. 4.8 Simulated impedance characteristics for different sizes of metallic platforms

Fig. 4.8 shows the simulated impedance characteristics of a compact tag antenna with different sizes of metallic platforms. It is noticed that the impedance variation is small without metallic platform and with various sizes of metallic platforms. Therefore, the impedance has very good tolerance for different sizes of metallic platforms. Fig. 4.9 shows the radiation efficiency versus frequency for various sizes of metallic platform. One can see that the radiation efficiency increases as the size of the metallic platform increases.

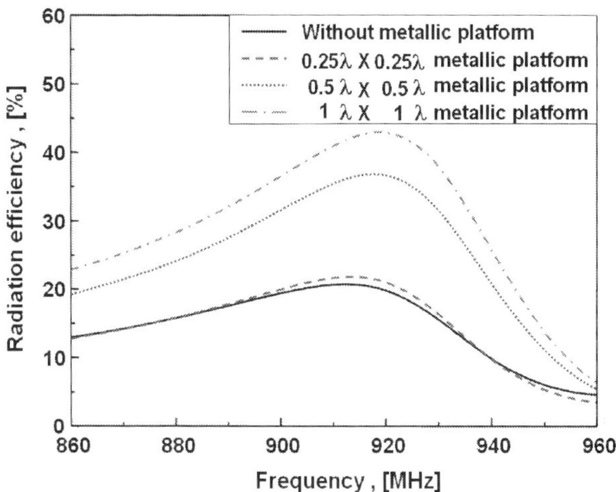

Fig. 4.9 Simulated radiation efficiency for different sizes of metallic platforms

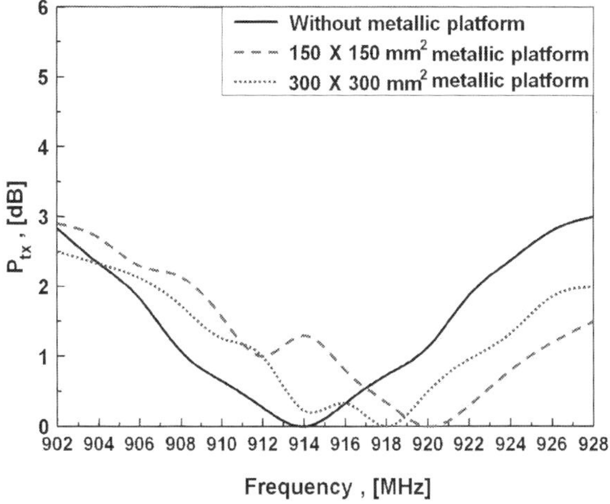

Fig. 4.10 Measured power bandwidth versus the different sizes of the metallic platforms

Fig. 4.10 shows the measured power bandwidth versus frequency when the tag is mounted on different sizes of metallic platforms. The bandwidth within 3 dB power variation for the square metallic platform of 150 ~ 300 mm length remains good. So, the bandwidth has a very good tolerance for the large sized metallic platforms. Fig. 4.11 shows the measured radiation patterns. It is shown that the direction of the antenna main beam does not vary with the size of the metallic platform, and its directivity is increased as the size of metallic platform increases. One can see that the proposed antenna gives a good performance when it is even mounted on various sizes of metallic objects.

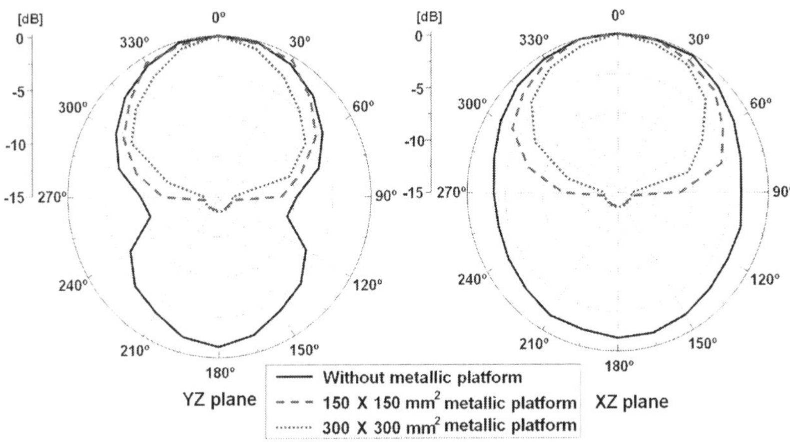

Fig. 4.11 Measured radiation patterns with different sizes of metallic platforms

4.3 Other RFID tag antennas

Two types of tag antennas which can be attached to metallic platforms have been introduced and discussed in earlier subsections. In addition to these, there are other types of tag antennas suitable for metallic platforms.

As mentioned in the previous section, incident electromagnetic waves totally reflects from metallic surfaces with a phase reversal. The metallic objects near an antenna change the antenna parameters and degrades radiation efficiency. Therefore, the metallic surface should be used as a ground plane of the antenna or as an energy-improving reflector. Both the patch with EBG ground plane and patch antenna with regular ground plane for a tag antenna attachable to metallic surfaces are analyzed (Ukkonen et al., 2005). According to their results, the patch antenna with EBG ground plane has higher radiation efficiency than the regular patch antenna. This is due to the suppression of surface waves when the EBG ground plane is used. However, the EBG structure needs a periodic structure. So it makes an antenna expensive, and its structure becomes larger.

According to the electromagnetic boundary conditions we mentioned, for magnetic field, there are only tangential components and no normal components of this field to the metallic surface. The tangential component of the magnetic field will be doubled when it is very near the metallic surface. The RFID tag antenna design (Ng et al., 2006) here exploits the fact above by having a loop antenna oriented such that the plane of the loop is perpendicular to the plane of the metallic surface where the RFID tag will be attached. With this orientation, the RFID tag antenna has improved performance when attached near a metallic platform, and this antenna has allowed better coupling to the magnetic components of the interrogation fields. Various types of loop antennas perpendicular to the plane of the metallic surface can be considered for the tag antenna. Although the circular loop antenna is the most common among all loop antennas, a rectangular loop is chosen to keep smaller height of a tag antenna.

Other types of tag antennas using a shorting plate (Hirvonen et al., 2004), a printed inductor (Son et al., 2006), and a U-shaped slot (Kwon & Lee, 2005) have been proposed to improve the antenna performance for metallic platforms.

5. Conclusion

The RFID is an emerging technology making ubiquitous identification possible. The potential applications of the RFID are numerous. A UHF (902-928 MHz) band RFID system becomes more attractive for many industrial services because it can be used for many applications such as security and access control, asset management, transportation, supply chain management, and baggage handling with high reading speed, capable multiple accesses, anti-collision, and long reading distance. Since RFID systems are applied in many fields, the technology used to realize the antenna without severe performance degradation for various types of platforms is perhaps the most important technology in improvement of the RFID system performance.

6. References

Balanis, C. A. (1997). *Antenna Theory: Analysis and Design 3rd edition*, John Wiley & Sons, ISBN 0-471-59268-4, New York

Hirvonen, M.; Pursula, P.; Jaakkola, K. & Laukkanen, K. (2004). Planar inverted-F antenna for radio frequency identification, *IET Electronics Letters*, Vol. 40, No. 1 (July 2004) ISSN 0013-5194

Iskander, M. F. (1992). *Electromagnetic Fields and waves*, Prentice Hall, ISBN 0-13-249442-6, New Jersey

Kwon, H. & Lee, B. (2005). Compact slotted planar inverted-F RFID tag mountable on metallic objects, *IET Electronics Letters*, Vol. 41, No. 1 (November 2005) 1308-1310, ISSN 0013-5194

Lee, B. & Yu, B. (2007). Compact structure of UHF band RFID tag antenna mounted on metallic objects, *Microwave and Optical Technology Letters*, Vol. 15, No. 1, (January 2008) 232-234, ISSN 0855-2477

Ng, M. L.; Leong, K. S. & Cole, P. H. (2006). A small passive UHF RFID tag for metallic item identification, *Proceedings of ITC-CSC2006*, pp. 141-144, ISBN 9749441877, Thailand, July 2006, ECTI, Chiang Mai

Pozar, D. M. (2005). *Microwave Engineering 2nd edition*, John Wiley & Sons, ISBN 0-471-44878-8, New York

Son, H.-W.; Choi, G.-Y. & Pyo, C.-S. (2006). Design of wideband RFID tag antenna for metallic surfaces, *IET Electronics Letters*, Vol. 42, No. 5 (March 2006), 263–264, ISSN 0013-5194

Ukkonen, L.; Sydänheimo, L. & Kivikoski, M. (2005). Effect of metallic plate size on the performance of microstrip patch-type tag antenna for passive RFID, *IEEE Antenna and Wireless Propagation Letters*, Vol. 4, (June 2005) 410-413, ISSN 0855-2477

Yu, B.; Kim, S. J.; Jung, B.; Harackiewicz, F. J. & Lee, B. (2007). RFID tag antenna using shorted microstrip patches mountable on metallic object, *Microwave and Optical Technology Letters*, Vol. 49, No. 2, (February 2007) 414-416, ISSN 0855-2477

11

RFID in Metal Environments: An Overview on Low (LF) and Ultra-Low (ULF) Frequency Systems

D. Ciudad, P. Cobos Arribas, P. Sanchez and C. Aroca
ISOM, ETSI Telecomunicación, Universidad Politécnica de Madrid (UPM),
Spain

1. Introduction

Depending on their frequency, radio waves can be absorbed by water and biological tissues (Bottomley & Andrew, 1978). They can also be shielded by metals due to the eddy currents. These effects constitute a huge problem for the implementation of radio frequency identification (RFID) systems in real environments.

RFID systems can be classified into three different groups depending on the physics involved: i) inductive coupling; ii) back-scattering; and iii) electrical coupling. Another classification can be done attending to the electromagnetic band used: i) LF (125-134.5 kHz); ii) HF (13.56 MHz); iii) UHF (~900 MHz); and iv) ISM (2.4 GHz). The last three systems (HF, UHF and ISM) require an environment without metals or water to work properly. On the other hand, LF systems are less affected by metals on the surroundings and can penetrate some materials like water, but they cannot work properly through metals due to the shielding or detuning of the electromagnetic signal. These problems could be avoided by reducing the working frequency. However, both LF and HF systems are based on the inductive coupling between tag and reader and this coupling become rapidly inefficient with the frequency reduction.

Some different solutions have been proposed to solve these problems. In this chapter we explain the basics of the inductive coupling method and the detuning and shielding effects due to metals. Additionally, a new system that is able to work at ultra-low frequencies ULF (1-100 kHz) and through a metallic shielding is proposed. Finally we compare the properties of the LF and the new ULF systems.

2. Inductive coupling based systems and metal environments

2.1 Basics on the inductive coupling method

Inductive coupling between two wires appears when the change in the current flow through one wire induces a voltage across the ends of the other wire. See Figure 1.

The reader (or transceiver) powers the tag (or transponder) through an AC electromagnetic field. The reader can also modulate this electromagnetic field to send information to the tag. Once the tag is powered, it can change the energy it takes from the electromagnetic field. These changes of energy give rise to some other changes in the powering of the reader. It allows the tag to send information. An accurate model of LF systems is given in (EM Microelectronic, 2002). Inductive systems can be modelled as the circuit showed in Figure 2 (EM Microelectronic, 2002).

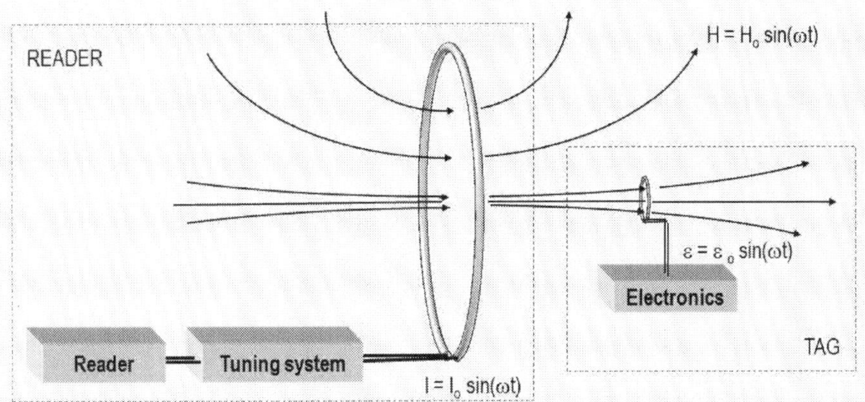

Fig. 1. Schematics of the inductive coupling.

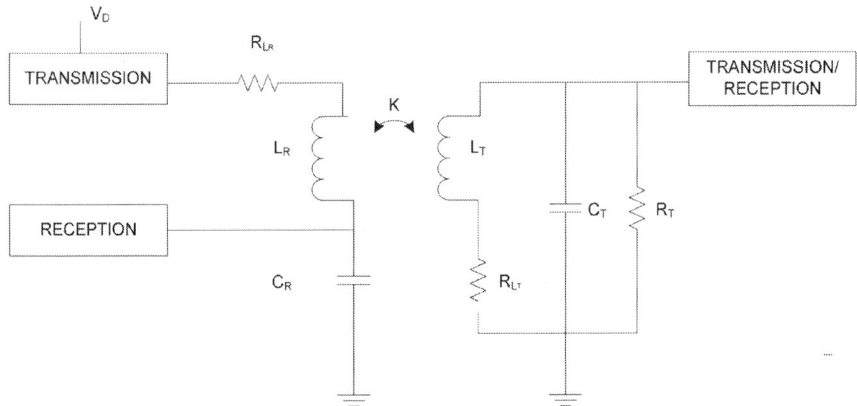

Fig. 2. Model of the LF systems. The circuit in the left is the reader. This one in the right is the tag.

This circuit allows the accurate calculation of the coupling factor k between reader and tag. It can be done by using the theory of transformers. See (EM Microelectronic, 2002).
Here we will not calculate the k factor. We will only analyse the resonance frequency. Both reader and tag must be tuned at the same resonance frequency to maximize k. The resonance frequency f$_0$ is given for both circuits by:

$$\text{(Reader)} \quad f_0 = \frac{1}{2\pi\sqrt{L_R C_R}} \tag{1}$$

$$\text{(Tag)} \quad f_0 = \frac{1}{2\pi\sqrt{L_T C_T}} \tag{2}$$

As introduced in Figure 2, L_R and C_R are the inductance and the capacitance of the reader, and L_T and C_T these of the tag.

2.2 Influence of metals in radio frequency sytems: distortion of the magnetic field, detuning, and shielding

The different effects produced by metals on RFID systems comes from the presence of eddy currents. These effects are: i) distortion of the electromagnetic field; ii) detuning; and iii) shielding. As a result, the readable area is reduced. See Figure 3.

The inductive coupling method is based on the resonance of the antenna-tag circuit. The metal produces a drift of the working frequency. This is the so called detuning effect. A proper design of the system can allow RFID systems to work with metals on the surroundings avoiding this effect. Some different techniques and methods have been proposed. They are based on introducing dielectric gaps in between the metal and the tag, or in the use of magnetic materials like ferrite in the tag. See for example (Dixon et al., 2007), (Dixon et al., 2008) and (Bovelli et al., 2006).

Shielding happens when working through metals. Sometimes it can be avoided by using a set of different antennas. However, if the tag is enclosed in metals it can only work in some particular geometries that include dielectric gaps (Finkenzeller, 2003).

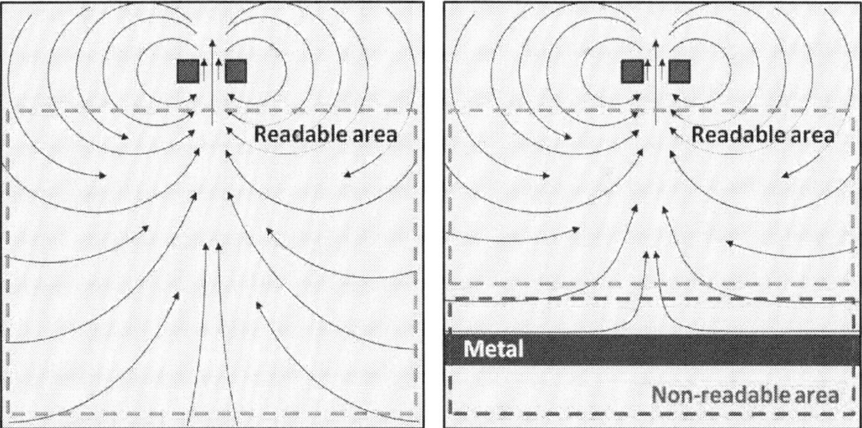

Fig. 3. Effect of a metal layer on an alternating magnetic field. On the left: the magnetic field produced by a wire in an environment without metals. On the right: the lines of the magnetic field are distorted by a metallic layer.

2.2.1 Distortions of the magnetic field

From Maxwell's equations and the appropriate boundary conditions, it is obtained that only normal electric fields to the surface and tangential components of the magnetic field are allowed in the surface of a perfect conductor. A metal is not a perfect conductor but it still produces a deformation of the magnetic field close to its surface. For mathematical details see (Balanis, 1997).

In addition, eddy currents appear in any conductive material in the presence of an AC electromagnetic field. These currents create a magnetic field perpendicular to the surface of the conductor.

As a result of these effects, a common planar RFID tag cannot work when it is in close contact to the surface of a metal. An appropriate design of the RFID system must be done to allow the tag to work close to metallic objects.

In (Dobkin & Weigand, 2005) it is shown a study on the read range of HF RFID tags as a function of the length of the gap between tag and a metal or water. This is not so critical is LF systems, but it still can cause some problems.

2.2.2 Detuning

Detuning happens due to eddy currents which produce a magnetic field perpendicular to the metal layer. This field opposes the original magnetic field applied. When metals are in between the reader and the tag or in the surroundings, they can be modelled as an additional parallel inductance, L_M, on the circuit of the tag (Hoeft & Hofstra, 1988).

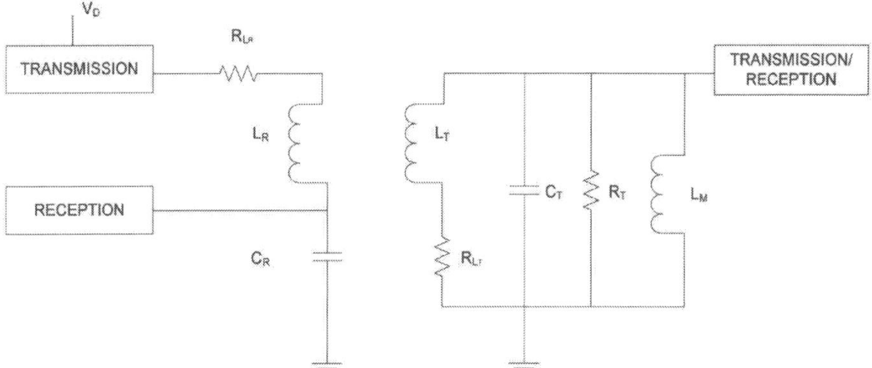

Fig. 4. Model of the LF system with metals on the surroundings of the tag.

The new inductance produces a change of this circuit resonant frequency from f_0 to f_1:

$$(\text{Tag + metal}) \quad f_1 = \frac{1}{2\pi\sqrt{L_{Teff}C_T}} \tag{3}$$

L_{Teff} is the effective inductance:

$$L_{Teff} = \frac{L_T L_M}{L_T + L_M} \tag{4}$$

The effect is usually a reduction of the total inductance (Bowler & Huang 2005). It increases the working resonance frequency f_{res}. Metals can also introduce a parasitic capacitance.

These parasitic impedances cause the detuning of the system since the resonance frequency of the reader and that of the tag are different. It hinds the tag from be properly powered.
If the metallic material is placed close to the reader, the impedance of the reader also changes. It can be used to measure the conductivity of materials through eddy currents (Bowler & Huang 2005). In RFID this effect also produces the detuning of the system.

2.2.3 Shielding and power loss

Shielding is also due to eddy currents. They allow the metal to absorb RF energy reducing the effectiveness of a RFID system. If an electromagnetic wave propagates through a metal a distance t_S, its amplitude B_0 is exponentially reduced according to the Skin's formula:

$$B'_0 = B_0 e^{-\frac{t_s}{\delta}} \tag{5}$$

B_0' is the amplitude of the magnetic field after having covered the distance t_S through the metal. The parameter δ is the penetration depth. This is the distance in which the amplitude of the magnetic field is reduced a factor e. For a good conductor, δ is given by:

$$\delta = \sqrt{\frac{2}{\mu_s \sigma \omega}} \tag{6}$$

μ_S is the permeability of the metal, σ its conductivity and ω the angular frequency of the electromagnetic field. Figure 5 shows the penetration depth δ for different metals.

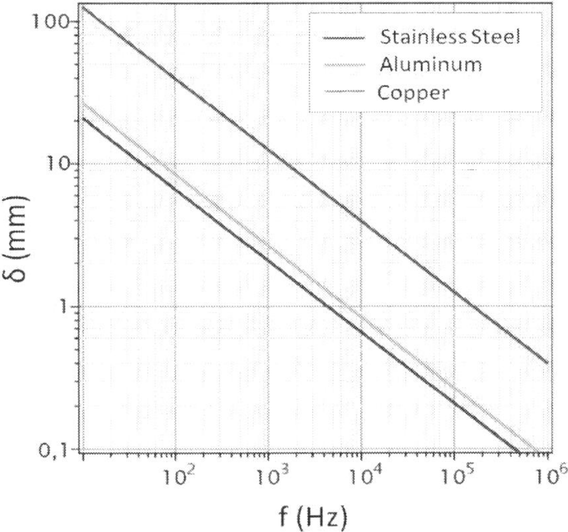

Fig. 5. Penetration depth versus frequency for different materials. $\mu \sim \mu_0$. The conductivity is 1.51 MS/m for stainless steel, 35.40 MS/m for aluminium and 58.00 MS/m for copper. Data from (Lide, 2009) and (Bowler & Huang, 2005)

Even if the detuning problems are avoided, the tag of a LF RFID system will only receive a fraction of the total energy without shielding. This is due to the reduction of the intensity of the magnetic field. Eddy currents and their related effects are explained in detail in (Ida & Bastos, 1997). Being H_0 the amplitude of the applied magnetic field when reaching the surface of the metal, the average power loss per unit volume in the metal due to eddy currents is:

$$p = \frac{1}{24}\sigma\omega^2 t_S{}^2 \mu_S{}^2 H_0{}^2 \qquad (7)$$

Inductive coupling-based systems can work through metals only with thicknesses bellow the penetration depth approximately. Thicker layers completely shield the tag. For example, attending to Figure 5 it can be roughly concluded that LF RFID systems cannot work through any aluminium layer thicker than 0.25 mm.

3. RFID through metals: ultra-low frequency ULF system

In this section we explain the new ULF RFID system. It is able to work in metallic noncleaned surroundings and even through metals. This is achieved by operating at ultra-low frequencies (1- 100 kHz). In the ULF range, inductive coupling cannot be used due to: i) the loss of the efficiency – in the inductive coupling it decreases squarely with the frequency-; ii) detuning problems; and iii) the high inductance and capacitance values needed to produce the resonance of the circuit. Instead, the capability of work at such low frequencies in a non-resonant system is achieved by measuring the change of the magnetization of a magnetic core integrated in the tag.

3.1 System description

The main idea is to measure changes on the magnetization of a magnetic material included in the tag. The ultra low-frequency RFID system is formed by three different elements:
1.   An antenna to produce an ultra low frequency magnetic field
2.   A tag with a soft magnetic core and a winding around it.
3.   A reader to detect any change on the magnetization of the tag.
The antenna produces an AC magnetic field that magnetizes the tag. The antenna must provide an area to place the reader without being magnetically saturated. This objective can be achieved by using different geometrical coil arrangements.

The tag has a low resistance winding around the magnetic core. When this winding is in open-circuit configuration, the magnetic field produced by the antenna magnetizes the magnetic material in the tag. At the same time in the winding it is induced an electromotive force (e.m.f.) able to feed a microcontroller. On the other hand, when the winding is in shortcircuit configuration, a current is induced in it. This current avoids the magnetization of the magnetic core. The micro-controller located in the tag can send information by opening and short-circuiting the winding. It produces some changes in the magnetization of the core that are easy to detect with a magnetic sensor.

The reader is a fluxgate in which a double demodulation in-phase technique is performed (Aroca et al., 1995). In the inductive systems, both the reception coil and tag must have large cross section areas and high turn number to work at low frequencies. An effective way to reduce the size of the antenna without losing effective cross section is to use soft-magnetic cores. This is one of the basics of the fluxgate magnetic sensors. Due to its magnetic core, the magnetic flux in a 3 cm³ fluxgate sensor is equivalent to a 10 m² cross section antenna for a 1kHz AC magnetic field.

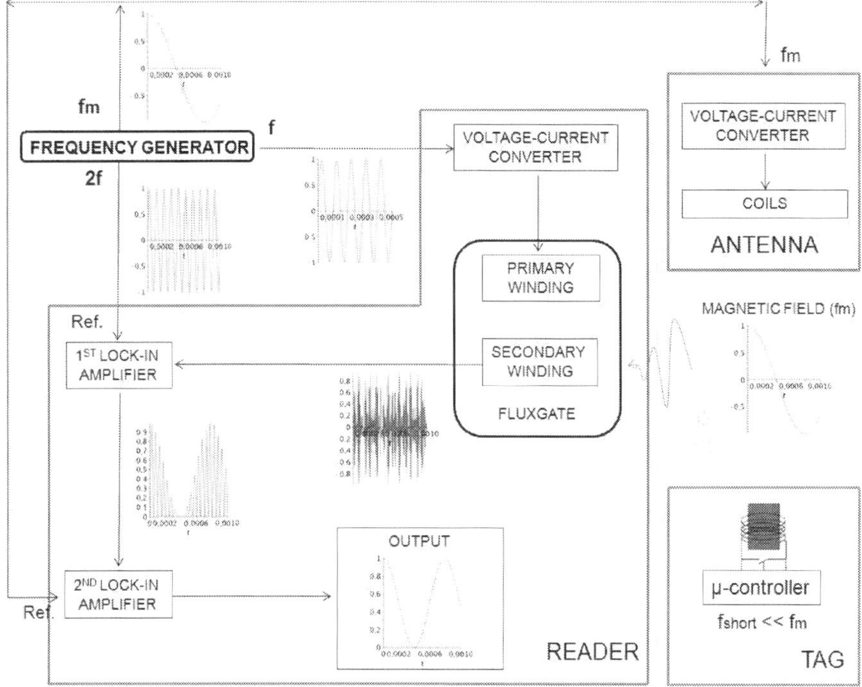

Fig. 6. Block diagram of the new ULF RFID system.

This system has been completely fabricated and tested in the laboratory (Ciudad et al., 2004) (Ciudad Rio-Perez et al., 2008). Different geometrical arrangements for the antennas have successfully been used. In addition, a theoretical model to develop the new RFID system for a particular application has been developed.

3.2 Fabrication
3.2.1 Tag
The tag is formed by three different elements: i) a soft-magnetic core; ii) a low-resistance winding; and iii) the electronics to send information.

Regarding their industrial production, the tags must be plastic or printed circuit board (PCB)-based and the magnetic material should be obtained by electrodeposition. We have done a PCB-based tag. Its dimensions are 5.5 cm x 8.5 cm x 1mm, like in a common credit card. The tag is fabricated by joining two PCBs. These PCBs have some copper tracks in one of their faces (Figure 7.A) and the magnetic material on the other (Figure 7.C). Both PCBs are joined with the magnetic material in the inner faces. The copper lines are soldered forming a winding all around the magnetic material (Figure 7.D). In our prototype the winding has 55 turns in the longitudinal direction.

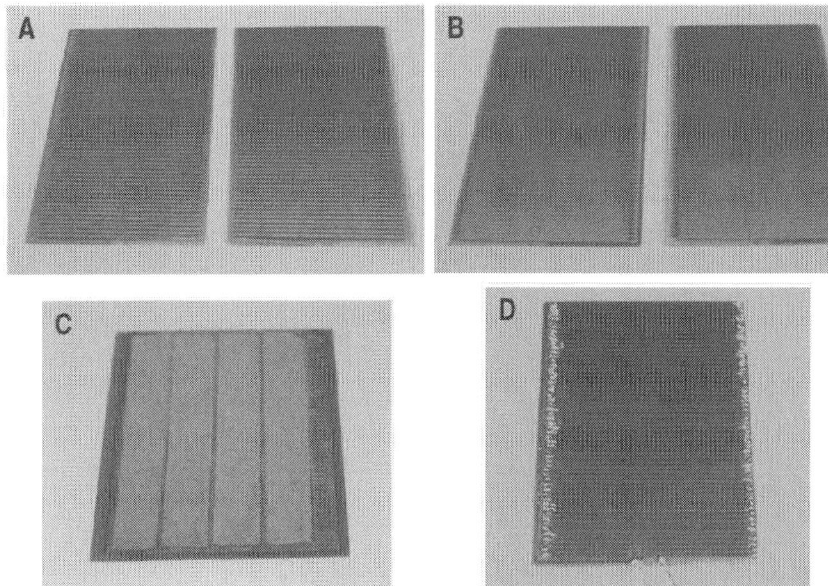

Fig. 7. Photographs of the PCBs used to fabricate the tag. A and B: before the electrodeposition of the magnetic material. C: one of the PCBs with the electrodeposited magnetic material. D: the card after soldering the tracks of both PCBs.

Magnetic core

We have electrodeposited CoP multilayers to fabricate the magnetic core of the tag. These multilayers are particularly useful due to their low coercivity and high permeability. CoP monolayers exhibit perpendicular to plane magnetic anisotropy. It means that the material is difficulty magnetized in plane. However, Co_xP_{1-x} multilayered films with different composition within the span $(0.74 < x < 0.86)$ show soft magnetic properties: i) in-plane magnetic anisotropy; ii) low coercivity; and iii) high permeability (Perez et al., 2000).

The total thickness of the multialyer was 40 μm. It is formed by the stacking $(Co_{0.74}P_{0.26}/Co_{0.83}P_{0.17})_N$ bilayers with each layer of 20nm. This two compostions were selected in order to have amorphous alloys covering all the substrate (Ciudad Rio-Perez et al., 2008).

$(Co_{0.74}P_{0.26}/Co_{0.83}P_{0.17})_N$ multilayered amorphous films were electrodeposited on Cu substrates under the conditions referred in (Perez et al., 2000). It was used an Autolab-PGSTAT30 potentiostat/galvanostat to control the current density. The electrolyte was: $CoCO_3$ (39.4 g/l – 0.33M), $CoCl_3.6H_2O$ (181 g/l – 0.76M), H_3PO_3 (65 g/l – 0.76M), and H_3PO_4 (50 g/l – 0.51M). The temperature of the electrolyte during the deposition was 80 °C. It was used a 99.9% chemical purity Co anode. The density of the electrical current needed to deposit $Co_{0.74}P_{0.26}$ was 100 mA/cm² whereas for $Co_{0.83}P_{0.17}$ was 500 mA/cm².

Electronics of the tag

In a first approach, the electronics of the tag was designed in two different modules: one for powering and another for communications. See Figure 8. The module for powering is an analog multiplier circuit. Some different capacitors are charged due to the induced e.m.f in the winding of the tag. The module for communications is formed by a microcontroller that provides control signals (0/1) to the gate of a MOSFET. This MOSFET acts like a switch to open and short-circuit the winding. The communication module is also similar to an analog multiplier circuit in which some capacitors have been removed in order to improve the response time. As microcontroller we have used a PIC16LF84A from MicroChip. It can be powered by only 2V and it can work with only 15 µA.

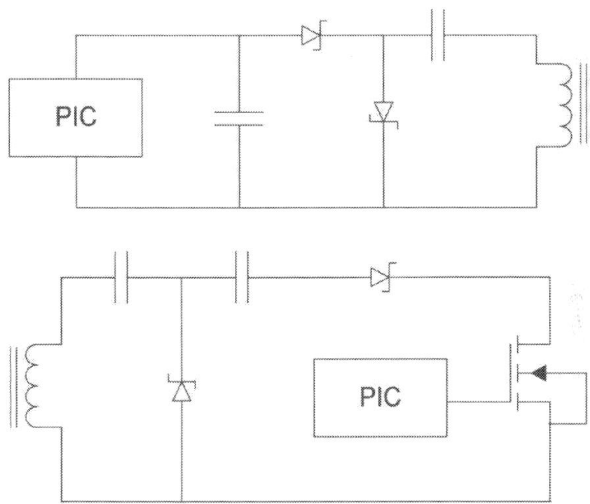

Fig. 8. Initial designs for the electronics of the tag. Up: module to power the PIC. Down: module for communications.

The inductive coupling-based RFID systems use to include a Zener diode 5V6 to protect the circuit of the tag from high voltages when approaching it to the antenna. In our case, this protection is not needed because the voltage is limited by the capacitors.
In a later stage, we simplified these circuits by unifying both modules as shown in Figure 9.

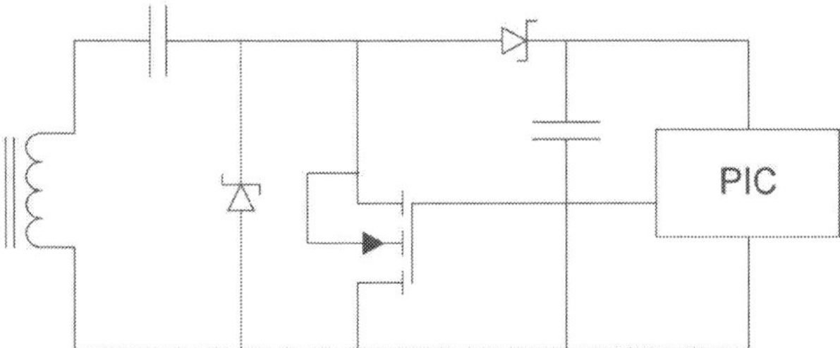

Fig. 9. Simplified electronics of the tag for powering and communications.

3.2.2 Reader

The reader is a tuneable magnetic field sensor. It detects the change in the magnetization of the tag due to the switching of the winding. The magnetic sensor must be tuned at the frequency of the magnetization of the tag. This frequency is that of the applied magnetic field by the antenna.

The best option is to use a fluxgate and the double demodulation in phase technique (Aroca et al., 1995).

Fluxgate

A fluxgate is formed by a soft magnetic core and two different windings. The primary or excitation winding generates a magnetic field able to saturate periodically in both polarities the magnetic core. In the secondary or pickup winding is induced an e.m.f. The amplitude of the second harmonic of the e.m.f. in the secondary winding is proportional to any external magnetic field. The sensitivity of the fluxgates spans in the range 10^{-10} to 10^{-4} T. A good introduction to fluxgate sensors can be found in (Ripka, 2001).

We have used two different fluxgates. A common one described in (Aroca et al., 1995) and a planar one described in (Perez et al., 2004). The last one is particularly interesting because it is fabricated in PCB technology and its magnetic core is electrodeposited.

Double demodulation in phase

Figure 6 shows the demodulation technique used in the RFID system. All signals for excitation and sensing are obtained from the same frequency source. The primary winding is excited by an AC current with frequency f. A first lock-in amplifier is used to select the second harmonic of the signal induced in the secondary winding of the fluxgate sensor (frequency 2f). The time constant must be carefully selected so that the amplitude of the second harmonic, as well as any change due to the presence of the tag in the surroundings, can be measured.

If the tag is magnetized with frequency f_m (f_m<<f) by the applied magnetic field, the amplitude of the signal after the first lock-in amplifier is modulated at this frequency. A second lock-in is used to demodulate the output signal of the first lock-in amplifier. The reference frequency for the second one is f_m.

3.2.3 Antenna

It produces the AC magnetic field to power the tag and magnetize the magnetic material in it. The antenna must also provide of an area in which the magnetic field is compensated. This area is needed to place the reader avoiding its magnetic saturations due to the applied magnetic field. Different antennas can be designed. We have successfully proved two different geometrical arrangements.

Antenna for a standard fluxgate

The first design of the antenna was a Helmholtz pair of coils with both coils connected in opposition. Figure 10. The fluxgate is placed centred in between the two coils. In this area the magnetic field is zero. We have used a standard fluxgate with a 12 kHz excitation signal. Applying an AC magnetic field of only 1 gauss of amplitude and 1.9 kHz of frequency, the reader is able to detect the signal of the tag even through an aluminium folder and in close contact to the tag and with a thickness up to 0.2 mm.

The magnetic field as a function of the distance for this antenna is given in (Ciudad et al., 2004).

Antenna for a planar-type fluxgate

The antenna for the planar fluxgate is also planar. Figure 10. This antenna can be fabricated in PCB technology. It is formed by two spiral and planar windings. They are connected to

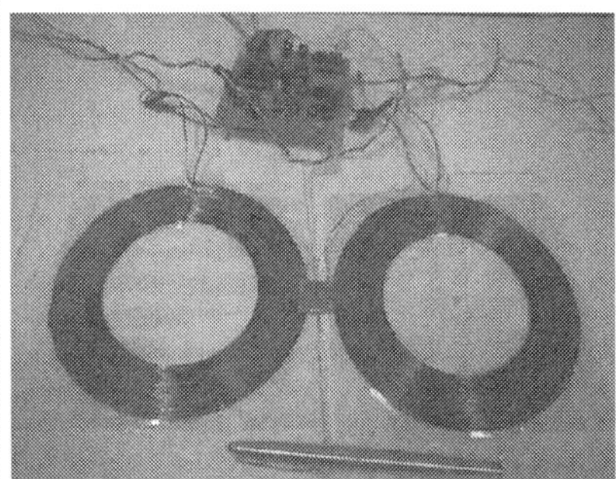

Fig. 10. Different designs for the antenna. Left: antenna for a standard fuxgate. The fluxgate is centred in the Helmholtz coils. Right: antenna for a planar one. The fluxgate is between the planar coils.

compensate the magnetic field they produce in the area in between them. This is the place to fix the fluxgate. Both windings have been done using a copper fill of 1 mm of diameter. Each spiral has 25 turns. The inner radius is 3.5 cm and the outer one 7.5 cm. The distance in between their centres is 18 cm. The magnetic field produced by this antenna as a function of the distance is given in (Ciudad Rio-Perez et al., 2008).

This RFID system works with an AC magnetic field of only 5 gauss of amplitude and 6 kHz of frequency.

3.3 Experimental testing

The system has been tested for the different fluxgates and antennas. It can work through metals at very low frequencies down to the 1 kHz order. This is two orders of magnitude bellow the working frequency of the inductive coupling-based systems.

Some of the signals produced in the different parts of the system are shown bellow. They have been taken from the system with planar fluxgate and antenna.

Figure 11 shows the voltage at the winding of the tag. Zero voltage is obtained when the microcontroller short-circuits the winding.

The output signal from the first lock-in amplifier is shown in Figure 12. This signal is proportional to the amplitude of the second harmonic of the induced e.m.f. in the pickup winding of the fluxgate. The amplitude of the second harmonic changes with frequency f_m due to the presence of the tag. In addition, this AC signal is modulated due to the change of the magnetization of the magnetic material when the microcontroller opens and shortcircuits the winding. Therefore, the low frequency modulation is produced by the switching of the winding of the tag. Compare Figures 11 and 12.

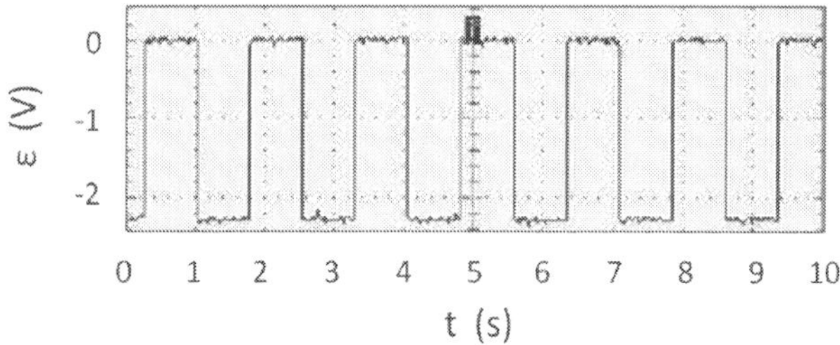

Fig. 11. Signal in the winding of the tag.

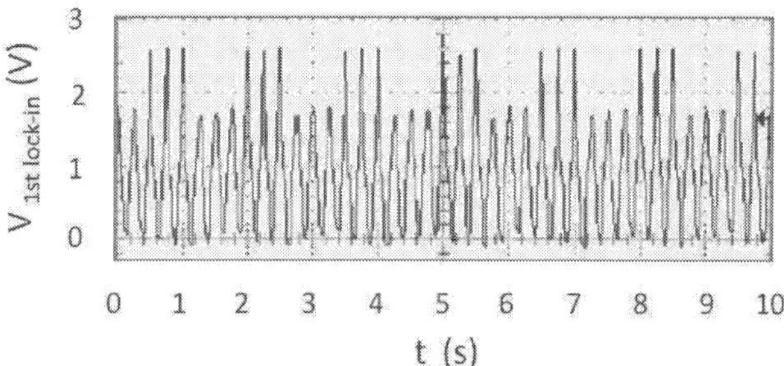

Fig. 12. Signal in the reader after the demodulation in the first lock-in amplifier.

Figure 13 shows the signal after the second lock-in amplifier. It has the same frequency that the signal produced by the microchip in the tag. It means that a signal with the same frequency that the produced in the PIC of the tag has successfully obtained in the reader.

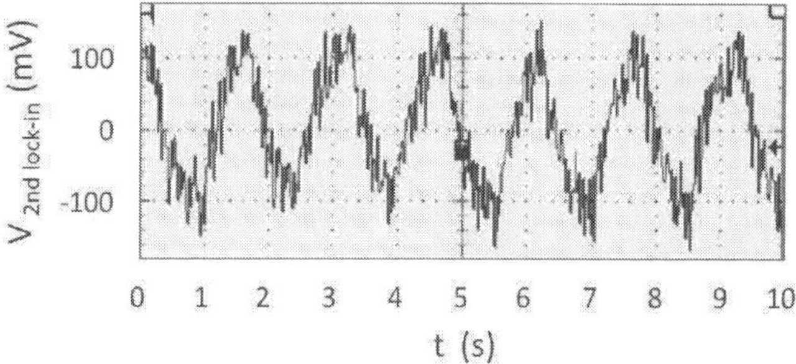

Fig. 13. Signal in the reader after the demodulation in the second lock-in amplifier.

These graphs (Figures 11, 12 and 13) clearly show that the microcontroller in the tag can be powered by a low frequency magnetic field and it can send information. They also show that the fluxgate with the second in-phase demodulation has successfully used as a reader.

3.4 Theoretical model
In (Ciudad Rio-Perez et al., 2008) it is given an accurate model to calculate the distance limitation of the ULF RFID system for a particular application. The model is also compared

with experimental data. This distance limitation can be due to failures in the detection or in powering the tag.

3.4.1 Detection of the tag: minimum sensitivity of the reader (fluxgate sensor).

The magnetic field in the tag position H_{ex} is assumed to be sinusoidal with amplitude H_0 and angular frequency ω:

$$H_{ex} = H_0 e^{j\omega t} \tag{5}$$

The magnetic flux through the tag and the induced e.m.f. in the winding are easily calculated. See (Ciudad Rio-Perez et al., 2008) for a detailed deduction. This e.m.f. is used to charge the capacitors that power the microcontroller. When the PIC in the tag short-circuits the winding, the induced e.m.f. gives rise to the flow of a current through the winding. This current causes a magnetic field. The total magnetic field (H_R) that magnetizes the magnetic core of the tag is the addition of the magnetic fields produced by the antenna (H_{ex}) and the winding (H_{tag}). The total magnetic field is:

$$H_R = \left(1 - \frac{1}{1 + \dfrac{R}{j\omega L}}\right) H_{ex} \tag{6}$$

R is the resistance of the winding of the tag and L its inductance. When the microcontroller opens the winding, $R=\infty$ and the magnetization of the magnetic core of the tag is given by:

$$M_{open} = \chi H_{ex} \tag{7}$$

χ is the magnetic susceptibility of the magnetic core. However, when this winding is shortcircuited, the magnetic core is not magnetized because $R=0$ and then $H_R = 0$. Therefore, being V the volume of the magnetic core, the change in the magnetic moment of the tag is given by:

$$\Delta m = \chi H_{ex} V \tag{8}$$

If a shielding layer of thickness t_s, conductivity σ and magnetic permeability μ_S is placed between the excitation system and the tag, the magnetic field is attenuated according the Skin's formula (5):

$$\Delta m = \chi H_{ex} V e^{-t_s \sqrt{\frac{\mu_S \sigma \omega}{2}}} \tag{9}$$

The tag is supposed to behave like a magnetic dipole. It implies that the magnetic field produced by the tag is reduced with the cube of the distance to the tag. This behaviour has been experimentally checked. See Figure 13.

The change of the magnetic field ΔH_{tag} when opening and short-circuiting the winding, at a distance r_t along its axis and at the other side of the shielding wall, is given by:

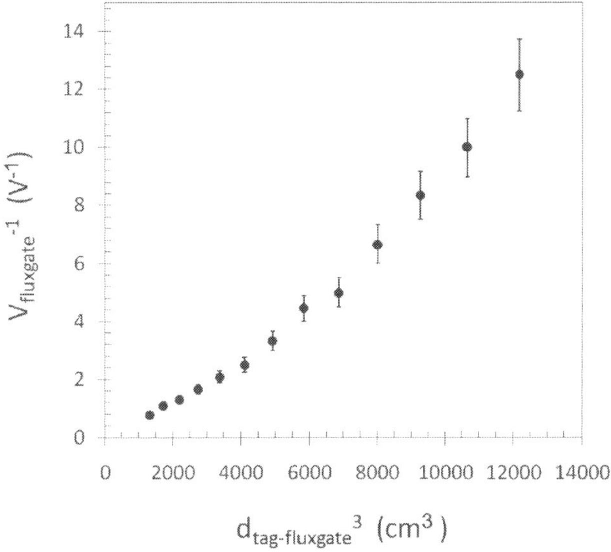

Fig. 13. Change of the signal in the pickup winding of the fluxgate $V_{fluxgate}$ when opening and short-circuiting the winding of the tag as a function of the distance between the tag and the fluxgate. Notice that $V_{fluxgate} \propto \Delta H_{tag}$. The relation $V_{fluxgate}^{-1} \propto d_{tag\text{-}fluxgate}^{3}$ is characteristic of the dipolar behaviour.

$$\Delta H_{tag} = e^{-t_s \sqrt{\frac{\mu_s \sigma \omega}{2}}} \frac{\Delta m}{2\pi r_t^3} = \frac{\chi H_{ex} V}{2\pi r_t^3} e^{-2t_s \sqrt{\frac{\mu_s \sigma \omega}{2}}} \tag{10}$$

This expression gives the minimum sensitivity of the fluxgate sensor that is needed in order to detect the tag at a distance r_t and through the shielding. This expression is in good accordance with our experimental measurements (Ciudad Rio-Perez et al., 2008).

3.4.2 Powering of the tag

Using any low-power microcontroller like a PIC16F84 from Microchip (working parameters: $\varepsilon = 2$ V and I = 15 µA at 32 kHz), the main limitation of the system is the maximum distance at which the induced e.m.f. in the tag is able to power its electronics. The r.m.s. value of the e.m.f. in the tag is given by:

$$\varepsilon_m = \omega \mu n S \frac{H_0}{\sqrt{2}} e^{-t_s \sqrt{\frac{\mu_s \sigma \omega}{2}}} \tag{11}$$

Formula (11) is in good accordance with the experimental values (Ciudad Rio-Perez et al., 2008). This simple model allows a proper design of the new RFID system for a particular application. Any particular arrangement of metals can be modelized by using an effective theoretical shielding.

4. Conclusions

Inductive coupling-based systems show different problems to work in the presence of metals. The low frequency (LF) systems can work with metals in the surroundings. However, they only can work through metals in some particular circumstances and designs. The different problems arising from metal non-cleaned surroundings have been showed in section 2. All these problems could be avoided if the working frequency is reduced. However, the inductive coupling becomes inefficient quickly.

We have developed and experimentally tested a new system to work through metals. It is shown in Section 3. It works at ultra low frequencies (1 - 100 kHz) and through metals. The new RFID system works without any resonant circuit. It is based on measuring changes of the magnetization of a magnetic core included in the tag. Different geometrical arrangements for the antenna and the reader have been designed. This is of importance since the magnetic fields produced by these antennas have different directions in the position of the tag. The characteristics of the antennas can be checked in (Ciudad et al., 2004) and (Ciudad Rio- Perez et al., 2008). A combination of those antennas will allow to avoid any directional problem. In addition, we give a theoretical model of the system. It allows a better design of the system for any particular application.

In section 3.4 it is explained a theoretical model of the system. According to this model and our experimental data, the work distance is bellow 0.4 m for a typical antenna and low intensity magnetic fields. The system has demonstrated to be able to work through aluminium layers with thicknesses up to 0.2 mm and in close contact to the tag.

Table 1 summarizes the characteristics of LF and ULF systems. The comments are relative to the different RFID systems. Some similar tables for other RFID systems can be found in (Wilding & Delgardo, 2004) and references therein.

5. References

Aroca, C.; Prieto, J.L.; Sanchez, P.; Lopez & Sanchez, M.C. (1995). Spectrum analyzer for low magnetic field, *Review of Scientific Instruments*, 66, (1995) 5355-5359

Balanis, C.A. (1997). *Antenna theory: analysis and design (2nd)*. John Wiley & Sons Publisher, 0- 471-59268-4, New York

Bovelli, S.; Neubauer, F. & Heller. (2006). C. A novel antenna design for passive RFID transponders on metal surfaces, Proceedings of the 36th European Microwave Conference, pp 580-582, September 2006, Manchester UK

Bottomley, P.A. & Andrew, E.R. (1978). RF field penetration, phase shift and power dissipation in biological tissue: implications for NMR imaging, *Phys. Med. Biol.* 23 (1978) 630-643.

Bowler, N. & Huang, Y. (2005). Electrical conductivity measurmement of metal plates using broadband eddy-current and four-point methods. *Measurement Scientific Technology.* 16 (2005) 2193-2200

Ciudad, D.; Perez, L.; Sanchez, P.; Sanchez, M.C.; Lopez, E. & Aroca, C. (2004). Ultra low frequency smart cards, *Journal of electrical engineering*, 55, 10/S, (2004) 58-61.

Ciudad Rio-Perez, D.; Arribas, P.C.; Aroca, C. & Sanchez, P. (2008). Testing thick magnetic shielding effect on a new low frequency RFIDs sytem. *IEEE Transaction on Antennas and propagation*, 56, 12 (December 2008) 3838-3843

Dixon, P.F.; Carpenter, M.P.; Osward, M.M. & Gibbs, D.A. (2007). *RFID Tags*, US Patent 7205898, April 17 2007

Dixon, P.F.; Carpenter, M.P.; Osward, M.M. & Gibbs, D.A. (2008). *RFID tags having improved read range*, US Patent 7378973, May 27 2008

	ULF System	LF Systems
Physical principle	Fluxgate magnetometry	Inductive coupling
Work frequency	1 kHz-100 kHz	125-134 kHz
Range	<0.4m (non-resonant configuration)	< 1m
Size issues	Small size (due to the use of fluxgates)	Large size
Data transfer rate	Very slow	Slow
Metals: in the surroundings	No problem	No problem (some design issues)
Metals: wrapping the tag	No problem. Distance range reduced	Only under very particular circumstances
Prize: Antenna	High	High
Prize: Tag	High (since it contains magnetic material)	Low
Sensors	The tag can power sensors connected to it as well as send the measurements.	Sensors cannot be powered by the RFID system
Applications	Any system having problems with metals and no high data transfer ratio requirements.	Animal tracking. Item tracking. Product indentification. Car key.

Table 1. Comparison of the characteristics of LF and ULF systems.

Dobkin,D.M. & Weigan S.M. (2005) Enviromental effects on RFID tag antennas, *2005 IEE MTT-S International Microwave Symposium Digest*, pp. 135-138, 0-7803-8845-3, June 2005, Long Beach-California, IEEE

EM Microelectronic. (2002). *AppNote 411: RFID Made Easy*. EM Microelectronic - Marin SA, September 2002

Finkenzeller, K. (2003). *RFID Handbook (2nd)*, John Willey & Sons Publishers, 0-470-84402-7, West Sussex

Hoeft, L.O. & Hofstra J.S. (1988). Experimental and theoretical analysis of the magnetic field attenuation of enclosures, *IEEE Transactions on electromagnetic compatibility*, 30, 3 (August 1988), 326-340, 0018-9375

Ida, N. & Bastos, J.P.A. (1997). *Electromagnetics and calculations of fields (2nd)*, Springer-Verlag, ISBN 0-387-94877-5, New York

Lide, D R. (ed.). (2009). *CRC Handbook of Chemistry and Physics*, 89th Edition (Internet version 2009), CRC Press, Taylor and Francis, Boca Raton, F.L.

Perez, L.; de Abril, O.; Sanchez, M.C.; Aroca, C.; Lopez, E. & Sanchez, P. (2000). Electrodeposited amorphous CoP multilayers with high permeability, *Journal of magnetism and magnetic materials*, 215-216 (2000) 337-339

Perez, L.; Aroca, C.; Sanchez, P.; Lopez, E. & Sanchez, M.C. (2004). Planar fluxgate sensor with an electrodeposited amorphous core, *Sensors Actuators A*, 109 (2004) 208-211

Ripka, P. (ed.) (2001). *Magnetic sensors and magnetometers*, Artech House Inc., 1-58053-057-5, Norwood

Wilding, R. & Delgardo, T. (2004). RFID demystified: Part 1 The technology, benefits and barriers to implementation, *Logistic & Transport Focus*, 6, 3 (2004) 26-31

Development of Metallic Coil Identification System based on RFID

Myunsik Kim[1], Beobsung Song[2], Daegeun Ju[2],
Eunjung Choi[2], and Byunglok Cho[2]
[1]*Sogang Institute of Advanced Technology(SIAT), Sogang Univ.*
[2]*Ubiquitous Gwangyang & Global IT Institute,*
Republic of Korea

1. Introduction

Recently, RFID gains increasing attention, since RF signal can eliminates the need an optical line of sight and transmits relatively large amount of information from several tens of tags in real time (Finkenzeller, 2003) (Landt. 2001). Based on these advantages, RFID is applied in various fields. For example, RFID is widely spreading on products identification in logistics and distribution fields instead of barcode (Chawla & Ha, 2007). The bus card and RF pass are famous applications of RFID. Also, the development of special tags such as metallic tag widens the applicable fields of RFID (Nikitin & Rao, 2006) (Kim et al, 2005). Among the RFID applications, this paper focuses on the RFID technique for the SCM (Supply Chain Management) regarding an iron and steel industry. Specially, the RFID based steel coil identification system during a crane operation is developed. Since the iron and steel industry is key industry providing material to other industries, it has no small effect.

The system is developed for two purposes as follows. Nowadays, many factories employ sophisticated machinery that automates many kinds of process. However, some processes such as the quality checking, packaging, loading / unloading products to freight vehicle, and so on are still dependent upon the workers, who encounters danger under the automated system. The more the industrial field becomes automated, the more the field is dangerous. Thus, the developed system ensures safety of workers by releasing them from the products identification and checking checking process. Also, the automated product identification system improves the efficiency of the manufacturing and distribution process by preventing missing or mixing of products.

One of technical challenges associated with the RFID based coil identification is to apply tho system to the existed automated system while sustaining the identification performance easily affected by environmental conditions such as reflection, refraction, and scattering of RF signal from metallic surface of coils, crane and equipments. To cope with the problem, two key techniques are proposed in this paper. First, the effective tag attachment method is proposed considering the shape and properties of metallic coils, and working environment. Second, robust reader antenna system is proposed to identify tag attached inside coil efficiently. An antenna case is developed to reduce the effect from the attached surface and improve tag identification performance by control beam pattern of the antenna.

To verify validity of the proposed system, simulation is performed using MWS 2008 EM simulator and test using various model coils in laboratory. The experimental results in real industrial environment in POSCO show that The coil is identified very successfully using the proposed system.

This paper is organized as follows. In chapter II, the necessity of metallic coil identification system in POSCO and first development is described. Experiment results using the developed system and its problems are shown in Chapter III. Chapter IV shows the further improvement of the RFID system and its simulation and experimental results are shown in Chapter V. Finally, conclusions are drawn in chapter VI.

2. RFID based coil identification system

2.1 Background of the research

In POSCO, the products such as metallic coil are packaged and banded after manufactured and stored until delivered to customer. Since the coil is heavy over several tons, cranes are used to move the coil as showing in Fig. 1. The crane is automated then it is important to manage the coil information correctly while it is moved. Currently, the coil Information is managed using the stored position in warehouse. In general, the information is correct, however, if there is error in the coil manufacturing schedule or sensed location of the crane, coils are lost or mixed. Thus, sometimes, wrong coil is delivered to customers, it cause problem in time, cost, and credit.

For the problem, a barcode label with product code, size, weight and etc is attached to a coil and workers check the information periodically. The barcode is printed tag with several vertical lines. In order to read the barcode, workers should come close and align reader and barcode for scanning the lines with laser light. It spends much time to read barcode one by one. Also, the printed barcode is easily stained or injured, it prevent from reading the stored data in the barcode.

For the problem, the RFID based coil identification system is proposed. An RF tag is attached to coil, which is identified using reader antenna installed to the crane and the information is transferred to MES (Manufacturing Execution System) server. Even though the coil storing map information is incorrect, it is fixed automatically when crane picks up the coil without any effort of workers.

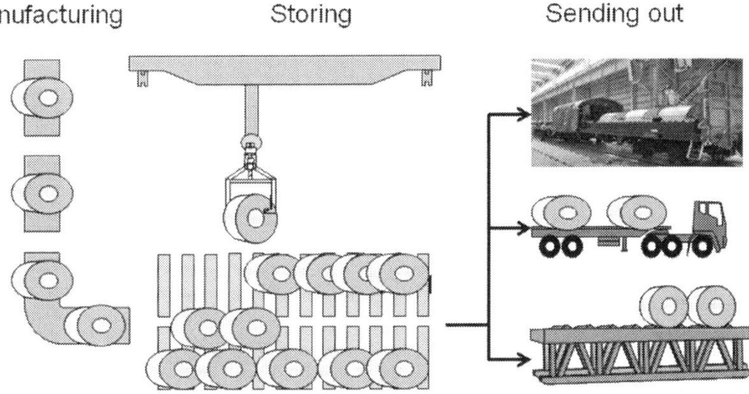

Fig. 1. The management of coil after manufacturing

2.2 Overview of developed system

Fig. 2 shows the overview of proposed system. RF tag is attached to inside of a coil, which is identified using reader antenna installed to crane shoe. The identified information is transmitted to MES server through TCP/IP interface then the real time sensing and tracking of a coil under the crane operation is available.

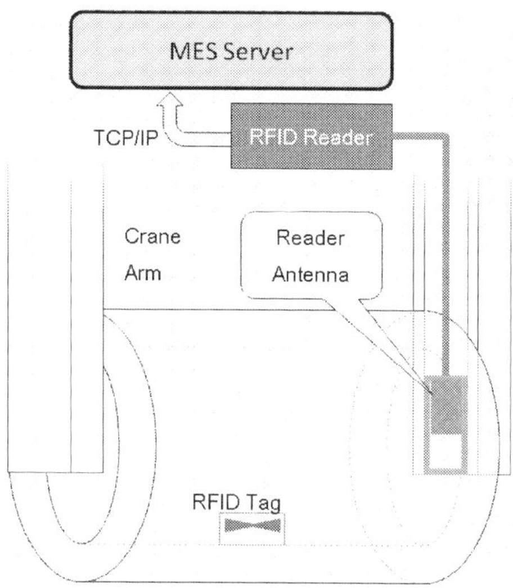

Fig. 2. Overview of developed system

However, since the coil and the neighboring equipments including crane are metallic object, the identification performance of the RFID system is lowered affected by the environmental effect. Also, in order to install the developed system in existing automated system without any changes, the system should be satisfy the conditions as follows.

1. The identification performance should be unchanged under the environment conditions surrounded by metallic object such as coil, crane, and other equipments.
2. The reader should read target tag only among neighboring tags.
3. The system is possible to be installed to current crane without any changes.
4. The tag should be cheap and light.

RFID system used in the developed system is shown in table 1. More detailed is described in following section.

Tag	UPM raflatac Dogbone Type
Reader	Alien ALR-9900 reader
Reader Antenna	Ceramic Patch Antenna
Interface to MES	TCP/IP
Tag on metallic surface	Flag tag technique

Table 1. RFID system applying in the developed system

(a) Nox-TM4 Metal Tag
SimplyRFID Corp.

(b) P0106AT Metal Tag
Sontec Inc

(c) Flag Tag
UPM Raflatac

Fig. 3. UHF RFID tag for metallic surface

2.3 Tag on metallic surface

Fig. 3 shows the tags can be used on metallic surface. Fig. (a) and (b) show metal tag, special tag that can be read, even though it is attached on metallic surface. Tag antenna is printed on a ferroelectric material such as ceramic with thickness of several millimeters. The basic principle of the metal tag is shown in Fig. 4. Wireless communication of RFID becomes possible by electromagnetic flux penetrating between two antennas of reader and tag as shown in Fig. 4-(a). However, when a metal is close to tag antenna, eddy current caused by reader's magnetic field is generated and it cancels the magnetic field necessary for communication as shown in Fig. 4-(b). When ferroelectric material is inserted between tag antenna and metal surface as shown in Fig. 4-(c), the material concentrates magnetic flux then the flux can flows without loss (Kim et al, 2005). Then the communication distance is improved as results. However, the price of the metal tag is much expensive than ordinary tag printed on film such as PI. Also, the metal tag is heavy then it comes off from the attached surface by vibration more easy comparing with ordinary tag while a tag attached object is moved. The cost and weight of the metal tag is chief obstacle to be applied.

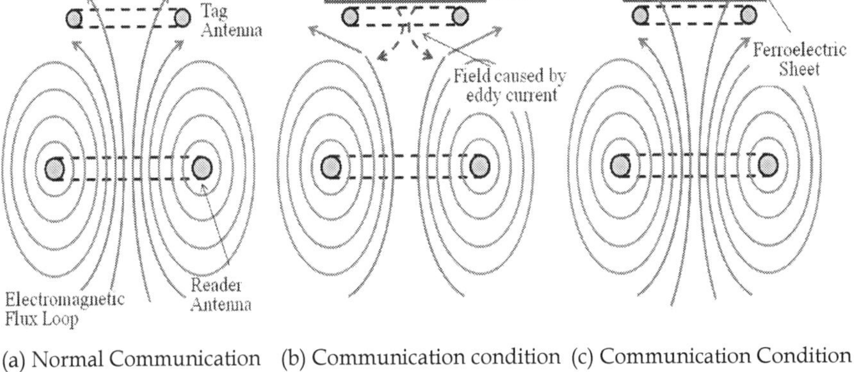

(a) Normal Communication Condition

(b) Communication condition with a metal surface in a vicinity

(c) Communication Condition with ferroelectric sheet present

Fig. 4. Basic principle of metal tag

Thus, the flagtag technique proposed by UPM is used in the developed system (Victor et al, 2006). Note that there is enough space between tag antenna and attached surface, the RF communication is available. Flagtag technique is very simple idea that makes space between tag and attaching surface. Fig 3-(c) shows the flagtag using label sticker. A tag is inserted in label and the tag is stood by folding the label as shown in the figure. Since general cheap film type tag can be used attaching on surface of various materials such as metal, paper, and so on with the flagtag technique, it has advantage in cost and applicability.

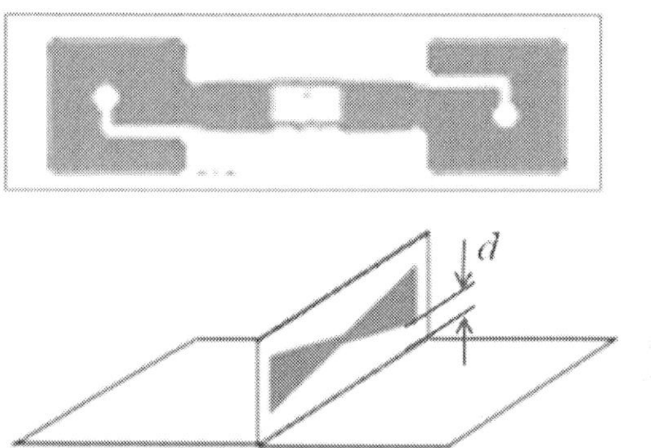

Fig. 5. Tag used in the system

Fig. 5 shows the tag used in the developed system. A UPM dogbone type UHF ranged RFID tag sized of 93 ×23 mm is used in paper label. The tag is erected by folding the paper label as shown in lower of Fig. 5. When the tag is attached, the identification performance is varied according to the distance d between the tag and the attached surface. Fig. 6 shows the strength of RF signal transmitted from the tag varied according to the distance d. As shown in the figure, the strength is almost same with normal condition when the distance is over 2 cm.

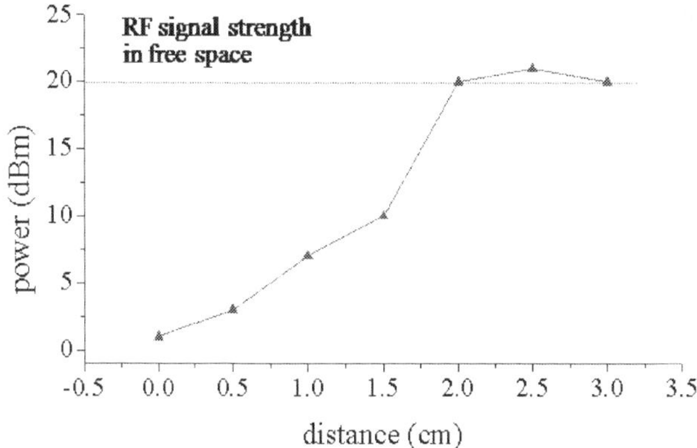

Fig. 6. Power of transmitted RF signal according to the distance d

2.4 RFID reader and antenna

ALR-9900 UHF RFID reader of Alien technology corp. is used in the developed system. In order to install reader in current crane, the smaller reader is better. Also, two more antenna port is required to install antennas to two crane shoe of a crane. Since the MES server is far from crane and it is hard to use wireless communication in factory environment, TCP/IP communication interface is required to transmit data without any loss. Table 2 shows the reader that can satisfy the above conditions.

Name	ALR-9900	SRU-FK0100
Photo		
Manufacturer	Alien Technology Corp.	Samsung Techwin
Tag Protocol	EPC Gen2 ISO18000-6c	EPC Gen2 ISO18000-6(TypeB)
Frequency	902.75~927.25 MHz	910-914 MHz
Size	190×160×40 mm	140×110×26 mm
Antenna Ports	4 Ports	4 Ports
I/F	RS-232, TCP/IP	RS-232, TCP/IP

Table 2. Specification of RFID readers

The specification of two readers is similar. Comparing with the Alien reader, SRU-FK0100 of Samsung Techwin has an advantage in size. However, the SRU-FK0100 reader affect to another sensor installed in crane then causes error in crane operation. After install the reader to crane and attach two antennas to crane shoe, the operation of folding and unfolding the shoe becomes unavailable. There is no reason to make the phenomena, since all sensors in the crane are shielded and the reader satisfies the standard of RFID reader specification.

Fig. 7 shows the noise transmitted from the antenna port of the two RFID readers. The noise level of two reader is under the standard. However, as shown in the figure, SRU-FK0100 reader has more noise than the ALR-9900 reader. It is regards as the reason that causes the crane to malfunction. Since the industrial field with many sensors for automation can be easily affected by any kinds of RF signal, the reader with less noise is better.

A Ceramic patch antenna sized of □80mm×7 mm is used with the ALR-9900 reader. Fig. 8 shows the antenna attached to crane shoe. Since the available width of the crane shoe is 12 cm only, the antenna is determined considering the required space for packaging. The antenna has gain of 2~2.5 dBi and can detect a tag of 6 m away with the ALR-9900 reader. To check the identification performance in real environment, we perform test in POSCO. Detailed experimental results are shown in following section.

909.5~910MHz   914~914.5 MHz   914.5~1000 MHz

ALR-9900

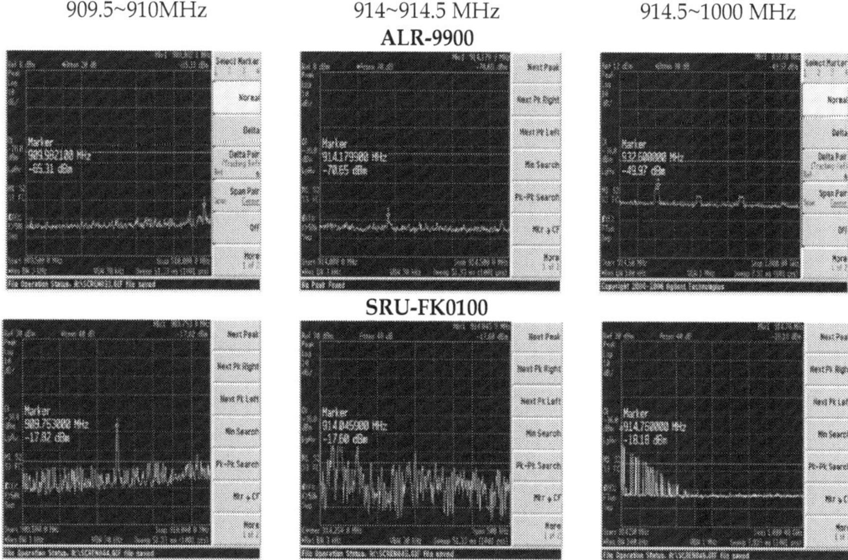

SRU-FK0100

Fig. 7. Noise according to various frequency range

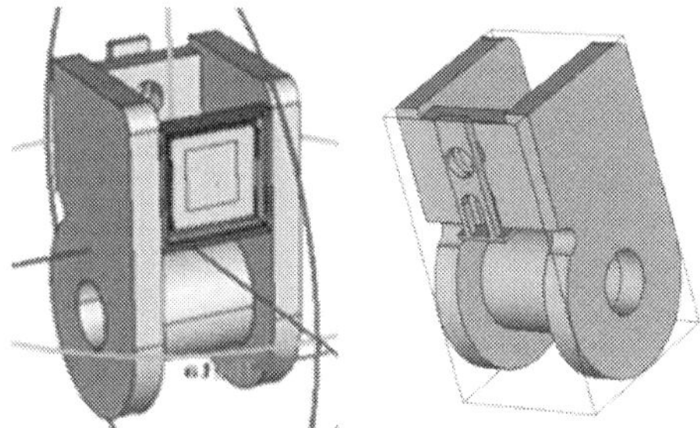

Fig. 8. Reader antenna attached to crane shoe

3. Experiment results

Fig. 9 shows the tag and antenna attached to coil and crane shoe. Label with tag is folded and attach inside of the coil parallel to the coil plane. In order to avoid damage, the label is put on 50 cm inner from the coil plane. Reader antenna is protected by plastic package and attached to crane shoe as shown in the figure. Fig. 10 shows the coil identification process when crane picks up the coil.

When the crane arm is down to pick up target coil, the tag inside the coil is identified using the antenna in the crane shoe. The identified information is transmitted to MES server and compared with the coil information in the storing map. If the two information are same, the crane lifts up the coil. Table 3 shows the experimental results using the developed system.

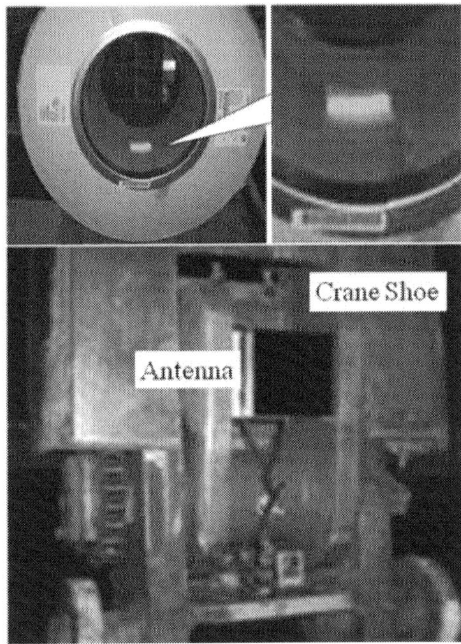

Fig. 9. (upper) Tag attached inside of coil, (lower) Antenna attached on crane shoe

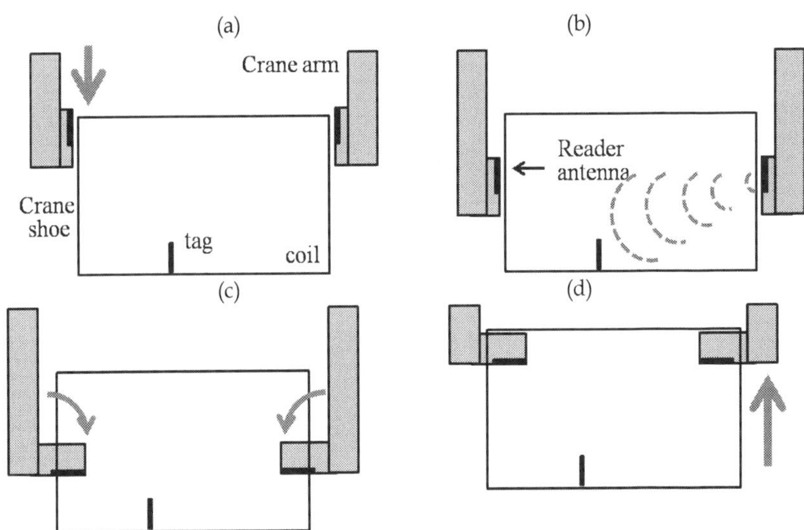

Fig. 10. Coil identification process using developed system while crane operation
(a) Crane arm is down to pick up target coil (b) Tag is identified,
(c) Unfolding crane shoe, (d) Lift up the coil

Day		1st	3rd	5th	7th
No. of Manufactured Coils with tag		440	429	511	505
Error	Identify wrong coil	4	9	9	6
	Missing	1	5	16	40
Error rate (%)		1.1	3.3	4.9	9.1

Table 3. Test result of coil identification

Tags issued following the production schedule is attached to a coil during packaging process and the coil is moved to warehouse using crane. Three cranes are selected for the test such as crane used for moving coils to warehouse, and in warehouse and to freight vehicle for shipment.

To test with as many tag attaced coils in the warehouse as possible, the test is performed during a week. As shown in the results, the fail rate increases. It is caused by the increasing the neighboring tags in the warehouse. First day, there are tags issued in the day then the reader can identify the target tag only. However, the error rate increases in 3d day. The tag behind the antenna is read by the back-radiation of RF signal. To solve the problem, we reduce the RF signal transmission power. It is resulted that the missing rate increases. When the power increases, the neighboring tags are detected together, and power decreases, the reader can not detect the target tag. Also, the tag is attached parallel to the coil plane, the tag lies down by the distortion caused by the effect of the curved surface of the coil. As the tag comes close to metallic surface of the coil, the identification performance is lowered, then the identification fail rate increases. For the problem, the developed system is improved in two directions. It is shown in following section.

4. Improvement of developed RFID based coil identification system.

In this paper, the developed RFID based coil identification system is improved in two directions: change of the tag attaching method and development of reader antenna package to control the beam pattern of RF signal transmitted from the antenna.

4.1 Tag attachment method.

First, the tag attachment method is changed. Fig. 11-(a) shows the previous method that the tag is attached parallel to the coil plane. In the case, since the lower paer of the label is straight, the tag is distorted by the curved surface of the coil. This problem can be solved by making the lower part of the tag curve fitted with the coil. However, the tag attachment process becomes complicated and the label should be made in various shapes according to the coil size. In addition, since the size of the tag shown in the coil plane is maximized with the attachment method, the tag is easily broken during the banding process and effect of wind passing through the coil.

For the problem, the tag attachment method is changed to be perpendicular to the coil plane as shown in Fig. 11-(b). The tag attachment surface is not curved and only the side of the tag is shown from the coil plane then the problem of distortion and damage of the tag can be minimized. However, when the tag is attached following the coil direction, the tag is at right angles with the reader antenna. Note that the RF signal transmitted from reader antenna to tag is maximized, when the reader antenna and tag is parallel (Stutzman & Thiele, 1999). Fig. 12 shows the power of the transmitted RF signal from reader antenna to tag according to the angle between them. Y-axis of the graph is the strength of transmitted RF signal to the tag from the reader antenna set 30 cm upper position of the tag reflecting the crane shoe position and the coil. The reader antenna send RF signal of 30 dBm. X-axis shows tag attached position from the reader antenna plane. The tag is attached at 20 and 50 cm, respectively. The results clearly show that the strength is lowered when the tag and reader antenna is perpendicular.

(a)   (b)

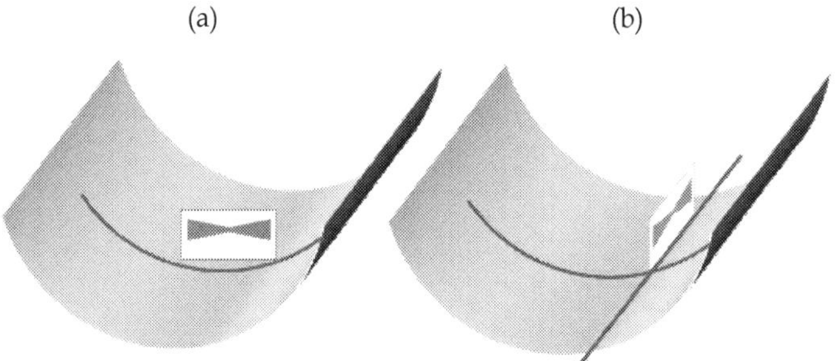

Fig. 11. Tag is attached (a) parallel to coil plane, (b) perpendicular to coil plan.

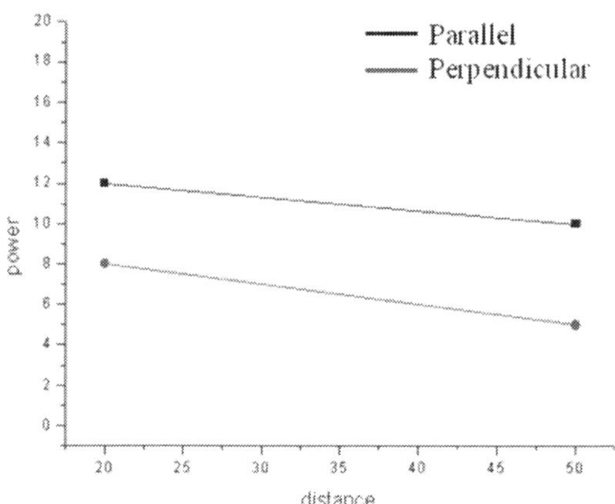

Fig. 12. Strength of the transmitted RF signal from reader antenna to tag according to the angle between them

However, the tag is attached inside of the coil then the RF signal transmitted from reader antenna propagates inside of the coil, which works as cylinder waveguide. Since there are various RF signal propagation modes according to the size of the coil (Pozar, 2005), (Balantis, 1996), the identification performance is different from that in free space. In order to test the RF signal propagation in coil, we make model coils using sheet zinc as shown in Fig. 13. The lengths of the model coils are 90 cm and 180 cm, the inner radiuses are 50.8 cm and 61 cm, respectively, and reader antenna is attached to position of 15 cm upper from the coil center reflecting the general coil size and the position of the crane shoe.

Fig. 13. Model coils used in the experiment.

The experiment conditions and the test results are shown in Fig. 14 and 15. Tag is attached to coil with the of 20 cm and 50 cm distance from the coil plane. The coordination of the graph is same with Fig. 12. When the coil radius is 61 cm, the strength of transmitted RF signal to the tag attached perpendicular to the coil plane is weaker than the parallel as same as in free space propagation. However, with the coil of 50.8 cm, the results are opposite. The reader can send more signal to the tag attached to the cylinder direction. It is caused by the effect of the RF signal propagation mode according to the coil inner radius.

The RF signal propagations in various coils are simulated using MWS 2008 EM simulator of CST AG. in same condition of previous experiment. The simulation results are shown in Fig. 16. The reader antenna and the tag are set as port 1 and port 2 and the blue line of the graph is the S12, the transmitted signal strength from port 1 to port 2. The initial strength of RF signal from port 1 is 30 dBm. As shown in the figure, the simulation results are same with the experimental results.

Note that the RF signal propagation in the coil is different from that in free space. The strength of the transmitted RF signal is affected by the propagation mode determined by the coil condition such as the radius of the coil. Since many kinds of coils are manufactured with various radiuses and lengths, it is impossible to make any standard for tag attachment method. However, the simulation and experimental results prove that even though the tag and reader antenna is orthogonal, tag inside a coil can be read. Based on the results, we choose the perpendicularly attachment method, because the method has more advantages such as easy to attach and with less possibility of damage.

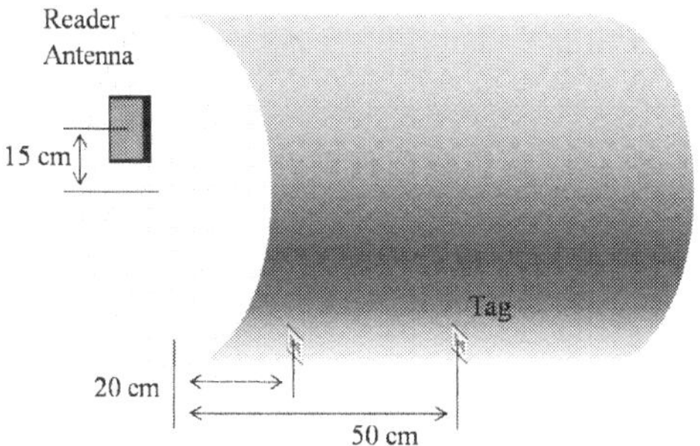

Fig. 14. The positions of tag and the reader antenna in the experiment

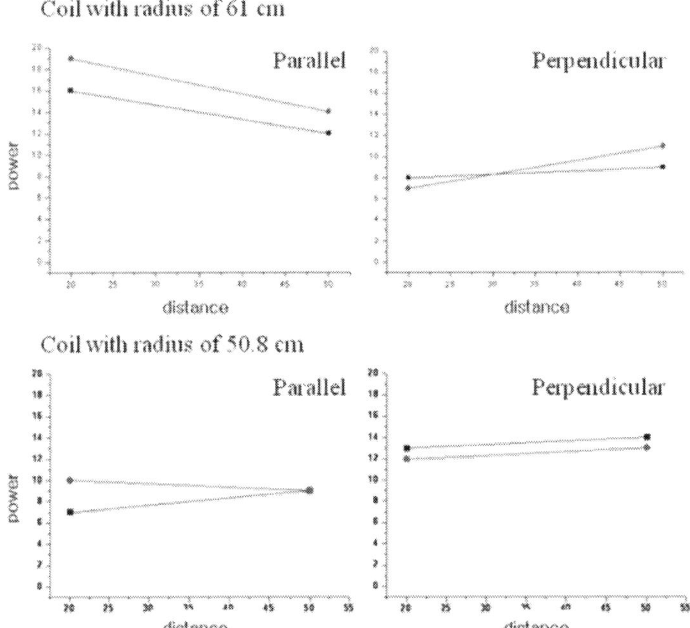

Fig. 15. Experimental results according to the tag attachment methods.

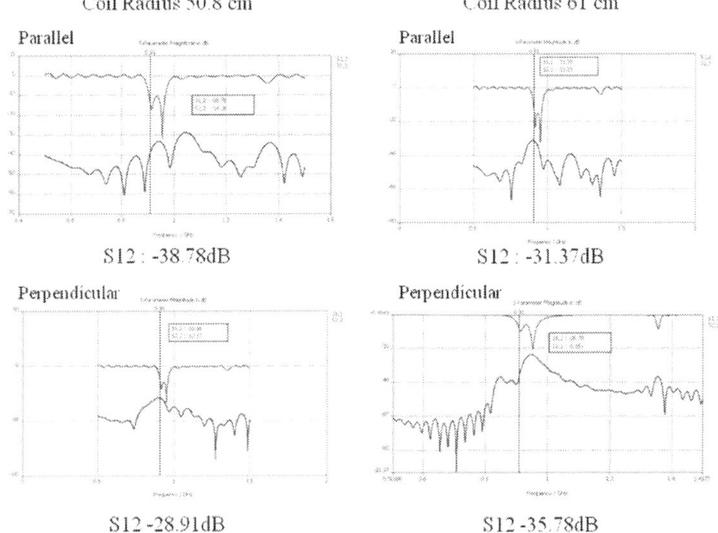

Fig. 16. Simulation results with same condition of fig. 15

4.2 Improvement of reader antenna

The above results show that there is limitation to improve tag identification performance by tag attachment method. When a RF signal is propagated in waveguide, there are weak and strong points of RF signal in the guide. Fig. 17 shows the RF signal propagation pattern in the coil simulated by MWS 2008 EM simulator. The signal is propagated irregularly and the weak and strong points are shown in the pattern. Also, there is radiation to back and side direction from the antenna. The radiation causes wrong detection of neighboring tags.

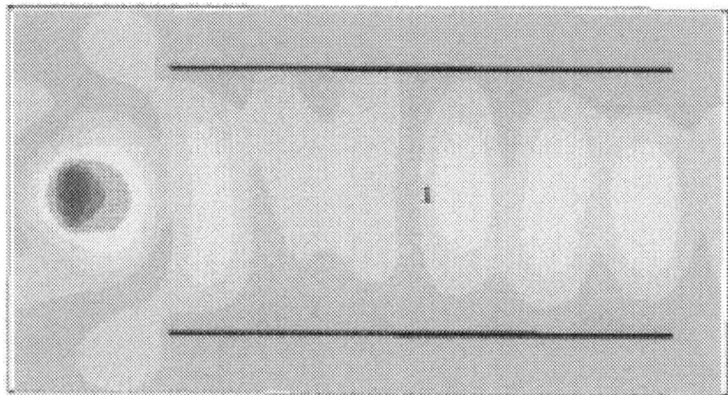

Fig. 17. RF signal propagation pattern in the coil

In the waveguide model, the reader antenna works as ports providing RF signal. Thus, the RF signal propagation pattern can be changed by controlling the RF signal radiation pattern from the reader antenna. The radiation pattern is controlled in two directions. First of all, the epi-radiation is restraint not to detect the wrong tags. Second, the beam-width should be widened as much as possible to keep the identification performance to the target tag with the high front to back ratio.

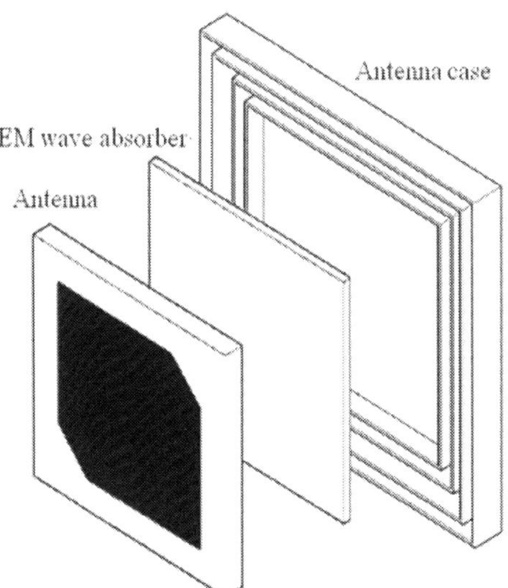

Fig. 18. Developed antenna case

Current available space for the reader antenna at crane shoe is □120 × 20 mm. Since the antenna should be small with enough durability to stand in the tough industrial environment, the ceramic path antenna is almost best solution. Thus, it is hard to improve the antenna.

For the problem, the antenna packaging technique is developed in this paper to control the beam pattern. Fig. 18 shows the antenna case design. The case is made using SUS (Steel Us Stainless) plate of 1 mm. The package is metal then the antenna is not affected by the material of the attached surface. The size of the case is □110 × 20 mm considering the available space in crane shoe.

When an antenna is close to metallic surface, the RF signal is diffracted following the metallic surface. This diffraction causes the radiation of RF signal to side and back direction. In order to reduce the diffraction, corrugated lines are inserted between antenna and the case. The corrugated line is generally used in horn antenna (Balantis, 1996) (Mentzer & Peters JR, 1976) (Pozar, 2005), by which the diffracted RF signal following the metallic surface is canceled. The RF signal enters and come out from slots then weaken by canceling each other as shown in Fig. 19. Also, electromagnetic wave absorber is inserted between the antenna and the case to prevent the RF signal from flowing follows the metallic surface of the case.

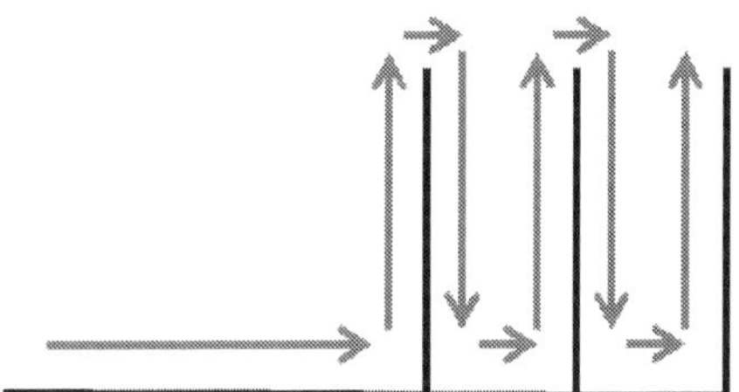

Fig. 19. RF signal is setoff by corrugated lines

To verify the validity of the proposed antenna packaging, the RF signal radiation pattern is simulated using MWS 2008 EM simulator. Fig. 20 shows the simulation results. The radiation pattern of a ceramic patch antenna, which is packaged in the case attached to crane shoe, is simulated. Upper figures shows the antenna model packaged in case and attached to crane shoe and 3 dimensional radiation pattern, lower graphs shows the 2 dimensional pattern. The inner green line shows the radiation pattern without case and red is with case. As shown in the figure, the beamwidth becomes wider and the epi-radiation decreases. It is expected that the identification performance will be improved with the developed case, specially, decrease of the wrong neighboring tags detection.

Fig. 21 shows the simulation results of RF signal propagation pattern in metallic coil with the case. The RF sginal regulary propagated in the coil with the proposed case. Also, the epi-radiation is almost disapeard.

Based on the simulation results, the antenna case is made and tested in electromagnetic dark room sized of L4.5 × W9 × H 4.5 m. Fig. 22 shows the electromagnetic dark room. In order to test more exactly, the model crane shoe is made using sheet zinc and the antenna packaged in the developed case is attached to the shoe.

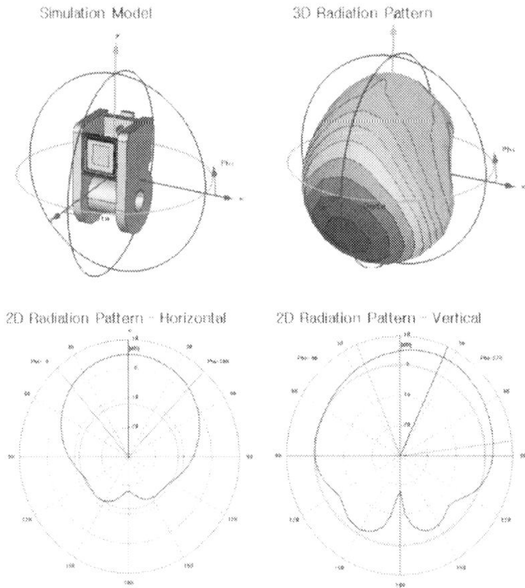

Fig. 20. RF signal radiation pattern varies according to the case existence

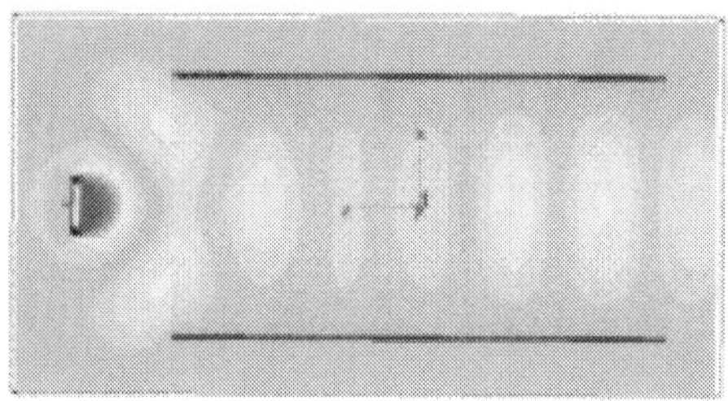

Fig. 21. RF signal propagation pattern in the coil when the case is exploited

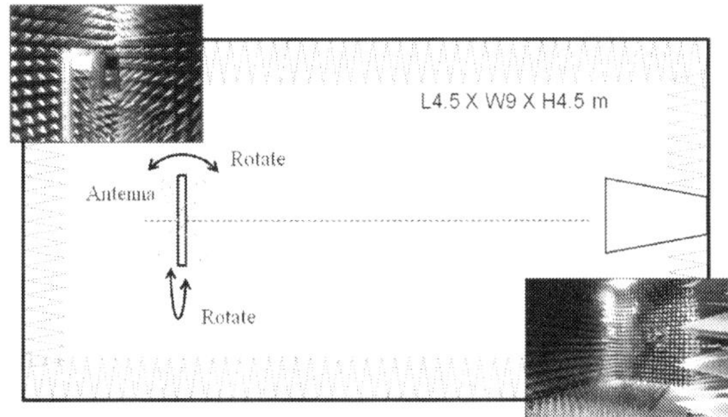

Fig. 22. (upper) overview of the electromagnetic darkroom (lower-left) antenna to be tested (lower-right) antenna to radiate RF signal

Fig. 23 shows the experimental results. The upper graph shows the results without antenna case and lower graph with case. And table 4 shows the measured data such as gain, beam width, and front-to-back ratio. As shown in the figure and table, the gains with case are better than without the case. However, the beamwidth becomes narrow with the case. It is caused by canceling the RF signal flows through metallic surface in the corrugated line of the case. However, the front-to-back ratio is dramatically improved, which reduces the wrong tags detection.

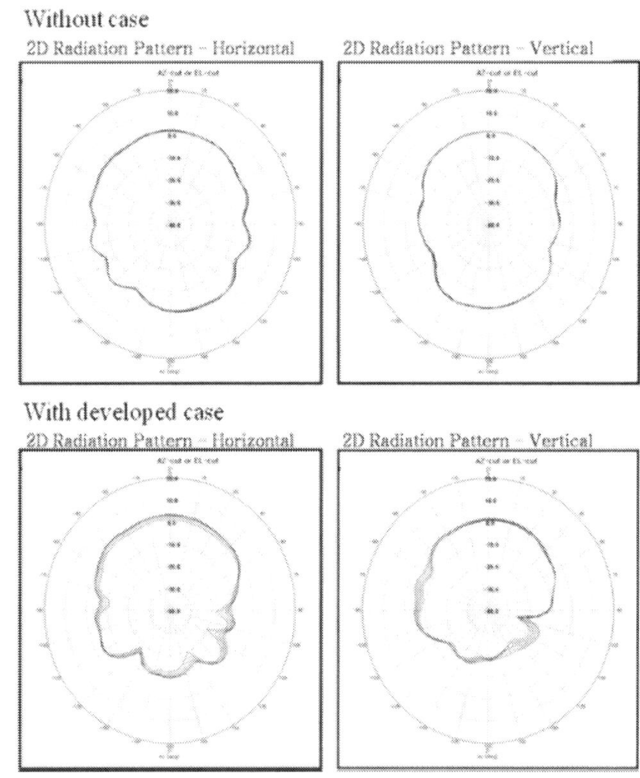

Fig. 23. (upper) Radiation pattern with case, (lower) without case

Property	Gain(dBi)		Bandwidth(ged.)		Front-to-Back Ratio (dB)	
	horizontal	vertical	horizontal	vertical	horizontal	vertical
Without case	2.10	1.48	**131.81**	**104.59**	1.66	3.49
With case	**3.98**	**1.49**	110.39	94.99	**13.31**	**16.29**

Table 4. Experimental results about RF signal radiation with / without case

5. Experiment results

To verify the validity of the developed system, the tag identification is tested using model cylinder coil. Two more coils with tags are positioned to check the effect of the neighboring coils and tags as shown in Fig. 24. The tag is attached inside in the coil head to the same direction of the cylinder with 50 cm distance from the coil plane. In order to whole range inside of the coil, the tag is attached from top of coil to bottom with an interval of 30 degrees in clockwise. The distances between coils are determined reflecting the position relation of stored coils in POSCO. The test results are shown in table 5. The tag position of 0 degree is the top of the coil. The sign of □ means that the wrong neighboring tags are detected, × the target tag is not detected. As shown in the table, the target tag are successfully identified with the antenna packaged in the developed case. Even though there are neighboring tags near from the antenna, the reader can detect target tag only.

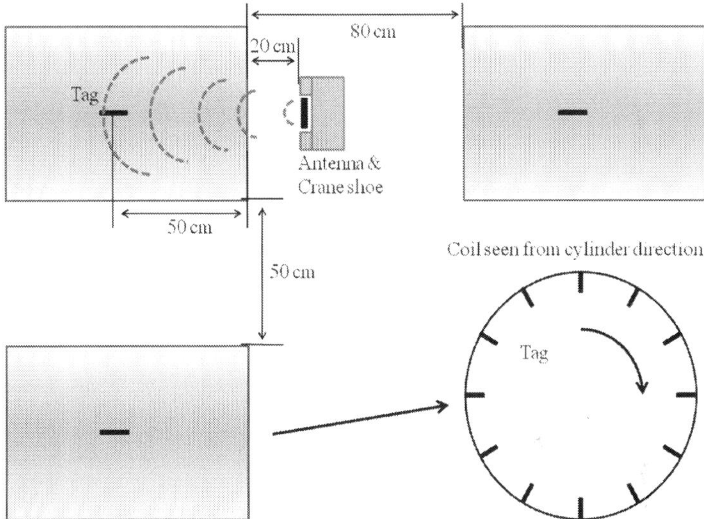

Fig. 24. Condition of experiment

Tag position (Degree)	0	30	60	90	120	150	180	210	240	270	300	330
Without Case	O	×	O	O	×□	O	O	□	×	O	O	O
With Case	O	O	O	O	O	O	O	O	O	O	O	O

Table 5. Experimental results of tag identification performance with / without case

Table 6 shows the experiment results in POSCO. The test is performed during two days, because enough tags to test the interference are already in the warehouse stored at previous test. The number in the parenthesis is the number of detected tag that is attached in previous test. The direction is parallel to the coil plane. As shown in the table, the error rate is dramatically decreases under the 0.5 %. The results satisfy the success rate of 99% that is required in the industrial filed then the system can be applied in the coil identification.

Day		1st	2nd
No. of Manufactured Coils with tag		405	392
Error	Identify wrong coil	2(1)	0(1)
	Missing	0	0
Error rate (%)		0.49 (0.25)	0 (0.26)

Table 6. Experimental results in real environment

6. Conclusions

This paper describes RFID based metal products identification technique for SCM of iron and steel industry. Specially, the coil identification system is developed. To cope with the falling off the tag identification performance affected by neighbouring metallic objects, the tag attachment method based on flagtag is proposed and the reader antenna packaging technique is developed to improve the performance of target coil identification. A Crane equipped with the developed system can detect the tag attached inside a target coil very successfully. Our main contributions can be summarized as: 1) The RFID based products identification system is developed for iron and steel industry, which is most difficult field to apply RFID. Thus, the system can be widely spread in other industrial fields. 2) The coil identification system during the process of manufacturing, storing, and shipping by crane is developed. Since the system is for managing the products information automatically, it can contribute the SCM in steel and iron environment. The future efforts includes the improvement of the developed system to cover another products such as steel plates and spreading the RFID technology to whole SCM systems that requires the products identification.

7. References

B. Victor, M. Otsuka, S. Stefan, T, Kumbayashi, H. Klaus (2006), A method for applying a RFID tag carrying label on an object, US patent, WO/2006/045395.

C. A. Balantis (1996), *Antenna Theory: Analysis and Design*, Wiley Text Books, ISBN-10: 9971512335.

C. A. Mentzer, L. Peters JR(1976)., Pattern Analysis of Corrugated Horn Antenna, *IEEE Transaction on antennas and propagation*, Vol.24 No. 3.

D. M. Pozar (2005), *Microwave Engineering 3/E*, John Wiley & Sons, INC. ISBN10:0471448788, Canada

J. Landt(2001), Shrouds of Time: The History of RFID

K. Finkenzeller (2003), *RFID Handbook: Fundamentals and Applications in Contactless Smart Cards and Identification*, John Wiley & Sons Ltd., ISBN-10: 0470955038, Canada.

P. V. NIikitin and K. V. S. Rao(2006), Theory and Measurement of Backscattering from RFID Tags, *IEEE Antenna and Propagation Magazine*, Vol. 48, No. 6.

S.-J. Kim, B.-K Yu, H.-J Lee, M.-J. Park, F.J. Harackiewicz, and B.-J Lee (2005), RFID tag antenna mountable on metallic plates, *Asia Pacific Microwave Conference*, Vol. 4, ISBN: 0-7803-9433-X.

V. Chawla and D. S. Ha (2007), An Overview of Passive RFID, *Communication Magazine*, Vol. 45, pp. 11 – 17, ISSN: 0163*6804.

W. L. Stutzman and G. A. Thiele (1999), *Antenna Theory and Design*, John Wiley & Sons Ltd., ISBN-10: 0471025909, Canada.

Virtual Optimisation and Verification of Inductively Coupled Transponder Systems

Frank Deicke[1], Hagen Grätz[1] and Wolf-Joachim Fischer[2]
[1]*Fraunhofer IPMS*
[2]*Dresden University of Technology*
[1,2]*Germany*

1. Introduction

RFID transponder technique is associated with various applications and usage scenarios. There are tags for wireless identification and tags that are able to store extended object information as well as including a data logging function, an actuator or a sensor. Besides passive UHF and microwave tags which provide long-range communication but only small energy transfer of some μW, inductively coupled passive tags can be better used for even sensor applications, today. In that case, the power consumption of the tag can be up to some mW (Finkenzeller, 2007) to power sensors as well as analogue and digital circuits for an extensive signal processing. A lot of physical parameters like acceleration, force, humidity, pressure or temperature can be measured. Whereby, many well known sensors and measuring principles can be implemented directly.

Such sensor tags but also others using LF and HF frequency range can be used in various industrial applications like process monitoring or automation. Just as complex and implantable diagnostic systems are available in medical engineering. Each of these RFID based applications need a customised design to optimise wireless energy transfer and data communication, because each application has different electrical, electromagnetic and geometrical demands. Therefore, antenna design is an important part of the whole system design. Both reader and tag antenna must be optimised taking into account a free air transmission channel or additional eddy current losses because of existing metals or fluids. Other important constraints considered are the specified maximum or minimum antenna dimensions, shape and used material, different link distances, arbitrary coaxial and non-coaxial antenna positions or tag rotation. Besides these mostly electromagnetic and geometrical properties, electrical system properties like power of the driver, demodulator sensitivity, approximated load resistance of the tag, used protocol, bandwidth or parasitics also influence the design process.

For a system designer an important question is if a particular RFID technique can be implemented and used successfully. Principally, an answer can be found using a lot of experiences, numerical simulations for antennas and extensive verification in the lab requiring prototypes and measurement setups. Thereby, system optimisation is done manually. But that standard design flow could be very time consuming and expensive because of producing many different prototypes and using complex measurement setups in the lab to characterise the transmission channel. Additionally, it is not sure that the best solution found is the optimum available, because the parameter space for optimisation and analysis is large, multidimensional and heterogeneous.

A first system success design approach based on software tools for system analysis and optimisation including automatic parameter variation and model generation seems to be more sufficient. Important questions like if a specific application would work using RFID technique or how to dimension and position antennas can be answered qualitatively and quantitatively on virtual level without doing prototyping. This design approach could be less time consuming and expensive as well as provide better results to work with.

2. System modelling

2.1 Transponder system

Transponder systems consist of different modules strongly dependent on application. The tag comprises for example a RF front end (Fig.1), a protocol stack with different complexity and different features, a state machine or a microcontroller, memory like EEPROM, RAM and flash or an analogue or digital interface to connect different actuators and sensors.

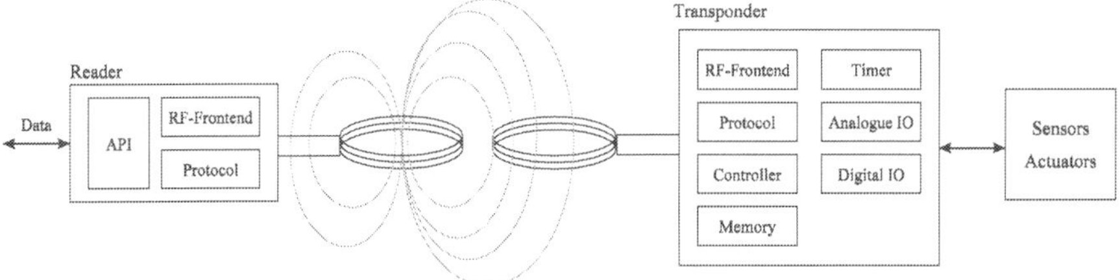

Fig. 1. Block diagram of a whole transponder system including reader and tag

On reader side, there is also a RF front end, a protocol stack and an application programming interface (API) to connect it to a computer or a middle ware. Furthermore, there are the antennas for both reader and tag ideally customised for each application.

In general, the goal of system design is to ensure a requested functionality on a specified link distance. On RFID level that means transferring enough energy from reader to tag wirelessly and to ensure an uni- or bidirectional wireless data communication. Hence, two objective functions, energy range and transponder signal range (Finkenzeller, 2007), can be derived. Energy range stands for a maximum distance, where the tag gets enough energy from the field generated by the reader. And transponder signal range means a distance between both reader and tag, where the reader receives data error-free sent by the tag. Both distances must exceed the requested link distance to get a working RFID system. For optimisation on electrical level, two important parameters, tag voltage and demodulator input voltage, are helpful for system evaluation.

2.2 Extracted parameters and parameter space

Principally, transponder system design is divided into different steps. These are the design of the transmission channel, the RF front ends, the digital protocol units and the application. There are many solutions for the RF front end and the digital protocol unit to meet different RFID standards. And there are various vendors providing powerful IPs, ICs or software packages. The design of these communication components is very challenging because of low device count and form factor. That implies using almost non-complex circuits and low power constraints in general. But mostly these demands are independent of particular applications, why these components can be reused in many different applications.

In contrast to ICs and protocol based software, the transmission channel depends directly on each application and must be customised for successful implementation. To do that, the kind of application or its implemented functions are not in foreground for optimisation. More important are derived system properties like variation parameters and constraints (Table 1) divided into transmission channel, electrical and protocol-dependent parameters.

Transmission Channel	Electrical Parameters	Protocol
Antenna	Reader	Carrier Frequency
• Size (Min, Max)	• Driver	Bandwidth
• Shape	• Demodulator	
• Material	Tag	
Antenna Configuration	• Power Consumption	
Environment	• Modulator	
	Parasitics	

Table 1. Important parameters and constraints for system optimisation divided into different categories

Antennas and its parameters size, shape and material belong to the transmission channel category as well as its configuration due to translation and rotation. Antenna size can be specified for example by inner and outer radius for round windings, antenna width, number of turns and used wire diameter with and without insulation. Another important point is the environment, in which the system should be implemented. There can be eddy current losses because of metals and fluids nearby the antennas influencing the behaviour of the transmission channel. The second category defines electrical system parameters for both reader and tag. It comprises for example the driver voltage, maximum driver current or demodulator input voltage of the reader and load, minimum and maximum voltage as well as modulation index of the tag. Parasitics like ohmic losses of resonance capacitors, antennas and input capacitance of the tag chip or internal resistance of the driver circuit are very important to get sufficient results. Besides geometrical, material and electrical properties, protocol specific characteristics like carrier frequency and bandwidth must be considered, too. Finally, transmission channel design, which is in the fore, is on low physical level where functions of upper protocol layers or application generally do not influence results directly. However, there is a heterogeneous and multidimensional parameter space with different parameter ranges as well as discrete or continuous parameter variation. Often objective functions with local or global extremes exist and the effort for detection could be high.

2.3 Electrical and electromagnetic model

To consider all important parameters during system design, the question now is which models can be used and how they should interact. Principally, there are two different models – electromagnetic and electrical. An idealised electrical model is shown in Fig. 2 for general discussions. It comprises a model for a reader with a voltage source and a series resonance circuit as well as a tag with parallel resonance circuit. The resistor R_L is the load of the tag.

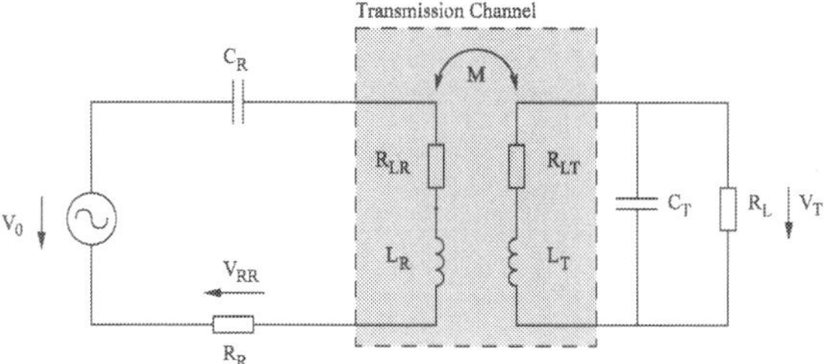

Fig. 2. Idealised electrical model of a transponder system using inductive coupling

The transmission channel can be described by the impedance matrix

$$\begin{bmatrix} V_R \\ V_T \end{bmatrix} = \begin{bmatrix} R_{LR} + j\omega L_R & -j\omega M \\ j\omega M & -(R_{LT} + j\omega L_T) \end{bmatrix} \begin{bmatrix} I_R \\ I_T \end{bmatrix}, \tag{1}$$

where V_R and V_T are the voltages over the antennas. I_R and I_T are the currents through the antennas.

For system design of passive tags, two objective functions are important. These are the energy range and the transponder signal range (Finkenzeller, 2007). Energy range means the maximal distance between reader and tag, where the tag can extract enough energy from the field. Transponder signal range means the maximal distance, where the reader can receive data error-free from the tag. The sensitivity of the demodulator is very important for the transponder signal range. The goal is that energy range and transponder signal range exceed the required minimum link distance after system optimisation.

To evaluate both energy range and transponder signal range, two objective functions can be used on the electrical level. These are the transponder voltage (2) and the demodulator input voltage (3) (Deicke et al., 2008a).

$$V_T = \frac{j\omega M I_R R_L}{\sqrt{2 R_{LT} R_L + \left(\dfrac{R_L R_{LT}}{\omega L_T}\right)^2 + (\omega L_T)^2}} \tag{2}$$

$$\Delta V_{RR} = V_0 R_R \left(\frac{1}{\mathrm{Re}[Z_R(Z_L)]} - \frac{1}{\mathrm{Re}\left[Z_R(Z_{L,Mod})\right]} \right) \tag{3}$$

The real part of the impedance Z_R is

$$\mathrm{Re}[Z_R] = R_R + R_{LR} + \frac{(\omega M)^2}{(-R_{LT} - Z_L)^2 + (\omega L_T)^2}(R_{LT} + Z_L). \tag{4}$$

Z_L is the parallel connection of C_T and R_L. $Z_{L,Mod}$ is the parallel connection of C_T and $R_{L,Mod}$. Whereby, $R_{L,Mod}$ is the load resistance during modulation. Furthermore, there are constraints on the electrical model. These are the quality factor of the reader (5) and the transponder (6). Generally, the quality factor is defined by the quotient of resonance frequency and bandwidth.

$$Q_R = \frac{\omega_0 L_R}{R_{LR} + R_R} \tag{5}$$

$$Q_T = \frac{2\omega_0 L_T R_L}{R_L^{\,2} + 2R_L R_{LT} - \sqrt{R_L^{\,4} - 4\omega_0^2 L_T^{\,2} R_L^{\,2}}} \tag{6}$$

Considering equation (1) to (6) and the discussion in previous sections, there are many different variables that influence V_T and ΔV_{RR}. On the one side there are electrical parameters characterising the transmission channel that depend on geometrical dimensions, antenna configuration, antenna material and eddy current losses due to fluids or metals nearby antennas. These parameters must be calculated by an electromagnetic model and forwarded to the electrical model. If magnetic materials like ferrite cores or ferromagnetic plates are placed inside or nearby antennas, magnetic field strength or antenna current must also be considered because of saturation. Then, there is an additional loop-back between electrical and electromagnetic model.

On the other side there are electrical components of resonance circuits like R_R, C_R and C_T that depend on antenna and transmission channel parameters as well as system constraints like bandwidth and quality factor. It follows that there is no closed solution available that takes into account both electrical and electromagnetic model. Because of that, manual optimisation is very difficult for experienced designers, as well. An exhausted search in that large multidimensional parameter space is not possible mostly because of considering a vast number of possibilities that would result in a lack of time. On manual optimisation only few solutions can be verified. And as a result, it is not really sure if the solution found, is the optimum for a particular application or not. That means the quality of the result can not be estimated in a sufficient way.

2.4 Current approaches and its bottlenecks

For RFID system dimensioning and analysis, different approaches had been discussed in literature. The selection of an adequate modelling approach depends on target-oriented use of variation parameters for design and optimisation. There are algebraic and numerical solutions in general. A well known work is (Grover, 2004) where many approximated formulas are collected to calculate self and mutual inductance for many different coil types. Using the approximated formulas for the electrical level from the application note (Roz, 1998) in combination with that work, simple system analysis can be done with an existing transmission channel including antennas and antenna configuration. Youbok introduced with (Youbok, 2003) a more detailed application note including formulas for most common antenna shapes and basic electrical circuits. For some standardised systems including co-axial antennas, no additional literature is necessary. Another interesting approach is discussed in (Finkenzeller, 2007) where a solution is presented to find the optimum antenna radius of reader for given read range and constant coil current. The reason is if the antenna radius is too large, the field strength is too low even at a distance of 0 between reader and tag antenna. And in the other way around, if the radius is too small, there is high field strength at distance 0, but it falls in proportion to x^3 from nearby the reader antenna. So, Finkenzeller explains that radius R and read range x should have the relation

$$R = x\sqrt{2} \; . \tag{7}$$

A question, that was not discussed, is if that formula is always true in free air independent of tag load or tag antenna size and shape. Mostly, it should not work in metal or fluid environments.

Another interesting approach also explained in (Finkenzeller, 2007) is to find the minimum field strength at tag side to power a passive tag. Therefore, the mutual inductance M in equation (2) is replaced by a simple approximation using magnetic field strength H. Then the equation is solved for H. After that the minimum magnetic field strength H_{min} can be estimated by defining a minimum tag voltage V_T and a load resistance R_L. With that result, the designer is able to dimension the reader of a system without any further relation to the tag side. An independent development of both reader and tag is possible if H_{min} is constant. That approach seems to be good for basic analysis and optimisation if the antennas and the antenna configuration are well known and less accuracy is accepted. If antennas are unknown at the beginning of system design like it is the case for many new industrial or medical applications, it is difficult to find an optimised system. One point, that would also impede the use of that approach, is, that electromagnetic and electrical model are mixed and used in one step. So, it is really hard to implement more model details in that closed formula even to increase accuracy. And there is also no numerical solver that can be used for such a mixed approach. It can not be considered in that way if data transfer works from tag to reader or not, because even formulas for simple models will be very complex and difficult to handle. Finally, that approach only helps to optimise energy range.

Besides these approaches with a reduced abstraction level, modelling using numerical methods is another way to increase model accuracy and to finally find better solutions. Therefore, specific computer-aided tools are used, like it is also done for many other problems in physics or engineering. But many specific tools such as ANSYS (ANSYS, 2007), FEMM (Meeker, 2006) or Spice (Quarles et al., 2005) only provide comprehensive functionalities for analysis and optimisation on particular modelling levels like mechanical, electrical or electromagnetic. Heterogeneous systems can not be analysed or optimised with one tool.

Another possibility for analysis and optimisation is the use of modelling languages like VHDL-AMS or Verilog-A. These are used to model physical behaviour such as acoustic, electrical, magnetic, mechanical, optical or thermal. Interactions between different modelling levels can be considered as well. Another advantage in comparison to numerical solvers like ANSYS is that modelling languages are standardised. Thus it can be used independent of a particular simulator. A disadvantage is that detailed models are very complex and handling these complex models is often not as good as using numerical solvers. Two approaches using standardised modelling languages are explained in (Beroulle, 2003) and (Soffke, 2007). Beroulle uses VHDL-AMS to model a transponder system on system level with a carrier frequency of 2.45 GHz to validate system performance. Soffke takes a similar approach for system analysis. He uses Verilog-A to model an inductively coupled transponder system. System optimisations are done manually. That means, found solutions can be close to an optimum, but it can not be evaluated easily if it is the case. Mostly it remains a big uncertainty.

All these approaches have in common not to be a good choice for system analysis and optimisation considering the whole transponder system and considering enough details in electrical and electromagnetic models to get sufficient results. Either it can be used for system analysis or to analyse parts of a whole system in detail without regard to interactions of other parts. Additionally, system optimisation is not described to find best solutions for different usage scenarios. Thus it is assumed to do it manually with all restrictions discussed above.

3. Virtual design approach

3.1 Objectives

From discussions above, objectives are derived for a virtual design approach that can be used for active and passive inductively coupled transponder systems. There, the focus is on antennas, transmission channels and its effects on the electrical level. Reader or tag antenna or both should be optimised dependent on application-specific requirements like geometrical, material or electrical properties and regarding whole system behaviour. Interactions between electrical and electromagnetic level should also be considered. During optimisation, a multidimensional and automatic parameter variation should be possible using adapted optimisation algorithm to get really optimised solutions and results with good quality. Besides pure optimisation, transponder systems should be analysed for different usage scenarios and different environments. Additionally, coaxial and non-coaxial antennas should be considered. That implies to move and rotate a tag in space for analysing operating range. To do that comprehensive analysis and optimisation, different model types should be selectable to choose between model accuracy, calculation time and possible model details like adding metal plates, for example. Finally, the goal is to make available a first system success design approach. This means that the first solution meets the requirements and can be used in practice without further extensive prototyping.

3.2 Design approach

These objectives were realised in a stand-alone software tool called Transponder Calculation Tool (TransCal) and introduced in (Deicke et al., 2008b). It was developed by the Fraunhofer IPMS. TransCal comprises different known solvers for electrical and electromagnetic models (Fig. 3.). These are closed formulas for electrical model and for ohmic losses of antennas including skin effect and proximity effect (Deicke et al., 2008a). Additionally, there is an adapted Neumann formula used for high speed calculation of self and mutual inductance for coaxial and rotated antennas. And there are links to external numerical solvers like FastHenry (Kamon et al., 1996) and Spice (Quarles et al., 2005). FastHenry is a 3D electromagnetic solver based on the Partial Element Equivalent Circuit method. It can be used to model arbitrary antenna configurations, 3D antennas or additional conductive structures nearby the antennas like metal plates or even metal rims to analyse transponder systems in a car or truck wheel, for example.

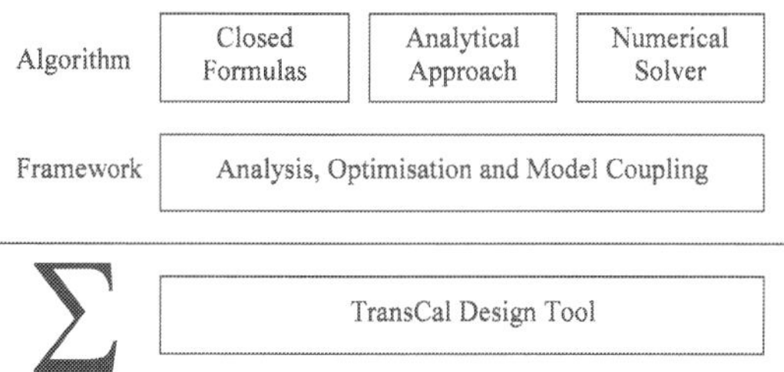

Fig. 3. Design approach for TransCal

Besides these algorithms needed for detailed modelling on both levels, a framework is used to implement algorithms for analysis, optimisation and model coupling to form a system simulation. The framework bases on C/C++ in connection with Microsoft Foundation

Classes to get a MS Windows compliant software tool with an appropriate graphical user interface. TransCal comprises five components like it is shown in the block diagram of Fig. 4. The analyser/optimiser module analyses and optimises different transponder systems using parameter variations and search algorithms. Furthermore, an automated model generator reorganises and adapts imported user defined netlists and generates antenna models as well as additional conductive structures nearby antennas. The model coupling module controls and synchronises different internal analytical algorithms and external numerical solvers selected by user.

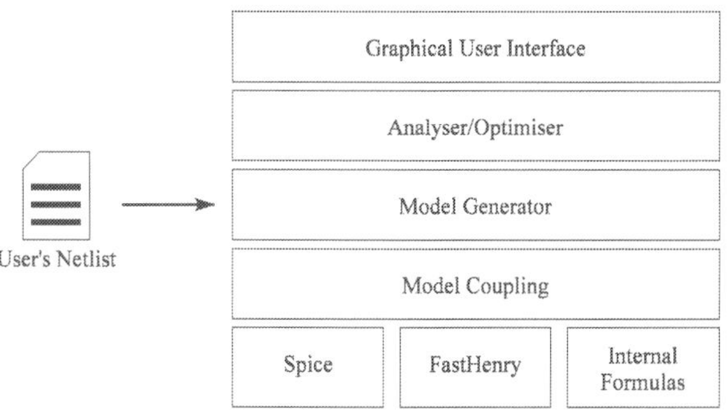

Fig. 4. Block diagram of TransCal

3.3 Input/Output parameters and Initialisation

The input of several user defined design tasks and the output of results are done with graphical user interface. Input parameters are geometrical and material properties of the antennas, variation ranges for optimisation and analysis as well as electrical properties. General settings for optimisation, analysis as well as used solvers can be made, too. The results are shown in a text-based output window and additionally stored in text files to provide the possibility for import in external data analysis and graphing tools. Fig. 5 shows a screenshot from TransCal with dialog-based input and text-based output. Each design is saved in a project file including all input settings and results to reopen and work on later.

For defining parameter space, constant and variable parameters have to be set. Dimensions such as inner radius, outer radius and width of antenna or antenna type, number of turns and link distance are variable parameters. Considering the optimisation of one antenna, there are five degrees of freedom (DOF). And considering an optimisation of two antennas, there are nine DOF. As a result, a five- or nine-dimensional parameter space must be used. That seems to be very complicated and time consuming for most optimisation algorithms.

An advantageous modification of that parameter space could be helpful to solve that optimisation task more efficiently. The reduction of variation parameters is to the fore. Principally, the variation of antenna geometry can be done by varying the number of turns if the antenna type is defined. Additionally, a constant fill factor has to be assumed. That is done by defining an outer diameter of the used wire including conductor and insulation. Using that substitution, the number of DOF can be reduced. Considering one antenna, the parameter space is reduced to one dimension assuming a constant link distance. And considering two antennas, the parameter space is reduced to two. If the link distance is variable, the number of DOF is increased by one. Considering objective functions V_T and

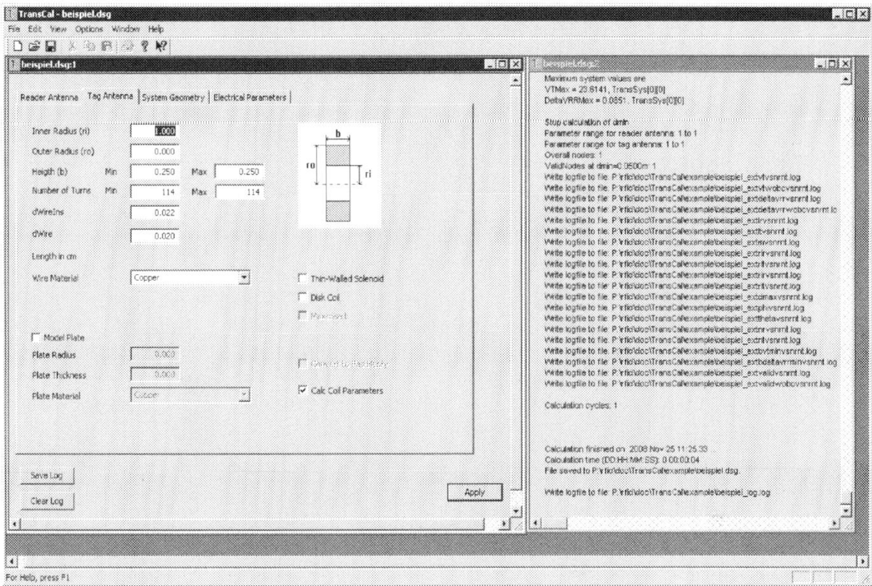

Fig. 5. Graphical user interface of TransCal with dialog-based input and text-based output

ΔV_{RR} versus the number of turns, the characteristic of these functions is concave. That can be used later to simplify optimisation process, too.

Subsequent to the definition of variation parameters, the generation of the n-dimensional mesh is done. The parameter space has discrete values excluding link distance. Each node of the mesh corresponds to a transponder system comprising a complete parameter set.

3.4 Optimisation and analysis

Looking from implementation side, optimisation and analysis are closely related to each other in general. The main difference is in generating new input parameters for system simulation. On system analysis, all defined nodes must be considered. Instead, there is a bigger parameter space on optimisation in general. Therefore, a more efficient algorithm is needed to consider as few as possible nodes during that process to reduce overall calculation time. The general flow, that was adapted from the well known simulation-based optimisation approach (Carson & Maria, 1997), is shown in Fig. 6. The analyser/optimiser module generates a new parameter set. First, electrical parameters of the transmission channel are calculated using an electromagnetic solver. These are the inductance and ohmic losses of the antennas as well as the mutual inductance. The impedance matrix is imported in the electrical circuit subsequently. After additional adaptations of resonance capacitors and quality factors, the objective functions V_T and ΔV_{RR} are calculated and imported in the analyser/optimiser module. There, the results are evaluated and a new parameter set is output. For optimisation, that loop is repeated until an optimised solution will be found for a given parameter space. Calculation of transmission channel and electrical circuit is controlled by the coupling module.

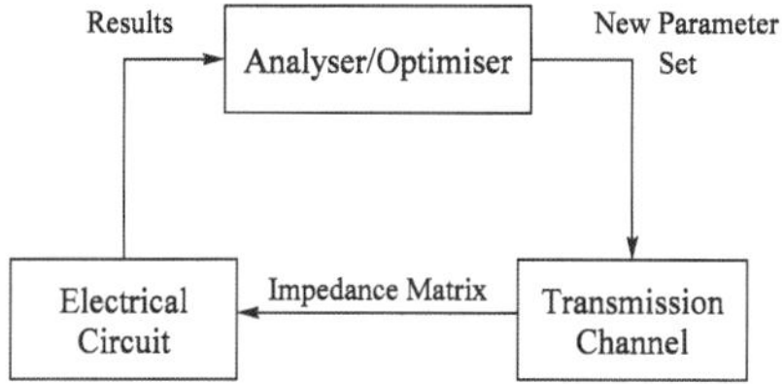

Fig. 6. Simulation-based flow for analyser and optimiser

During system optimisation the goal is to find an optimal parameter set for any particular application with a minimum amount of computing resources and time. Therefore, it is desired that not all possibilities are evaluated explicitly. Especially with complicated and heterogeneous systems, optimisation could be a challenging part. Using simulation-based optimisation, all nodes must be found that meet defined constraints for objective functions. Fig. 7 depicts an example, where all systems are looked for that met constraints for objective functions at a constant link distance. With that contour plot objective functions V_T and ΔV_{RR} are shown over the number of turns for reader N_R and tag N_T. The dash-point line is the equipotential line for the minimum transponder voltage $V_{T,min}$. All nodes on the left hand side have a tag voltage that is at least the minimum value. The dashed line is the equipotential line for the minimum demodulator input voltage $V_{RR,min}$. All nodes enclosed with that line have a demodulator input voltage that is at least the minimum value. The intersection of both areas shows all transponder systems that fulfil requirements for energy

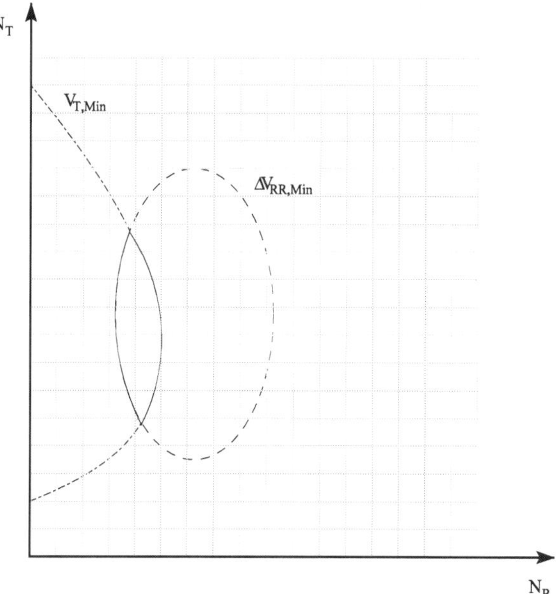

Fig. 7. 2-dimensional mesh for an optimisation task and marked sections where objective functions V_T and ΔV_{RR} meet requirements

range and transponder signal range at a given link distance. For that design example, these systems are optimal solutions. If the intersection comprises only one node, that node defines the system with maximum link distance.

Applying introduced simplifications for variation parameters, the objective functions are concave. Because of that, robust and simple gradient based search algorithm and logical operations are used to find the intersection. To find the node with maximum link distance, an additionally root finding algorithm was implemented. So, the overall optimisation task is divided into different steps using different simple and robust algorithms. In addition to that advanced optimisation method, a brute force method was implemented that considers all nodes available. Many design examples had shown that calculation time of the advanced method is less than 4% of the brute force method.

3.5 Model coupling

The coupling module controls and synchronises different calculation types selected by user before starting analysis or optimisation. There are internal closed formulas and additionally external numerical solvers that can be selected for each modelling level to adjust used model accuracy and calculation time (Fig. 8). On the one side, it is possible only to use internal closed formulas and analytical algorithms to speed up calculation. Thereby, less accuracy is accepted. And on the other side, external numerical solvers can be used for both electrical and electromagnetic model to get best accuracy. The communication between TransCal and these external solvers are done using command and result files. At the moment, FastHenry and Spice can be used. But if necessary, other simulators can be connected, too. A third way is to mix internal algorithm and external solvers like it is shown in an example later.

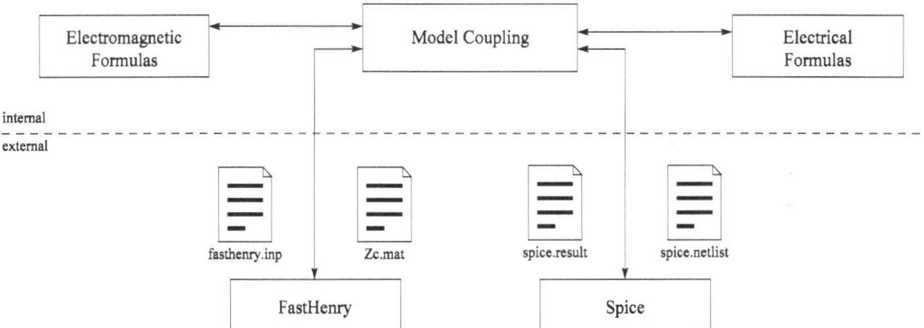

Fig. 8. Model coupling module and connected solvers

The model for FastHenry simulator is generated by model generator of TransCal for each simulation step, because of changing antenna geometry. Besides calculation of coaxial antennas in free air, FastHenry additionally has the possibility to model non-coaxial antennas concerning translation and rotation as well as 3D antennas. Furthermore, metal plates inside or under each antenna as well as car or truck rims can be modelled regarding its influence on transmission channel. If other conductive structures should be used in models, model generator must be extended before. That can not be done by user.

Using Spice, different user defined netlists can be imported by TransCal to provide the possibility to add particular components as well as to analyse different circuit concepts for reader and tag. The electrical model is focused on low level such as transmission channel, parasitic elements and dependencies of the whole system primarily. More complex components are replaced by basic equivalent circuits. That concern to, for example, ICs of reader and tag. Mostly, detailed descriptions of such ICs are not available for system design and of course not needed for that task in most cases. Often, basic properties extracted from datasheets can only be used. If internal IC behaviour is available, it can be used for modelling and simulation, too. However, effort for modelling and simulation should be considered in comparison to the gain on accuracy. Because of that, a good approach is to consider different resonance circuits, parasitic elements and different possibilities to connect the demodulator input as well as the tag IC. Fig. 9 shows an example for an extended electrical model. There, additional parasitics are considered like R_{CT} as ohmic losses of the resonance capacitor of the tag and C_L as the input capacitor of the tag IC. During analysis or optimisation the imported netlist is parameterised again for each simulation step dependent on changes of transmission channel.

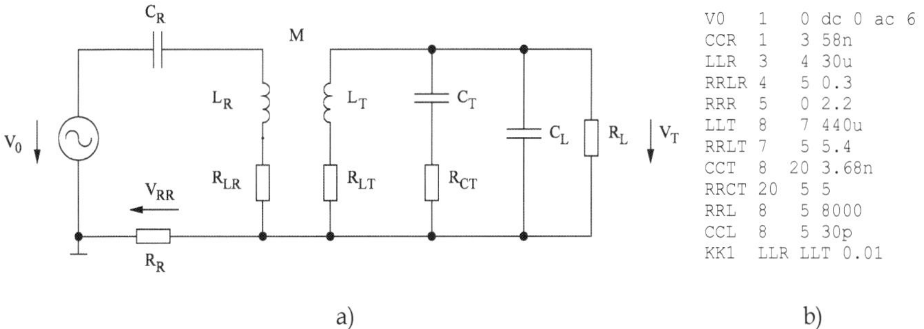

```
V0    1     0 dc 0 ac 6
CCR   1     3 58n
LLR   3     4 30u
RRLR  4     5 0.3
RRR   5     0 2.2
LLT   8     7 440u
RRLT  7     5 5.4
CCT   8    20 3.68n
RRCT 20     5 5
RRL   8     5 8000
CCL   8     5 30p
KK1   LLR LLT 0.01
```

a)   b)

Fig. 9. Circuit example a) and netlist b)

4. Examples

4.1 Basic optimisation

The first example shows basic optimisation functions of the introduced approach with more details. There, reader and tag antenna geometry must be optimised regarding to a given input parameter set (Table 2). It describes a passive transponder system for a standard ID and sensor application compliant to ISO 18000-2 protocol. The reader antenna is a disc coil and defined by inner radius, used wire including insulation as well as minimum and maximum number of turns. Additionally, electrical parameters are defined for driver and demodulator input as well as carrier frequency and bandwidth. Unlike reader antenna, a maximum winding space is defined for tag antenna like it is often defined in applications. It includes inner radius, outer radius and maximum antenna width. The tag antenna is a multi-layer coil.

Additionally, the used wire is not defined. The wire diameter varies between 0.08 to 0.25 mm regarding to IEC 60317. The tag includes a front end IC IPMS_RFFE125 (FhG IPMS, 2007), a microcontroller MSP430F123 (Texas Instruments, 2004) and additional application-specific components. The estimated load is 11 kΩ approximately.

With that input parameter set, a parameter range is defined to vary antenna geometry of both reader and tag. In a first step V_T and ΔV_{RR} are calculated for different antenna geometries at minimum link distance (Fig. 10). The wire diameter of tag antenna is 0.2 mm in that case. Maximum values for V_T and ΔV_{RR} are not in the same region of parameter space. So, the system with maximum transponder voltage or maximum demodulator input voltage is not the system with maximum link distance.

	Reader		Tag	
Antenna	$r_{i,min}$	135 mm	$r_{i,min}$	21.5 mm
Parameters	$r_{o,max}$	-	$r_{o,max}$	23.5 mm
	b_{max}	-	b_{max}	4 mm
	N_{Min}	10	N_{Min}	-
	N_{max}	30	N_{max}	-
	DWI	1.2 mm	DWI	variable
	DW	0.8 mm	DW	variable
	Type	DC	Type	MLC
Electrical	V_0	6.0 V	$V_{T,min}$	3.3 V
Parameters	ΔV_{RR}	0.01 V	$V_{T,max}$	3.6 V
	$R_{RR,min}$	4.4 Ω	R_L	11 kΩ
	f_0	125 kHz	$R_{L,Mod}$	30 Ω
	b_f	10 kHz		
Link Distance	d_{min}	15 cm		

Table 2. Input parameter set for reader and tag

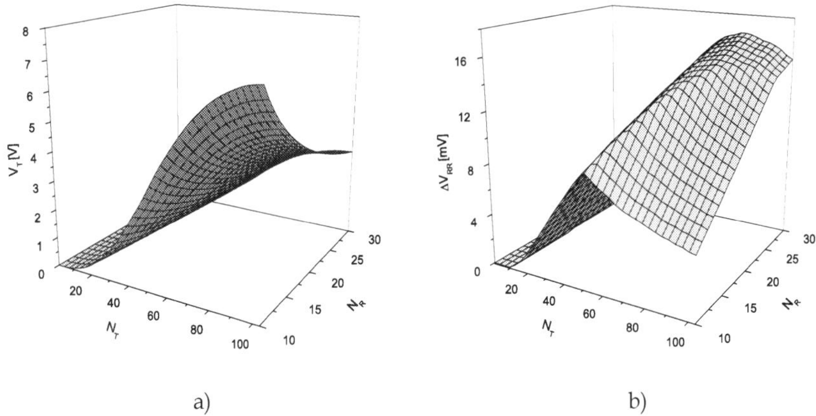

a)   b)

Fig. 10. Tag voltage V_T a) and demodulator input voltage ΔV_{RR} b) vs. number of turns for reader and tag antenna at a fixed link distance

In a next step both antennas and system setup are optimised for maximum link distance considering different wire diameters for tag antenna. The number of windings varies between 21 and 22 for reader antenna, but there is no direct influence from wire diameter. Fig. 11 a) shows the impedance of the tag antenna over the wire diameter for a constant maximum winding space. The impedance for maximum link distance decreases with increasing wire diameter, whereby ohmic losses decreases faster than inductance. As a result, the quality factor increases, too. Mutual inductance has the same behaviour as self inductance. So, the bigger losses because of smaller wire diameters are compensated by increasing self and mutual inductances to an appropriate value. In the second diagram (Fig. 11 b) the maximum link distance and the outer diameter of the tag antenna is shown. Both parameters increase with wire diameter. The increase of the link distance from 0.08 mm to 0.25 mm is 34% approximately. But the increase of the outer diameter of tag antenna is 6% only. Because of that, increasing wire diameter for a given winding space is a good approach to maximise link distance without making tag antenna much bigger. But finally, antenna optimisation is always a compromise between maximum antenna dimensions, wire diameter, electrical system parameters and link distance.

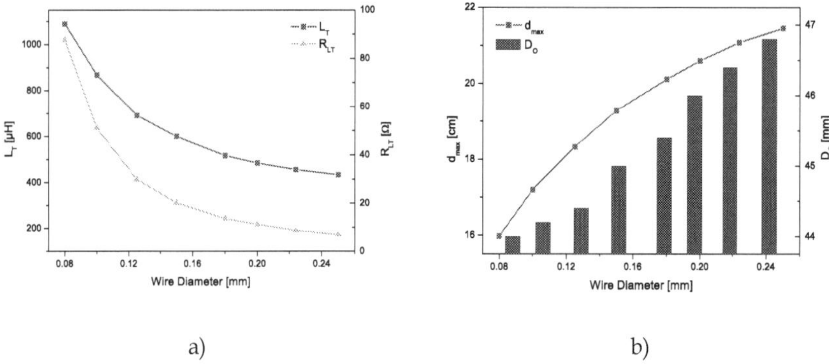

a)   b)

Fig. 11. Impedance of tag antenna a) as well as maximum link distance and outer diameter b) vs. wire diameter of tag antenna

In a further discussion, the model accuracy and calculation time are compared for introduced combinations of used solvers. Therefore, prototypes were made for reader antenna and two tag antennas with wire diameter of 0.1 mm (MLC01) and 0.2 mm (MLC02). The whole system was tuned to finally measure maximum link distance. During modelling, the difference between link distance for MLC01 and MLC02 is 16.5%. In practice, difference is 20.4%. If absolute values are considered, the maximum link distance differs between 12.1% for MLC02 and 16.3% for MLC01, whereby external simulators FastHenry and Spice are used. Fig. 12 a) shows the error for calculating maximum link distance in relation to the measured values of MLC02 over possible combinations of used solvers. On the one side, all calculated values are bigger than the measured one. And on the other side, the error is reduced by increasing model accuracy like mentioned before. But for coaxial antennas, using Spice including a more detailed electrical circuit is more important than using FastHenry instead of analytical formulas for the transmission channel.

If calculation time (Fig. 12 b) is also considered, the most efficient way is to use analytical formulas for transmission channel and Spice for electrical circuit. That is a good compromise between calculation error and time. Finally, the introduced approach of TransCal is good to make a virtual design and to use the results qualitatively and quantitatively.

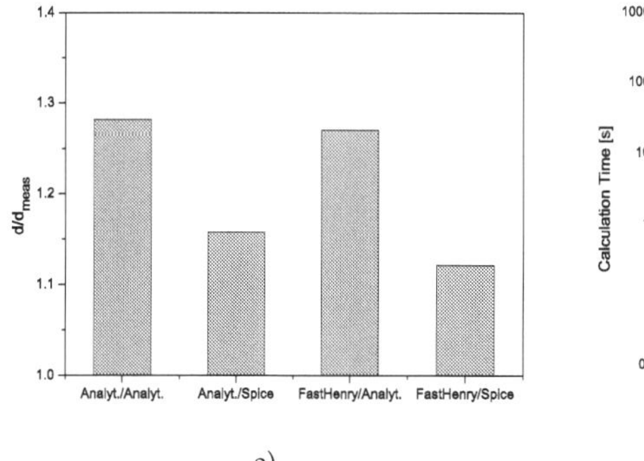

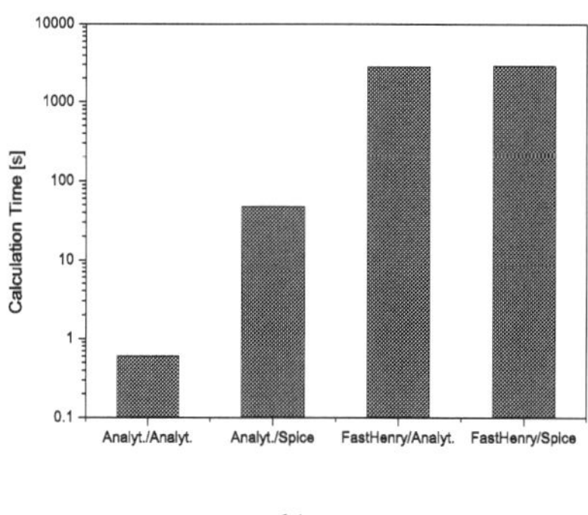

a)   b)

Fig. 12. Calculated link distance in relation to measured a) and calculation time b) vs. possible combinations of used solvers

4.2 3D antenna – dimensioning and analysis

Inductively coupled transponder systems are mostly analysed and optimised using coaxial antennas. That approach is simple to use, but it is not sufficient for many applications and its operating range or coverage. Because of that, analysis of tag rotation and translation is important, too. In that section, analysis of rotation is in the fore. Standard antennas have a limited coverage. However, 3D antennas with three coils assigned in perpendicular to each other (Fig. 13), seem to have a better one. But a question is if 3D antennas will provide power supply and communication for arbitrary rotation or not. Answers can be found by making some analysis using TransCal.

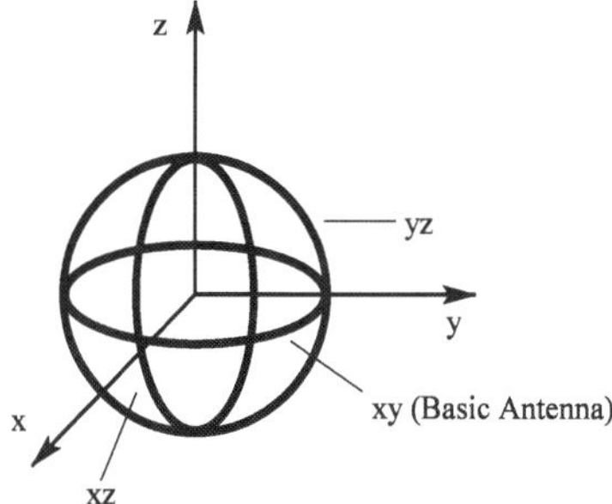

Fig. 13. 3D antenna with perpendicular coils in the xy plane, the xz plane and the yz plane of a Cartesian coordinate system

For modelling and optimisation of practical 3D antennas, different design constraints can be used. The primary goal is to find a good approximation for geometrical and electrical properties in comparison with a theoretical approach where three equal coils are used. Therefore, design constraints are constant maximum link distance or constant coil impedance. If the coils are optimised for constant impedance, the resonance circuits are equal and as a consequence production process simplifies. Additional constraints are overall dimensions of a 3D antenna that also influence the winding space of each coil.

If 3D antennas are generated using constant maximum link distance, V_T and ΔV_{RR} must have the same values at the same link distance and same rotation approximately. Starting from the inner coil that is also called basic antenna, the coil in the xz plane must have an inner radius that is bigger than the outer radius of the basic antenna. In a second step, the coil in the yz plane is generated, whereby its inner radius is bigger than the outer radius of the coil in the xz plane. 3D antenna generation and optimisation is done by TransCal automatically before starting further analysis. For that example a 3D antenna is derived from MLC02. A model for FastHenry simulator is shown in Fig. 14.

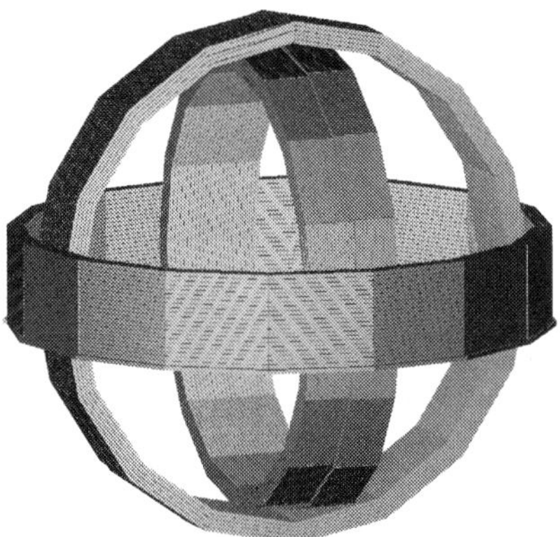

Fig. 14. 3D antenna model for FastHenry simulator

The simulated coverage of the single antenna MLC02 and a derived 3D antenna are shown in Fig. 15 over the rotation about the x axis (φ) and y axis (θ). Coverage means, where the functionality of the tag is ensured. There, the transponder voltage V_T and the demodulator input voltage ΔV_{RR} exceeds the defined limits at a given link distance. The coverage of the

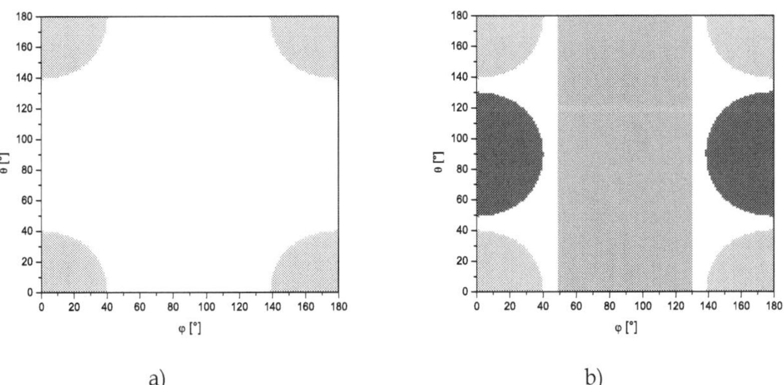

a)   b)

Fig. 15. Coverage of a single antenna a) and a 3D antenna b) vs. rotation about x axis (φ) and y axis (θ)

single antenna is symmetric and represents 20.8% of the considered parameter space approximately. In comparison, the 3D antenna has a coverage of 91.4% approximately. It is more than 4.5 times bigger than of the single antenna. If link distance is very small, it could be 100%. Finally, 3D antennas can be used to increase coverage if tag position is not defined or if tag is rotating during use. But in most cases, 3D antennas are not able to get full coverage during rotation.

4.3 Tags in truck or car tyres

The identification of a car or truck tyre with respect to manufacturer, type and mileage as well as the measurement of physical parameters like pressure, temperature or stress during use to warn against overstraining or damage (Lehmann, 2004) has been discussed intensively for some time past. A passive and inductively coupled transponder system is one possibility to realise that functionality, because it provides wireless energy transfer and wireless data communication, less maintenance and the ability to integrate different sensors. The properties of the transmission channel and the functionality of such a transponder system depend on the rim material and shape as well as the dimensioning and the positioning of both reader and tag antenna or using an additional steel cord. These dependencies will be analysed in the following on a virtual level to evaluate system usability and to derive important design rules concerning this usage scenario.

To realise the introduced application, a tag must be integrated in a tyre. The reader is outside the tyre. Therefore, different types of antennas and antenna configuration had been published in literature. Some examples are (Benedict, 2003), (Lehmann, 2004) and (Pollack, 2000). The used antennas and antenna configurations must ensure the functionality of the tag independent of the size of the tyre, the speed of rotation and the position of the tag. Because of that, an annular tag antenna is assumed to be for example under the tread

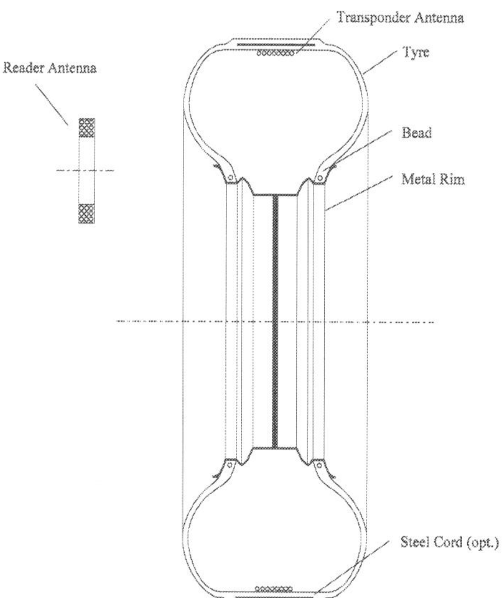

Fig. 16. Cross section of a wheel with integrated tag antenna and external reader antenna

(Fig. 16). The wheel and the tag antenna are coaxial. The annular antenna can be placed in the middle of the tyre, nearby the sidewall or the bead as well as on the rim or directly on a run-flat system (Continental, 2005; Rodgard, 2008). A steel cord can be placed inside the tyre optionally. The rim model can be generated by TransCal automatically using different variation parameters for a comprehensive analysis (Deicke et al., 2009).

The transponder circuit can also be placed under the tread or inside the tyre. The reader antenna is outside and in parallel with the wheel. It is smaller than the tag antenna. So, the reader antenna can be positioned on the axle or on an advantageous position in the wheel case. A passive tag is used for identification and the integration of different sensors. Therefore, the ST1 transponder IC is used in a test setup. It provides an analogue 125 kHz RF front-end, a hard coded module to support low level ISO 18000-2 protocol functions, a 16-bit microcontroller core, RAM, flash, EEPROM, timers and several peripheral components to connect, for example, different sensors with an analogue or digital interface. Additionally, analogue and digital signal processing can be done on the tag side (Grätz et al., 2007). In the test setup, a combined pressure and temperature sensor MS5541 (Intersema, 2008) is connected to the ST1. So, the power consumption of the passive tag can be estimated to use it for analysing and optimisation of the whole system in the following. For system setup, the general input parameters of Table 3 are used.

	Reader		**Tag**	
Antenna Parameters	r_i	31 mm	r_i	243 mm
	b	10 mm	b	11 mm
	N	56	N	33
	Type	MLC	Type	TWS
Electrical Parameters	V_0	6.0 V	$V_{T,min}$	5.0 V
	ΔV_{RR}	0.01 V	$V_{T,max}$	5.5 V
	f_0	125 kHz	R_L	14 kΩ
	b_f	10 kHz	$R_{L,Mod}$	30 Ω

Table 3. Input parameter set for reader and tag of a tyre application

Both reader and tag antenna are optimised for maximum link distance in free air. Then, this optimised non-varying antenna setup is used to analyse and compare different rim configurations with and without steel cord. Later, both antennas can be optimised dependent on particular rim shape, size and material. The reader antenna is a multi-layer coil (MLC). It is smaller than the tag antenna that is a thin-walled solenoid (TWS). Additionally, Table 3 lists important electrical parameters of the reader and the tag.

Two parameters are very important for system evaluation if a transponder system is analysed within a tyre. The first parameter is the minimum load resistance $R_{L,min}$ with a defined transponder voltage and link distance. Therewith, the maximum power consumption can be calculated to estimate which sensors and signal processing functions for measuring physical parameters can be implemented into the tag. The second parameter is the demodulator input voltage to determine if data communication is possible from tag to reader.

For the first analysis below, an antenna configuration is used without a steel cord. Thereby, the rim width is varied to analyse different aspect ratios. Additionally, the reader antenna is moved in radial direction starting from a coaxial position. With this example, the diameter of the rim is 40 cm and the round plate of the rim is centred. The rim material is aluminium. The diagrams of Fig. 17 a) and b) show the minimum load resistance and the demodulator input voltage of the current setup with a link distance of 7.5 cm. The rim width is varied from 1 cm to 10 cm. That corresponds to an aspect ratio from 4.3 to 0.43 including the range of most used car and truck tyres. The reader antenna is moved in radial direction beyond the tag antenna. The extremes of $R_{L,min}$ and ΔV_{RR} are not with a coaxial antenna configuration. They are at a radial shift of 19.3 cm approximately. If the reader antenna is moved beyond the tag antenna, the load resistance increases sharply. ΔV_{RR} has the same behaviour in the opposite direction.

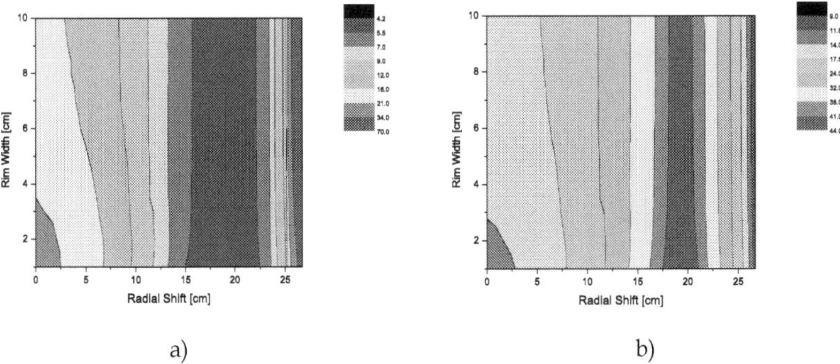

a)   b)

Fig. 17. $R_{L,min}$ [kΩ] a) and ΔV_{RR} [mV] b) vs. rim width and radial shift of the reader antenna

The diagrams also show, that $R_{L,min}$ and ΔV_{RR} do not depend strongly on high rim widths. As a result, low aspect ratios are not really an influencing factor with that configuration and it can be assumed that the system would also work on lower values and flatter tyres respectively. If a low rim width is considered, the distances between the round plate and the antennas are smaller. So, the influence of the plate is bigger and for example on a coaxial antenna configuration, the load resistance is bigger as on larger rim widths.

The diagram of Fig. 18 a) depicts the minimum load resistance versus the radial shift of the reader antenna with different link distances and a constant rim width of 5 cm. The absolute minimum value decreases and moves radially outwards with reducing link distance. So, it is above the rim flange for very small distances. For example, considering a link distance of 2.5 cm, the minimum load resistance is at a radial shift of 21.8 cm approximately. If the link distance increases, the absolute minimum also increases and moves towards the centre of the rim. At a link distance of for example 15 cm, the minimum is at a radial shift of 13.3 cm. At very high distances, the curve corresponds to a free air configuration without a metal rim qualitatively. If the curve with a link distance of 2.5 cm is considered, the minimum of $R_{L,min}$ is 1.3 kΩ. For a link distance of 15 cm, the minimum is 11.3 kΩ. Thus the maximum power consumption of the tag is 19.1 mW and 2.2 mW respectively if a transponder voltage of 5 V is considered. This is sufficient to power the introduced passive tag for measuring pressure and temperature. Considering the difference of $R_{L,min}$ between the centre position of the reader antenna and the position of the absolute minimum value, it increases with reducing the link distance. If the link distance is small, the correct position of the reader antenna is more important than at higher link distances. The load resistance increases sharply if the reader antenna is moved beyond the tag antenna.

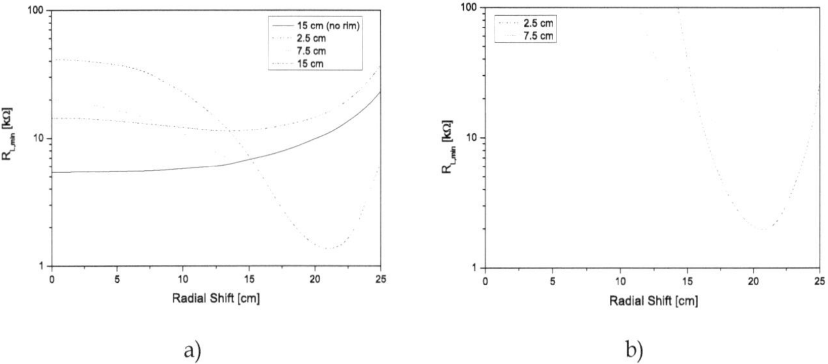

a)   b)

Fig. 18. $R_{L,min}$ vs. radial shift of the reader antenna for different link distances and a constant rim width without a) and with steel cord b)

With a second analysis, the same configuration is considered with a steel cord, like it is shown in Fig. 19. The steel cord is as wide as the rim and is 1 cm above the tag antenna. The thickness is 2 mm. The diagram of Fig. 18 b) depicts the minimum load resistance versus the radial shift of the reader antenna with different link distances, like it was done in the analysis before. In comparison to Fig. 18 a), the curves are steeper and the range of radial shift, where the tag would work, is smaller. Additionally, the minimum load resistance is higher than without a steel cord. With a link distance of 15 cm, the tag can not be powered from the RF field of the reader.

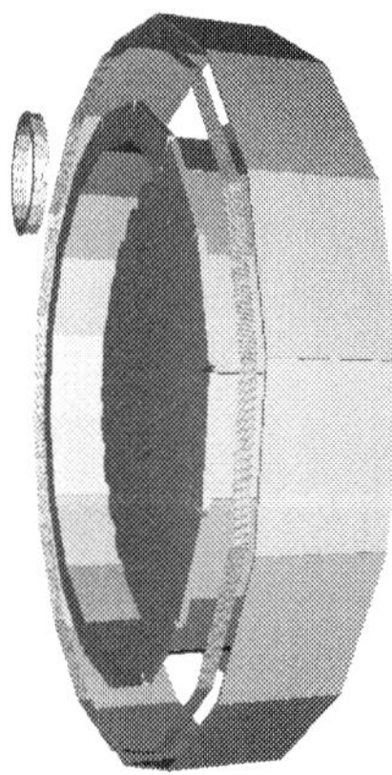

Fig. 19. FastHenry model of the transmission channel including the rim, a steel cord and the antennas. The tag antenna is over the left rim flange.

Finally, the behaviour of an inductively coupled transponder system within a wheel setup can be discussed before doing any prototyping. Whereby, analysis can be done for different rim sizes and shapes, different reader and tag antennas, different antenna positions as well as different electrical properties to find best solutions.

5. Conclusion

A first system success design approach for inductively coupled transponder systems was introduced. It bases on a software tool called TransCal. This tool can be used for system analysis and optimisation including automatic parameter variation and model generation. Many different customised designs including RFID technique can be solved in an easy and convenient way. Whereby, the focus is on transmission channel analysis, antenna design and its effects on the electrical level. During design process, the system is divided in an electromagnetic and an electrical model to consider necessary details in an appropriate way. Therefore, analytical algorithms are implemented in TransCal. Additionally, external numerical solvers can also be used to increase model accuracy. A model coupling module controls and synchronises internal and external solvers for these modelling levels to provide a system simulation finally. That functionality is used by an adapted simulation-based optimisation algorithm to find optimised solutions in a large, multidimensional and heterogeneous parameter space. In comparison to manual optimisation that only bases on human experience, better results concerning quality and quantity can be found within less amount of time. Using this approach, different usage scenarios can be considered such as coaxial or non-coaxial antennas including rotation and translation, different environments including eddy current losses as well as 3D antennas to increase tag coverage.

Besides the theoretical introduction, three design examples are presented to show the advantages and limits of that approach. The first example introduces the basic optimisation process considering accuracy and calculation time, too. The second example is an analysis of 3D antennas versus single-coil antennas to compare system behaviour during rotation. In a third example a transponder system was implemented in a tyre for analysing different antenna positions. Finally, important questions like if a specific application would work using RFID technique or how to dimension and position antennas can be answered qualitatively and quantitatively on virtual level without doing a lot of prototyping. Additionally, it was shown that this design approach is less time consuming and expensive as well as provide better results for LF and HF systems to work with.

6. References

ANSYS Inc. (2007). ANSYS 11.0 User Manual. www.ansys.com

Benedict, R.L. (2003). EPO Patent No. EP 1384603. European Patent Office

Beroulle V., Khouri R., Vuong T. & Tedjini S. (2003). Behavioral Modelling and Simulation of Antennas: Radio-Frequency Identification case study. *BMAS Conference 2003*, pp. 102-106, San Jose, USA, October 2003

Carson, Y., Maria, A. (1997). Simulation Optimization: Methods and Applications. *Winter Simulation Conference of the INFORMS Simulation Society*, Atlanta, Georgia, December 1997

Continental AG. (2005). ContiSupportRing and Continental SSR. www.conti-online.com

Deicke, F., Grätz, H. & Fischer, W.-J. (2008a). Combined System Analyses and Automated Design of RFID Transponder Systems. *IEEE International Conference on RFID*, pp. 328-335, Las Vegas, USA, April 2008

Deicke, F., Grätz, H. & Fischer, W.-J. (2008b). Computer-Aided Design of Antennas, Transmission Channels and the Optimisation of Transponder Systems. *RFID SysTech 2008*, pp. 32-41, June 2008

Deicke, F., Grätz, H. & Fischer, W.-J. (2009). Analysis of Antennas for Sensor Tags Embedded in Tyres. *RFID SysTech 2009*, pp. 23-28, June 2009

FhG IPMS (2007). RFFE125 Datasheet. http://www.ipms.fraunhofer.de/en/ applications/sas-info.shtml

Finkenzeller, K. (2007). RFID Handbook. Fundamentals and Applications in Contactless Smart Cards and Identification. John Wiley & Sons Ltd.

Grätz, H, Heinig, A., Deicke, F. & Fischer, W.J. (2007). ST1 – ein freiprogrammierbarer Schaltkreis für 125 kHz Transponderlösungen. *Mikrosystemtechnik-Kongress*, Dresden, Germany, October 2007.

Grover, F. W. (2004). Inductance Calculations: Working Formulas and Tables. Reprinted in Dover Publications

International Electrical Commission (2005). IEC 60317 Specifications for particular types of winding wires. www.iec.ch

Intersema Sensoric SA (2008). MS5541-30C Datasheet. http://www.intersema.ch/ products/guide/calibrated/ms55341-30c/

Kamon, M., Smithhilser, C., White, J. (1996). FastHenry User's Guide. http://www.fastfieldsolvers.com

Lehmann, J. (2004). WIPO Patent No. WO 2004/108439. World Intellectual Property Organization

Meeker, D. (2006). Finite Element Method Magnetics User's Manual http://femm.foster-miller.net

Pollack, R.S. (2000). EPO Patent No. EP 1037755. European Patent Office

Quarles, T., Pederson, D., Newton, R., Sangiovanni-Vincentelli, A. & Wayne, C. (2005). The Spice Page. http://bwrc.eecs.berkeley.edu/Classes/IcBook/ SPICE/

Rodgard (2008). Pneumatic Tire Run-Flat Systems. http://www.rodgard.com/ mobility.htm

Roz, T. & Fuentes, V. (1998). Using low power transponders and tags for RFID applications. EM Microelectronic Marin SA

Soffke O., Zhao P., Hollstein T. & Glesner M. (2007). Modelling of HF and UHF RFID Technology for System and Circuit Level Simulations. *RFID SysTech 2007 - ITG-Fachbericht Vol. 203*. Duisburg, Germany, July 2007.

Texas Instruments (2004). MSP430x12x2 Datasheet. www.ti.com

Youbok, L. (2003). Antenna Circuit Design for RFID Applications. Microchip Technology Inc.

14

Fabrication and Encapsulation Processes for Flexible Smart RFID Tags

Estefania Abad[1], Barbara Mazzolai[2], Aritz Juarros[1], Alessio Mondini[2],
Angelika Krenkow[3] and Thomas Becker
[1]Fundacion Tekniker, Eibar,
[2]Scuola Superiore Sant'Anna, CRIM lab, Pisa,
[3]EADS Deutschland GmbH, Innovation Works, München,
[1]Spain
[2]Italy
[3]Germany

1. Introduction

RFID tags are often envisioned as a replacement for the current barcodes. These systems are simple wireless transponders with integrated memory chips. Nowadays the challenge in this field is the integration of sensors on board and there are some examples of tags in the market including temperature and humidity sensors (Opasjumruskit et al. 2006). However, there are no commercial labels containing chemical sensors. In this chapter book, we present an integrated process flow for the integration of gas sensors onto flexible substrates together with a RFID transponder to get a Flexible Tag Microlab (FTM) innovative system for food logistic applications (see figure 1). In the proposed scenario, the FTM is designed to be handled by a specifically designed reader with onboard sensing capabilities (Vergara et al. 2007). RFID technology in the 13.56 MHz band was chosen since it is the best compromise for integration on a flexible tag. Furthermore this band is very suitable for the food logistic application, considering possible constraints such us the surrounding environment (e.g. humidity) and range of communication. In order to be compliant with recent RFID developments the ISO 15693 standard has been selected.

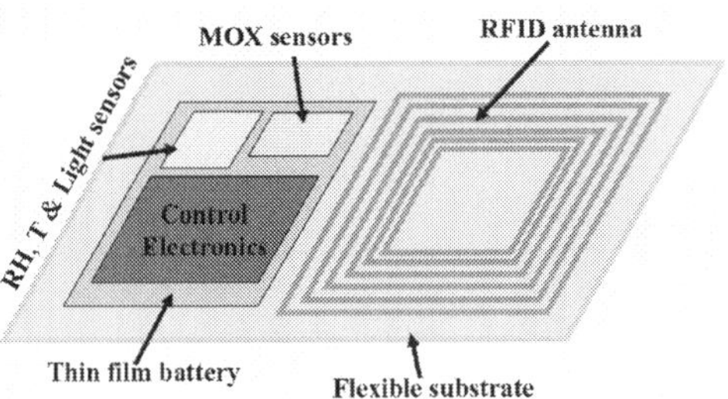

Fig. 1. Main functional blocks of the FTM inlay for food logistics.

This visionary application involves both the fabrication of the so-called inlay, which is the flexible substrate, acting mainly as a passive interconnect structure, with all components needed for the FTM assembled on it, and the development of particular assembly and packaging issues for the new ultra-low power consumption substrates for gas sensor integration.

Flexible substrate, microcomponents assembly and encapsulation technologies have been used throughout the electronics industry and continue to play a major role in new designs and applications. Flexible substrate technologies refer to a group of processes for the construction of multi-layer flexible circuits, commonly based on the use of a polyimide as raw material. Typical fabrication of this type of circuit involves the masking and etching methods similar to those employed by printed circuit board manufacturers. The component assembly techniques have been developed to perform a hybrid integration of flexible substrates with integrated electronic circuitry on bare dice. Flip chip bond, wire bond, electrically conductive glues and tapes are some examples of connection methods. Envisioned miniaturized systems could be assembled by combining these techniques and solder steps. There is virtually no limit to the types of terminations possible for flexible circuits. Wire, cable, contacts, printed circuit boards, chips and microstructures can be connected to flex circuitry with these breakthrough technologies.

The process flow employed for the two metal levels interconnect fabrication will be described in detail. The material used is the DuPont™ Pyralux® AP 8525R double-sided copper-clad laminate, formed by a Kapton foil with a copper layer on each side. The vias and windows openings are performed by femtosecond laser ablation. The copper interconnections are realized by photolithography and wet chemical etching.

The MOX sensors hotplates specially developed to fulfil the FTM constrains in terms of low power consumption has been used to prove two integration technologies into the flexible substrates: Chip on Flex (COF) wire bonding and Anisotropic Conductive Adhesive (ACA) flip chip bonding. Both technologies will be compared and benchmarked for future product developments.

2. RFID flexible inlay fabrication

Flexible substrate and component assembly technologies (Numakura 2001) for the FTM have been developed and/or optimised. Flexible circuit technology refers to a group of additive or subtractive processes for the construction of multi-layers flexible circuits, commonly based on the use of a polyimide (PI) as substrate. Specifically, two different materials for substrates can been considered: DuPont Pyralux flexible composites and photosensitive polyimide Pyralin PI2730 products.

Flexible composites technology uses Pyralux copper-clad laminated composites[1], constituted by DuPont Kapton polyimide film and copper foil on one or both sides, as flexible substrate. The copper interconnections can be generated by standard photolithography (using either DuPont adhesive photoresist coverlay that works as a negative photoresist or a positive liquid photoresist) and wet etching. On the other hand, the vias definition in Kapton can be performed either by photolithography and dry etching, or directly by femtosecond laser ablation.

---

[1] http://www.dupont.com/fcm/products/pyralux.html

The main advantages of this technology are:
- Easy and quick to fabricate
- Good mechanical and electrical properties
- Low price

And the main drawbacks:
- Multilayer circuits need bonding and electrical contact trough vias.

On the other hand, there is the polymer thin film technology based on the use of photosensitive polyimide. The polymer thin film represents an extension of the conventional thin film technology. In this case, thin (< 20 μm) polymer dielectric films are deposited over a substrate such us silicon. Then, a thin (< 2 μm) conductor layer, usually copper, is deposited (PVD or CVD) and processed photolithographically. Vias can be easily achieved by using a photopatternable polymer, as for example the Pyralin PI2730 products[2]. The Pyralin PI2730 series are photosensitive negative working polyimides. Thin films of this product can be applied by spin coating.

The main advantages of the thin film polymer technologies are:
- Narrow lines and vias
- Very high conductor a package density
- Very good mechanical and electrical properties of cured polyimide films
- Multilayer construction

And the main drawbacks:
- High cost
- Immature technology

A multiple spin steps process with the polyimide Pyralin PI2730 represents a powerful solution for the fabrication of multilayer high density integrated circuits. The efficiency and performance of this approach have been tested, comparing with the results obtained by an approach based on the use of Pyralux double sided copper, in terms of feature size, time and easiness of process. The results of this comparison activity (including advantages and drawbacks of both the approaches) are briefly reported in the following:
- Pyralin PI2730, in a multilayer process configuration, allows a better integration of complex circuits in a flexible tag.
- Pyralin PI2730 can be also used for developing the passivation layer.
- The approach based on Pyralin PI2730 shows a lower reproducibility in realizing planar and homogeneous surfaces, when the flexible tag dimensions increase.

On the basis of the above mentioned results experimentally obtained, and considering the low complexity of the tag circuits to be realized together with the usual dimensions of a flexible Tag (credit card), the Pyralin based process is not necessary at this, and therefore the Pyralux double sided copper was selected for developing the flexible tag.

A straightforward process flow for the fabrication of flexible substrates has been implemented. The outline of this process is presented in the left part of Figure 2. The material employed is the DuPont™ Pyralux® AP 8525R double-sided, copper-clad laminate (Kapton), which is an adhesiveless laminate for flexible printed circuit applications. The Kapton has a thickness of 50 μm and the copper layer has a thickness of 18 μm on each side.

---

[2] Liquid Polyimide: Dupont Pyraline PI2730. http:://www.hdmicrosystems.com

In this procedure, the vias definition in Kapton was performed directly by femtosecond laser ablation. Then, the copper interconnections of the two metal levels necessary for the substrate were generated by standard photolithography and wet etching. Finally, contacting through the vias was also implemented. Further details of this procedure are given elsewhere (Abad et al. 2005). An example of the double sided flexible circuit (a) and antenna (b) fabricated using this process is presented in the right part of Figure 2.

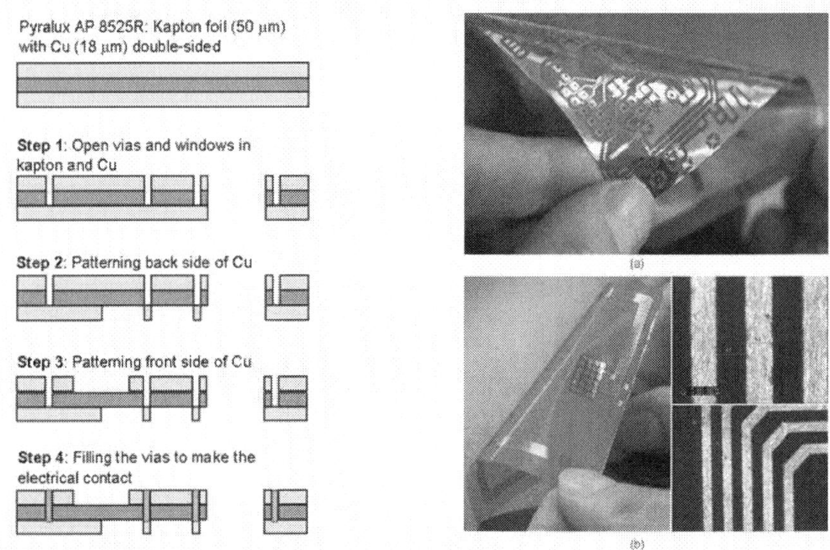

Fig. 2. Process design for flexible substrates fabrication, left image. Photographs and microscope images of the flexible circuit (a) and the flexible antenna (b). Detail of the copper tracks of the inductor.

Figure 3 shows a prototype of the developed FTM. The implemented system is a semi-active tag with a passive read-out and a battery powered sensing part, as reported in (Zampolli et al. 2007). The main functional blocks include a flexible antenna, a microcontroller for sensor control and signal acquisition, a RFID front-end and a complex programmable logic device (CPLD) for signal modulation/demodulation, commercial sensors (relative humidity, temperature and light), an EEPROM memory and a thin film flexible battery. For this prototype packaged chips were integrated on the flexible circuit using conventional assembly technologies.

3. MOX sensors integration

The integration of MOX sensors on a flexible tag has several critical aspects, mainly due to mechanical reliability and power consumption and requires specific assembling methods and protection of the chips from the environment. The power consumption issues were addressed in the design of Ultra-Low Power Hot Plates (ULPHP) but mechanical aspects

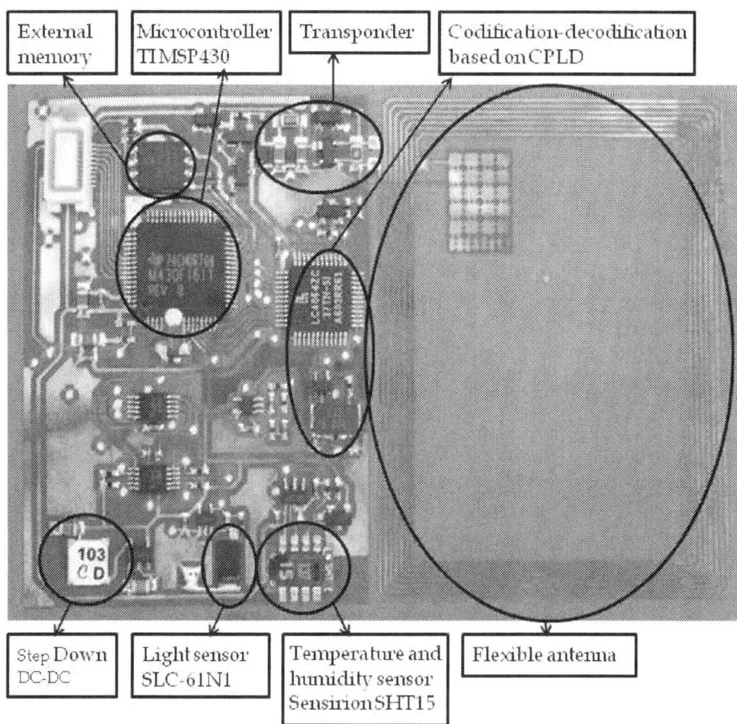

Fig. 3. FTM prototype including relative humidity, temperature and light sensors and a flexible antenna. A credit card size was chosen for the tag.

were considered as well. Details of the design and fabrication of the chips are given elsewhere (Elmi et al. 2008; Abad et al. 2007).

To reach the extremely low power consumption necessary for flexible tag operation, MOX sensors are designed to be used in discontinuous mode as reported in (Sayhan et al. 2008). The gas sensors, based on MEMS structures, need to be electrically connected to the rest of the tag electronics, and subsequently mechanically encapsulated and protected from damaging. While for the electric connections the same strategies as for other dies can be adopted, gas sensors have some mechanical peculiarities: the sensing layer must be exposed to the air sample being analysed. Therefore, assuming to have the sensing layer on the same side as the contact pads, the typical flip-chip underfilling techniques cannot be applied, since they would cover and damage the sensing layer.

The suspended membrane and the sensing layer must be protected from damaging, being generally very fragile they could brake if they get in contact with particulate or water drops. Considering the above issues, two MOX sensor encapsulation strategies were followed in parallel, aiming at an overall risk reduction of this activity.

ACA flip chip bonding

Flip chip technology utilizing ACAs has been proved to be a possible solution for MEMS packaging (Pai et al. 2005 and Johansson et al. 2006). Using this technology, a special procedure has been designed for the integration of the ULPHP, involving the following main steps illustrated in Figure 4 (a):

1.   Window opening by femtosecond laser ablation.
2.   Patterning of the electrical contacts.
3.   ACAs flip-chip for assembly.
4.   Polymer casting and curing for encapsulation.

The anisotropic conductive adhesives can provide uni-directional conductivity, which is always in the vertical, or Z axis. The directional conductivity is achieved by using a relative low volume loading of conductive filler. The low volume loading, which is insufficient for inter-particle contact, prevents conductivity in the plane of the adhesive. The Z-axis adhesive is placed between the surfaces to be connected and pressure and/or heat is applied to form the bond, as illustrated in Figure 4 (b). This type of products is now being used in flexible circuit interconnection, especially in copper/polyimide circuits. Due to their anisotropic conductivity, these adhesives can be deposited over the entire surface, thus facilitating the material application and avoiding the use of a dielectric layer and the formation of bumps onto the chip pads.

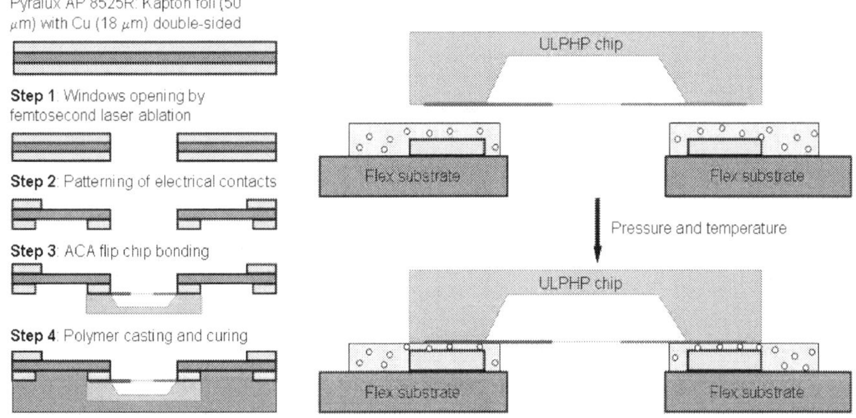

Fig. 4. (a) Process design for the MOX sensor integration. (b) Schematic view of the thermo-compression bonding using ACAs. Conductive particles are trapped between the flexible circuit and the chip pads.

A simple flexible substrate layout was designed in order to prove the viability of ACA flip chip assembly of the ULPHP. For this initial trial, the single sided flexible substrates shown in Figure 5 (a) were produced. Each test substrate includes a 1.15 mm x 0.4 mm laser ablated window and 12 copper pads (140 µm x 100 µm size and 60 µm spaced) linked to connectors. The ACA bonding can be tested by measuring the resistance at the corresponding connectors with a multimeter.

The adhesive employed to assemble the ULPHP onto the flexible substrate was the Z-axis film 5460R from 3M™. The 5460R ACA film is a 40 µm cyanate ester and epoxy/thermoplastic blend loaded with 7 µm size gold plated nickel particles. The film is attached to a liner to facilitate the handling and supports a maximum current of 100 mA / 0.1 mm². Using this film it should be possible to achieve an interconnect resistance of less than 0.05 Ω.

Electrical interconnections in ACA flip chip bonding are formed by a thermo-compression cycle. The procedure for the flip chip assembly includes several steps: (1) heat pre-tacking the film to the flexible circuit, (2) removal of the release liner, (3) ULPHP flipping and alignment to the substrate and (4) bonding by a thermo-compression cycle. All these operations are performed in a Dr. Treski AG 8800 flip chip station with alignment errors below 10 µm. Manufacturer's recommendations were used for the pre-tacking and thermo-compression bonding steps. The values employed for these steps are gathered in Table 1. The heating was realised by placing the test flexible substrate on the hot chuck of the flip chip system. The temperature control over the film and the applied pressure are the key parameters to achieve a good bonding. For this reason, a thermocouple was employed during the thermo-compression cycle to assure that the required film temperature was reached. The pressure was applied by pushing the ULPHP towards the substrate using the vacuum gripper of the system with a force previously calibrated.

Figures 5 (b) and (c) show the images of an ULPHP chip assembled on the flexible substrate using the Z-axis film 5460R following the procedure described previously. The mechanical reliability of the assembly was tested by bending the flexible substrate after ACA flip chip bonding of the ULPHP die chip. The assembly supports sharp bending without any damage.

Step	Temperature	Pressure	Time
Tacking	80 °C	0.1-1 MPa	5 s
Bonding	170 °C	3 MPa	20 s

Table 1. Pre-tacking and bonding parameter for the ACA 5460R from 3M™.

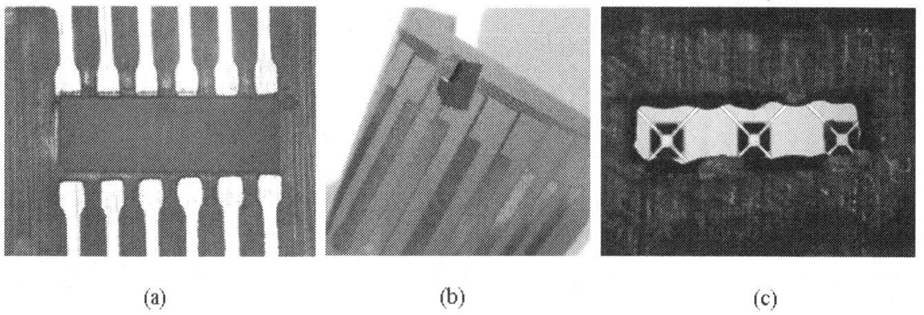

(a)   (b)   (c)

Fig. 5. (a) Flexible substrate fabricated to prove the viability of ACA flip chip assembly of ULPHP. (b)ULPHP chip assembled on the flexible substrate (c) View of the sensor membranes exposed to the air throw the ablated window.

The electrical behaviour of the bonding connections was characterized using some preliminary ULPHP test dies integrating 6 heater resistors. At first, the series resistance formed by the heaters and the ACA bonding connections was compared to the resistance of the heaters of a new, not encapsulated die. The resistance was measured with an I-V ramp in the range from -10 mA to +10 mA, which is higher than the 7 mA maximum current expected during the ULPHP sensor operation. The comparison of the average curves of 6 heaters is shown in Figure 6 (a), and no significant difference can be discerned between the resistance of the stand-alone ULPHP and the ULPHP integrated on the flexible substrate.

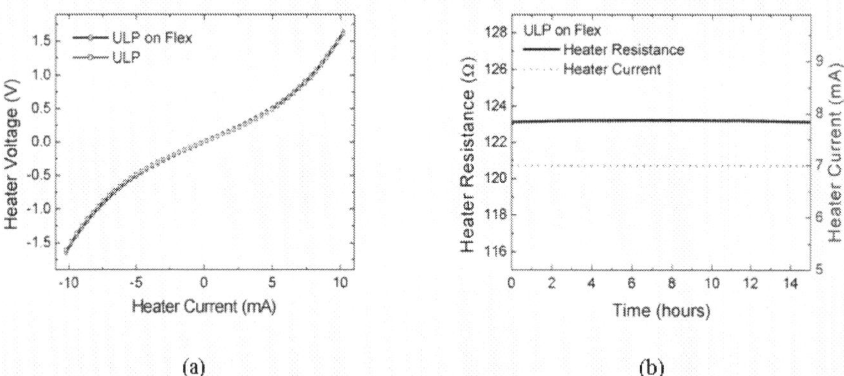

(a)   (b)

Fig. 6. (a) Comparison of resistance measurements between ULPHP (red) and encapsulated ULPHP (black). Each curve is the average of 6 different heaters within the test die. (b) Electrical short-term stability of the ACA flip chip bonding. The heater resistance does not change significantly during 15 hours of hotplate operation at 7 mA.

The characteristic non-linearity of the I-V curve is due to the temperature increase of the hotplate heater resistors while dissipating power.

To evaluate the short-term stability of the electrical connections, a 7 mA current was applied continuously for 15 hours and the heater resistance was acquired. As can be seen from Figure 6 (b), the heater resistance of 123.1 Ω does not change during the 15 hours of operation.

These measurements experimentally confirm that the 7 mA current supply necessary for powering the ULPHP and the pad dimensions of 0.012 mm² are compatible with the specifications of the ACA film, resulting in approximately half of the maximum specified current density of 100 mA / 0.1 mm².

3.1 Chip on flex wire bonding

Wire bonding is a valid and well-established method for attaching chips directly to flex circuits for both low and high volume applications. A rigid support, called stiffener, is necessary to obtain a more reliable connection between the MOX sensor and the flexible circuit. Stiffeners are very important in wire bonding to flex. The most common stiffener materials for COF are aluminium and stainless steel, but other materials like ceramic are also used.

These materials keep the bond pad area rigid, preventing ultrasonic energy from being absorbed by the flex circuit. Moreover, by cutting a hole in the flex circuit by laser ablation, the die can be mounted directly on the metal stiffener, which can act also as very efficient heat sink. A perforated cap, put on the top, allows gas flow to the sensing layer and protects the bonding wires. The cap can be equipped with an air filter; the filter allows protecting the sensor from the atmospheric particulate and water drops.

Figure 7 shows a schematization of the COF structure.

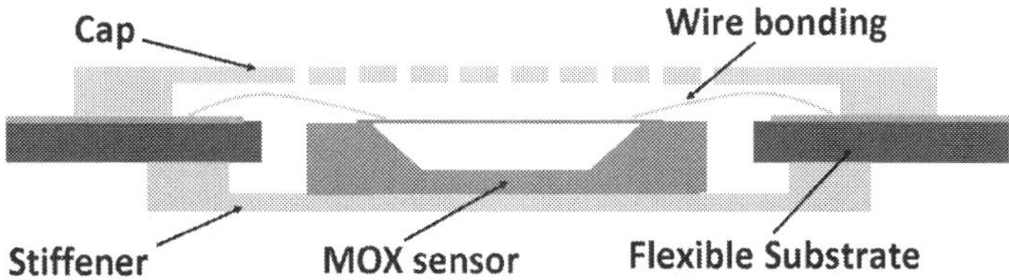

Fig. 7. Schematization of the chip-on-Flex structure.

For this work, two types of hot plates have been integrated on the flexible tag:

- a 5x5 mm² hotplate;
- a 1.0x1.5 mm² hotplate (also called Ultra-Low Power Hot Plate).

The 5x5 mm² hotplate has been used to test and to validate the Chip on Flex process for the integration of the MOX sensors. The successive step, the integration of the miniaturized sensor, has been a more challenging work, because of the dimension of the hotplate.

For these sensors, two kinds of stiffener have been realized:

- a stiffener of 10x10 mm² fabricated with a milling machine, for the first version of hotplate (the bigger one) with a central opening of 6x6 mm², to lodge the sensor die (Figure8(a));
- a stiffener of 2.5x2.5 mm² realized on aluminium cylindrical support with a Kern HSPC micro CNC, which is a 5 axes CNC micro-milling and drilling machine with a tolerance of 1 μm. All the components fabricated are cut with a Sodick AP 200L WEDM, a fine Wire high precision 6 axis Electrical Discharge Machine (Figure 8(b)).

Moreover, for the integration of the ULPHP, the cap for protecting the membrane of the sensor has been realized (Figure 8(c)), by means of the same technology used for the fabrication of the stiffener.

(a)

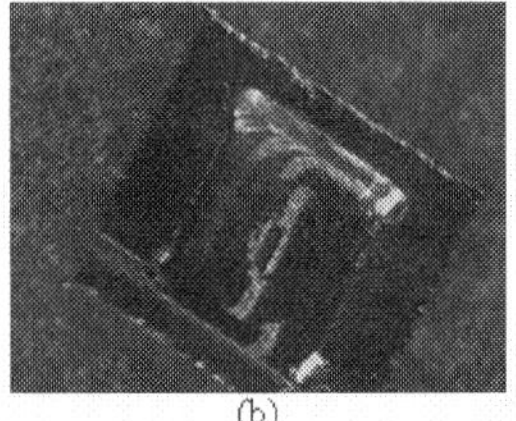
(b)

(c)

Fig. 8. (a) Stiffener 10 x 10 mm²; (b) Stiffener 2.5 x 2.5 mm²; (c) Cap 2.5 x 2.5 mm².

The wire bonding has been realized with a Kuliche&Soffa 4523 digital wire bonder with a 25 μm aluminium wire for both the hot plates versions. As previously described, an important function of the stiffener is the role of rigid support during the bonding step. In fact, the flexible circuit has been designed in order to have the copper pads along the stiffener board (Figure 9(a)). In this way the bonding is most efficient and it permits to obtain more reliable connections, which are very important during the tag manipulation. Regarding the flexible substrate, a high resolution has been necessary for the realization of the bonding pads. In particular, pads of 140 µm x 100 µm spaced by 60 µm (figure 9(b)) are required.

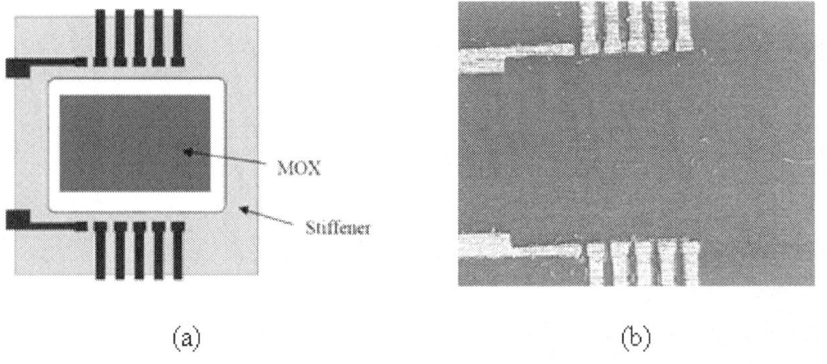

(a)   (b)

Fig. 9. (a) Assembling scheme; (b) Flexible substrate fabricated to prove the viability of COF wire bonding.

Figure 10 shows the first prototype of COF gas sensor integration, realized with the 5 x 5 mm^2 hotplates. The integration has shown good results concerning the electrical connection and the reliability of the whole structure. Regarding the mechanical property, this realisation leads to a quite large rigid portion (100 mm^2) on the flexible substrate, which affects the flexibility of the tag and makes it relatively weighty.

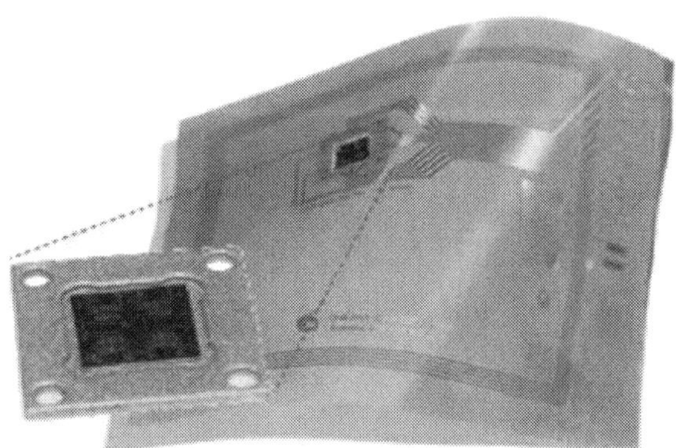

Fig. 10. First prototype of COF gas sensor integration.

The second prototype integrates the ULP array die of 1.5x1.0 mm² by COF wire bonding technique on a circuit realized on Kapton. Figure 11(a) shows the wire bonding between a ULP 4-sensor-array and the copper tracks on the Kapton substrate, and Figure 11(b) shows the protective cap on top of the sensor array. The mechanical bending of a prototype and the stiffener on the back-side of the sensor array can be easily disclosed. In this case the rigid surface is 6.25 mm², 16 times less than the first prototype, and this rigid surface does not affect the flexibility of the tag (Figure 11(c)).

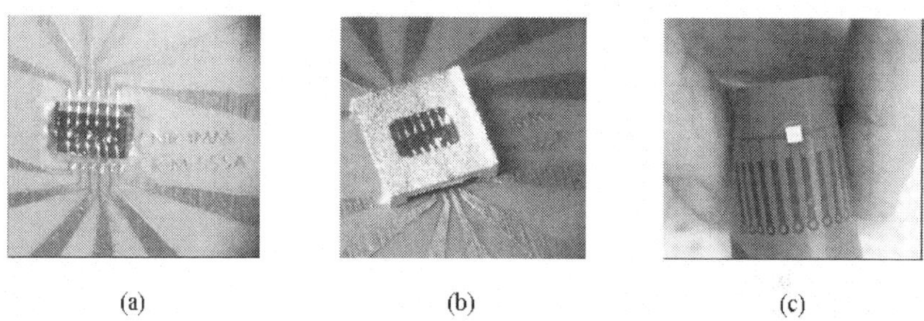

(a)(b)(c)

Fig. 11. Final prototype of ULP gas sensor array die integrated on flex. (a) wire-bonded array on stiffener. (b) array covered by protection cap. (c) effect of the stiffener during bending of the flex tag.

The whole height of the structure is about 0.5 mm without cap and 1 mm with the cap. The reliability of the electric connections was tested, and a 100% yield was found in the first 3 test arrays fabricated. The electric connections did not show any damages, even after a sharp bending of the substrate, with a bending radius of less than 1 cm as shown in Figure 11(c).

Test Array 1			Test Array 2			Test Array 3		
91.89	92.18	91.17	92.34	91.89	91.95	91.45	91.78	91.78

Table 2. Measured heater resistance values in Ohms (Ω).

The ULPHP used for integration with the COF technique were characterized by a heater resistance of 91.5 Ω (average value specified by the supplier). The 3 test arrays encapsulated shown the heater resistance values reported in Table 2. These measures show an average value of 91.83 Ω that confirms how the wire bonding does not introduce a significant resistance series.

3.2 Comparison of both technologies and outlook

In the previous sections, we have presented the two specific strategies developed in this work for MOX sensor assembly on flexible substrates. Using both technologies ULPHP for MOX sensors have been successfully integrated on test substrates.

In the first case, an Anisotropic Conductive Adhesive is used in a flip-chip-like technique. This approach has been proved to be very simple, low cost and mechanically and electrically very reliable. The use of ACAs present several advantages, these adhesives can be deposited over the entire surface, thus facilitating the material application and does generally not require underfilling layers and bumps formation. Furthermore in this process a small window is opened in the flexible substrate by laser ablation, exposing the sensing layer to the air on the opposite side of the tag without using any mechanical part. Being very small, the hole mechanically protects the sensing layer and the suspended membranes. If more protection of the sensing layer were needed, the window could be perforated instead completely ablated.

The second approach is based on the Chip-On-Flex wire bonding technique. In this case, a very reliable mechanical structure is used to protect the MEMS die, based on a metallic "stiffener" on the backside and on a perforated cap on the front-side of the gas sensor. The electric connections are realised by wire-bonds, which is a very mature bonding technology, and the whole gas sensor area, though being only 2 mm small, is in fact a stable zone within the flexible tag, ensuring high mechanical reliability. The COF technique is already widely used in industrial fabrication of mechanically critical flexible circuits, though it was never applied to gas sensor integration before.

Despite the higher complexity of the COF implementation with respect to the ACA flip-chip technique, which does not require caps or stiffeners, the COF technology has one main advantage over ACA: the conductivity of the bonding wires is superior in terms of low series resistance and maximum allowed current compared to the specifications of the ACA materials. Although at the present status the maximum current to be provided to the hotplates is compatible with the ACA specifications, for some applications higher currents may be necessary, and the COF technique could be exploited there.

In summary, the ACA flip chip bonding process represents a cost effective way to manufacture tags with MOX sensors using reel to reel production lines. If highly reliable tags are needed for harsh environments, the robust COF solution is advisable. However, a silicon based solution for the stiffener and the cap instead of the metallic one is envisaged in order to realise a "classical" microsystem. By doing so the integration of a micro reactor (Becker et al. 2000) could be realised as well. Figure 12 shows the cross section of a micro reactor integrated into a flexible tag.

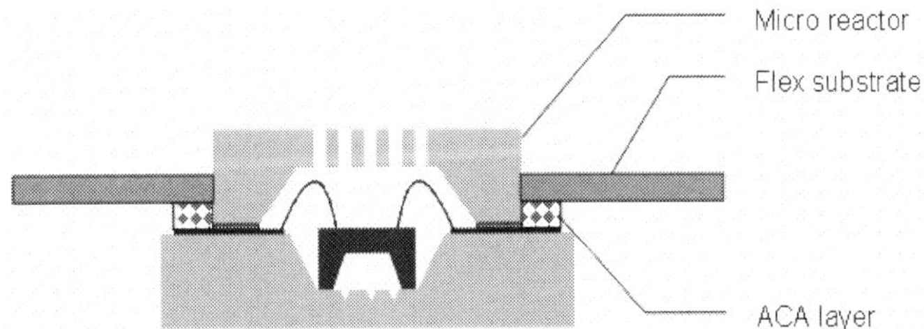

Fig. 12. Proposed all silicon MST solution using a silicon micro reactor chamber embedded into the flexible substrate. The bonding can be done by ACA flip chip bonding.

4. Conclusions

Flexible substrate technologies and assembly of components, in particular of the new ULPHP specifically developed for MOX sensor integration using special bonding technologies have been accomplished with the aim of developing a FTM inlay for food logistic control. A prototype of an ISO 15693 compliant semi-active tag, including low power control electronics, RFID antenna, commercial sensors, memory and a thin film battery, has been presented. The assembly of the ULPHP on flexible substrates using ACA flip chip technology and COF bonding has also been proved. Considering that the HP is the mechanically critical structural component of the gas sensors, these achievements will allow the integration of chemical sensors in the RFID tag that represents the major innovation of this FTM realisation.

5. Ackowledgements

This work was co-financed by the EU Frame Program VI within the IST priority, through the Integrated Project GoodFood (FP6-IST-1-508744-IP). The authors wish to thank CNR-IMM Bologna to have supplied the MOX sensor hotplates. The authors appreciate the technical support with the flip chip station by Mr. Javier Berganzo and Mr. Rafael Salido. The authors also wish to acknowledge 3M for providing ACA samples.

6. References

Abad, E.; Raffa, V.S.; Mazzolai, B.; Marco., S.; Krenkow, A. & Becker, Th. (2005). Development of a flexible tag microlab. *Proceedings of SPIE*, Vol., 5836, 599-606.

Abad, E.; Zampolli, S., Marco, S. Scorzoni, A., Mazzolai, B., Juarros, A., Gómez, D., Elmi, I., Cardinali, G. C., Gómez, J.M., Palacio, F., Cicioni, M., Mondini, A., Becker, T., Sayhan, I. (2007). Flexible tag microlab development: Gas sensors integration in RFID flexible tags for food logistic. *Sensors and Actuators B*, 127, 2-7.

Becker, Th.; Mühlberger, S.; Bosch, V.; Braunmül, C.; Müller, G. & Benecke, W. (2000) Gas mixture analysis using silicon micro-reactor systems. *J. Microelectromech. Sys.*, Vol., 9, 478-484.

Elmi, S. Zampolli, E. Cozzani, F. Mancarella, and G.C. Cardinali, (2008) Development of Ultra Low Power Consumption MOX Sensors with ppb-level VOC Detection Capabilities for Emerging Applications. *Sensors and Actuators:* B. 135 342-351.

Johansson, A.; Janting, J.; Schultz, P.; Hoppe, K.; Hansen, I.N. & Boisen A. (2006) A SU8 cantilever chip interconnection. *J. Micromech. Microeng.*, Vol., 16, 314-319.

Numakura, D. (2001) Flexible circuits. *Coombs' Printed Circuits Handbook*. 5th ed. 1205-1344, McGraw Hill, New York.

Opasjumruskit, K.; Thanthipwan, T.; Sathusen, O.; Sirinamarattana, P.; Gadmanee, P.; Pootarapan, E.; Wongkomet, N.; Thanachayanont, A. & Thamsirianunt, M. (2006) Self-Powered Wireless Temperature Sensors Exploit RFID Technology. *IEEE Pervasive computing*, Vol., 5, 1536-1268.

Pai, R.S. & Walsh, K.M. (2005) The viability of anisotropic conductive film as a flip chip interconnect technology for MEMS devices. *J. Micromech. Microeng.*, Vol., 15, 1131-1139.

Sayhan, I., Helwig, A., Becker, T., Müller, G., Elmi, I., Zampolli, S., Padilla, M., Marco, S. (2008). Discontinuously Operated Metal Oxide Gas Sensors for Flexible Tag Microlab Applications. *IEEE Sensors Journal*, 8, 176-181.

Vergara, A., Llobet, E., Ramírez, J.L., Ivanov, P., Fonseca, L., Zampolli, S., Scorzoni, A., Becker, T., Marco, S., Wöllenstein, S. (2007) An RFID reader with onboard sensing capability for monitoring fruit quality. *Sensors and Actuators: B.,* 127, 143-149.

Zampolli, S., Elmi, I., Cozzani, E., Cardinali, G.C., Scorzoni, A., Cicioni, M., Marco, S., Palacio, F., Gómez-Cama, J.M., Sahyan, I., Becker, T. (2007). Ultra-low-power components for an RFID Tag with physical and chemical sensors. *Microsystem Technologies,* 14, 581-588.

15

Conduct Radio Frequencies with Inks

Rudie Oldenzijl, Gregory Gaitens and Douglass Dixon
Henkel Corporation
United States

1. Introduction

The objective of this chapter is to give additional insight into the inks which are being used for printing RFID antennas; their properties, how to apply them, their performance, benefits and drawbacks and future developments for these inks. In addition, some attention will be given to adhesives, which are necessary to bond the die or die strap to the antenna.

Conductive inks have their place in the HF-RFID market, with proven performance in the field. But, the UHF-RFID market is a relatively new application for inks and, while much testing has been conducted, the technology is still in its infancy.

Using conductive inks for printing RFID antennas has several benefits: printing is an additive process and generates much less waste than certain subtractive processes, investment in the technology is minimal (especially for companies that are already active in the printing industry), and inks are relatively environmentally friendly. Certain inks do contain solvents, but the quantities are limited and, if one chooses not to use solvent-containing inks, there are water-based and UV-curable inks available, which are environmentally friendly.

One may think that the RF-performance of an ink is inferior to that of aluminum or copper, but experiments have proven that the opposite is true. Ink performance in RFID applications has been studied extensively and the general consensus is that inks perform as well as copper or aluminum.

2. Theoretical background

2.1 Skin depth

To evaluate the use of various materials for RFID applications, it is important to understand the basic function and historical background of RFID technology. RF radiation generates an AC current in conductive antennas. These AC currents tend to flow on the outside of an antenna and this is called the "skin effect", see Figure 1. This skin effect increases at higher frequencies, in this case the radiation goes even more to the outside and the skin depth decreases. The skin depth is the thickness of the outer layer of the material, which transfers 63 % of the RF signal. This skin depth is determined by the permittivity and the conductivity of the antenna material. Since most antenna materials have a permittivity of 1, the conductivity is the most important variable.

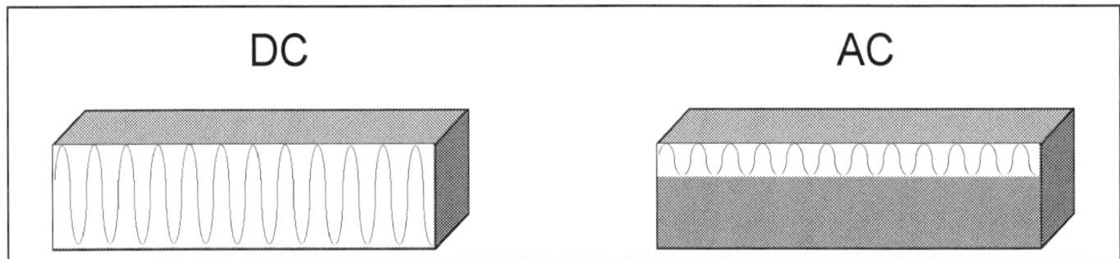

Fig. 1. Skin effect with AC-currents

A more conductive material will have a thinner skin depth. In other words, the radio waves are transferred by less material. For a less conductive material, the skin depth is thicker. Ideally, for an antenna to work properly, 100 % of the radiation should be transferred. As a rule of thumb, the antenna should be around three to five times the thickness of the skin depth so that it can transport all the RF-radiation.

A comparison of the RF-performance between different antenna materials was done in (Syed et al., 2007) and (Nikitin et al., 2005). Bulk copper (18 μm), thin deposited copper (2 μm), bulk aluminum (10 μm) and two conductive inks (5 and 8 μm) were compared with each other. Despite the fact that the conductivity of the silver ink was roughly ten times less conductive than the bulk copper, the performance of the inks was consistently   80-99 % of the performance of the bulk copper antenna. Since conductive inks are less conductive than bulk copper or bulk aluminum, the skin depth is also larger. As mentioned in (Syed et al., 2007) the skin depth for the ink, at 915 MHz, is in the 7-8 micron range, while bulk copper has a skin depth of 2.18 micron. By printing a thicker layer of the ink, this difference can be overcome. Results in (Nikitin et al. 2005) indicate that, for certain antenna designs, a silver ink antenna performs as well as a copper antenna for HF-RFID. For lower frequencies, the skin depth is higher and thicker layers must be printed.

2.2 Electrically conductive inks

Electrically conductive inks are used in a wide variety of applications. The conductive tracks in printed circuit boards are a mature and well-known application. Other examples are membrane touch switches, heating elements, sensors and displays. RFID and medical applications are more recent and emerging, while photovoltaics and printed transistors present future opportunities for electrically conductive ink applications.

Simply put, a conductive ink consists of a polymer, a solvent, a pigment and additives. The conductivity or resistance values of inks are normally given in Ohm/square/25μm. A high conductive ink has a low resistance.

The polymer or binder will keep the pigment particles together and determine the mechanical properties – such as adhesion, flexibility and hardness of the final ink. To make the ink processable, the binder is dissolved. Application technique determines which solvent is used. For a typical application method such as screen printing, high boiling point solvents are used. High-speed printing methods like rotogravure, require low boiling point solvents.

Most important is the pigment, as it determines the conductivity of the final ink. Most of today's materials employ silver as the pigment. Usually the silvers are flake- type materials, as these tend to provide the highest conductivity (Figure 2). Characteristics like the surface area, particle size and the type of lubricant have great influence on the final conductivity of the dried ink.

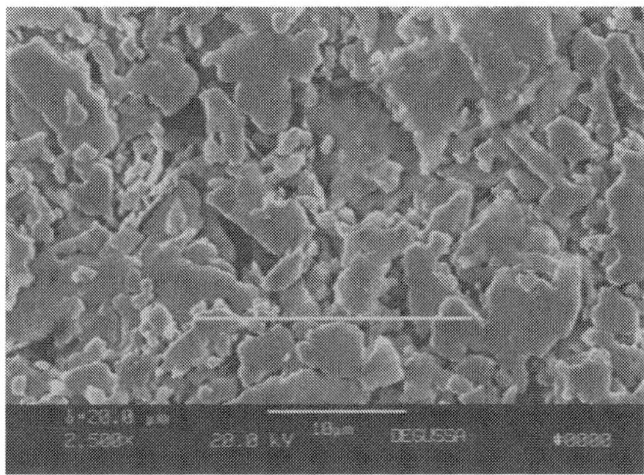

Fig. 2. An example of silver flake (source: Website *www.Ferro.com*)

In order to optimize the properties of the ink, several additives can be used in a formulation. These can improve wetting, flow, adhesion, flexibility or the rheology of the ink. Some additives can be used to promote the conductivity.

3. Application methods

Application methods are of key consideration for RFID inks and it is, therefore, important to understand the primary printing techniques, which include screen printing, flexography and rotogravure. Screen printing is very well-known application method for conductive inks . Flexography and rotogravure are relatively new for RFID, but very established in the packaging industry.

Table 1 below illustrates some typical values of the different printing techniques

	Flexography	Rotogravure	Screen printing
Cost	Relatively inexpensive	Requires long runs to be cost effective	Relatively Inexpensive
Speed	15->150 mpm	30->400 mpm	3-60 mpm
Print quality	Good	Excellent	Excellent
Print thickness	2-5 µm	2-5 µm	5-25 µm
Boiling point range solvents	80-140°C	80-140°C	130-210°C

Table 1. Comparison printing techniques

The printing technology selected will have a large impact on the process speed for RFID tags.

3.1 Screen printing

Screen printing is a porous printing process. Distinct from flexography and rotogravure, the image and the non-image are in the same plane. A mesh stencil carrier -- usually nylon, polyester or stainless steel -- is stretched tightly over a frame. An image is created by covering the non-image parts by an adhesive foil. Another common method is covering the mesh with a photopolymer and subsequently exposing the mesh to a photographic positive and polymerizing the non-image parts. After removing the un-polymerized parts, an image remains. During printing, the frame is loaded with ink and a squeegee presses the ink through the image parts of stencil and onto the substrate, see figure 3.

Screen printing is used for displays, signs, instrument panels, textiles and posters. It is a relatively slow printing process, by comparison. A totally automatic flatbed screen printing machine has an output speed around 1500 impressions per hour.

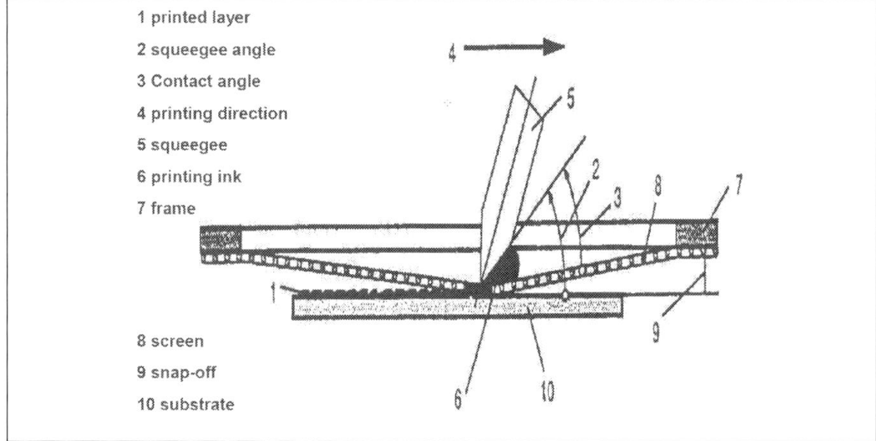

Fig. 3. Schematic figure of the screen printing process

A modification on screen printing is rotary screen printing, where an ink supply and the squeegee are fitted inside a rotary screen cylinder. This process is a competitor for flexography and gravure in both output and costs.

3.2 Flexography

The original name of flexographic printing was aniline printing, but the name was changed in 1952 for the following reasons:
- Flexible and elastomeric printing plates were used
- It was the most widely used technique for printing flexible materials
- The process is flexible and enables printing of flexible, non-flexible, rigid materials
- The process may be used for applications other than printing (coating, sizing and laminating, for example)

Flexography is used in a wide variety of printing in packaging, publication, business, and consumer products. Flexographic printing usually starts with a fountain roller which takes ink from a container. Subsequently, ink is transferred to the anilox roll. In this step, the transferred amount is controlled by the pressure of the doctor blades. The anilox roller delivers the ink to the image plate, usually a soft rubber or a photopolymer, which transfers it to the substrate (figure 4).

The anilox roller is the most important part of the flexographic printing system. It is engraved with tiny, uniform cells and transfers the ink with a precise thickness to the printing plate. Cell volume and the number of cells per inch are important features of the roller (figure 5). Figure 6 illustrates typical data of an anilox roller.

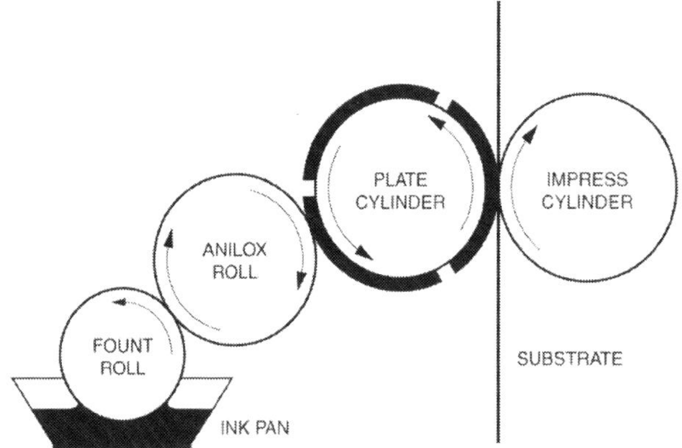

Fig. 4. Schematic representation of a flexographic press

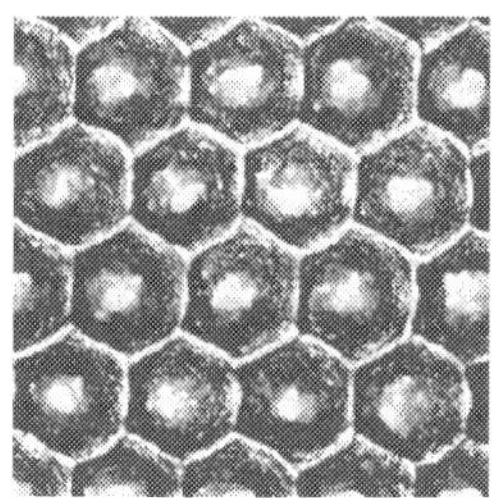

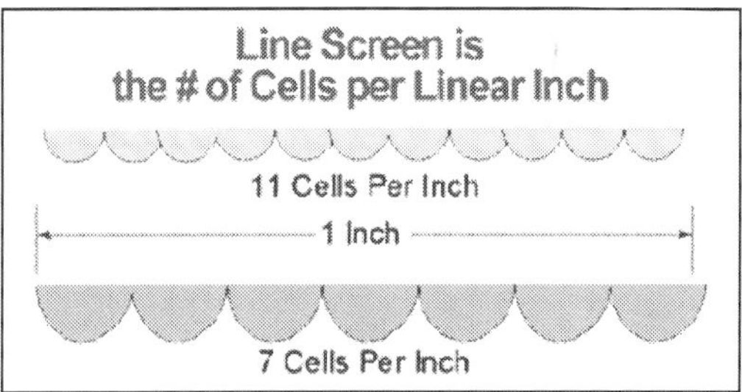

Fig. 5. Typical anilox shape and schematic of lines/inch (Figure Courtesy of Harper Corporation of America)

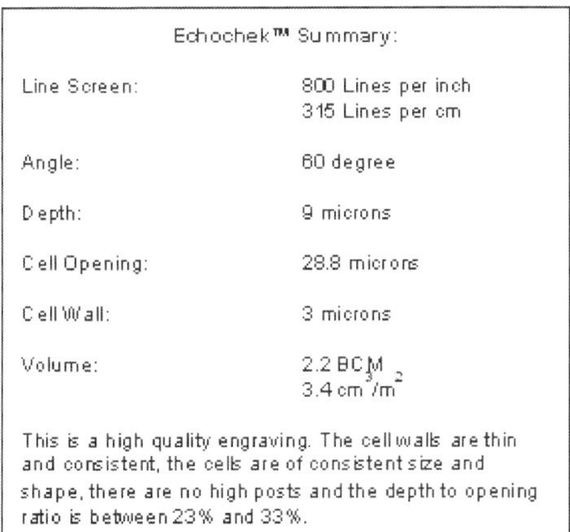

Echochek™ Summary:

Line Screen:	800 Lines per inch
	315 Lines per cm
Angle:	60 degree
Depth:	9 microns
Cell Opening:	28.8 microns
Cell Wall:	3 microns
Volume:	2.2 BCM
	3.4 cm³/m²

This is a high quality engraving. The cell walls are thin and consistent, the cells are of consistent size and shape, there are no high posts and the depth to opening ratio is between 23% and 33%.

Fig. 6. Typical parameters anilox roller (Figure Courtesy of Harper Corporation of America)

In order to maintain good printing properties, most flexographic presses are equipped with an automatic viscosity control system. When necessary, the viscosity can be adjusted.

3.3 Rotogravure printing

Rotogravure printing is a form of intaglio printing (figure 7). The print is recessed in a metal plate or cylinder in the form of discrete cells. These printing cylinders are usually made of copper which are chromium plated for protection and comparable to the anilox roller used in flexography. During rotating of the cylinder, ink is transferred from a container to the cells of the printing plate and subsequently applied to the substrate. A doctor blade removes the excess ink. The amount of ink transferred depends on the depth of the cells.

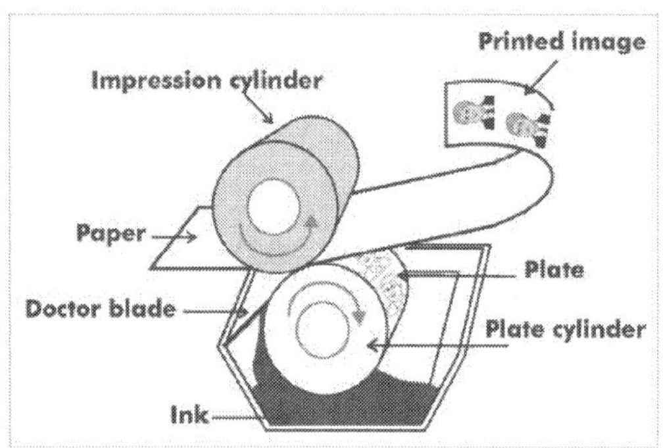

Fig. 7. Schematic representation of a rotogravure printer

Gravure printing is used for publication and package printing and delivers high-quality output at high speeds. Publication printing is undertaken on very expensive, custom made, presses. The cylinders usually contain eight A4 pages and print approximately 35000 impressions per hour.

Package printing by gravure is managed on standard-sized machines, but with a wide variety of cylinder widths to suit the product. Speeds of 300 meters per minute are normal. The cylinders are stored for reprinting. This requires additional space and increases the production costs.

4. Rheology (Rheology and drying of RIFD ink)

4.1 Rheology

To make the RFID antenna, inks are applied by one of the different techniques noted previously. In order to print effectively, the inks require a certain rheology. Rheology characterizes the deformation and flow of a material. An ink must have the proper deformation and flow properties to be applicable by certain printing techniques.

Screen-printable inks are usually high viscous pastes. They must stay on the screen, but once a force is applied, they should thin down and flow through the screen. Once pressed through the screen, the materials cannot flow out, but must remain in place in order to achieve line definition (Hoornstra et al., 1997).

Figure 8 illustrates the different stages during the screen printing process and some typical viscosity values of screen inks during these stages.

The same is done for the flexo and gravure processes in (Geiger & Henderson, 2001). This is schematically shown in figure 9.

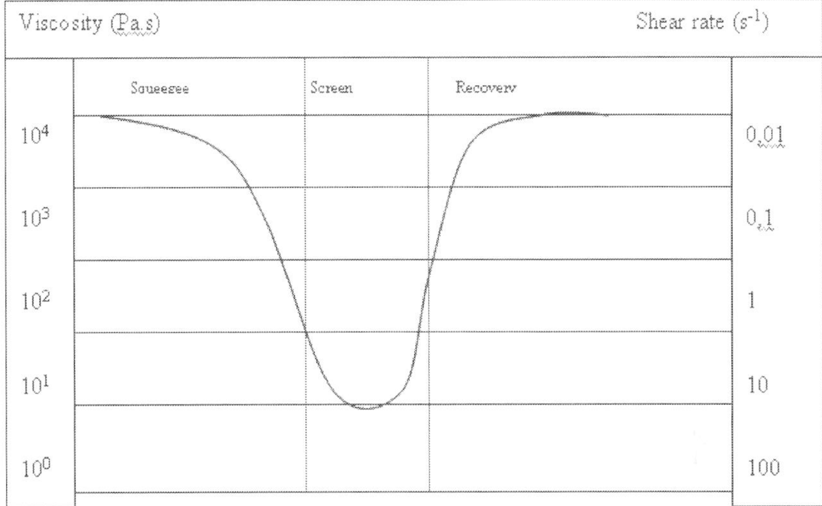

Fig. 8. Typical shear rates and viscosities during the screen print process

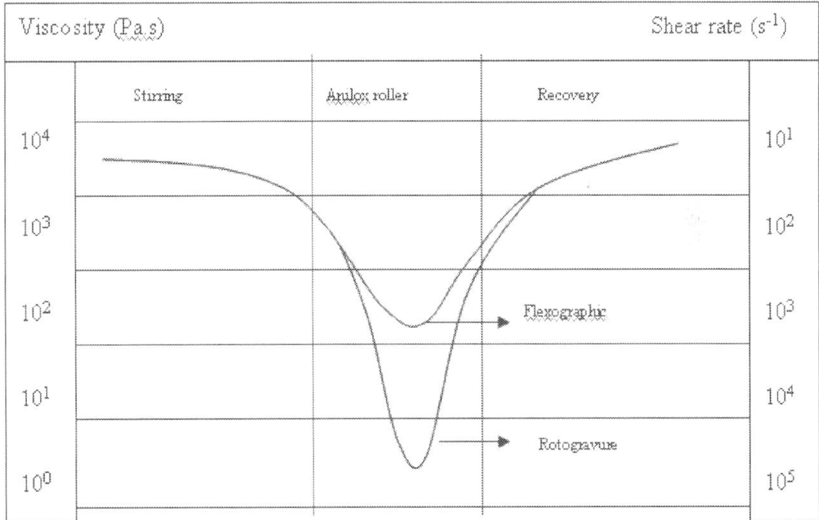

Fig. 9. Typical shear rates and viscosities during flexo and gravure process

There are differences in flexo printable graphical inks and conductive inks. A closer look at a graphical print shows the typical "dot" structure, as seen in figure 10. Usually relatively low volume anilox types are used for this application and, in this case, all dots are not connected, but different colors are printed on top of each other to yield the desired color. If the same anilox volumes were used for conductive inks, the printed structure would most likely not be conductive. For conductive inks, higher cell volumes are used to create a conductive trace. Figure 11 shows the influence of the cell volume on the trace morphology.

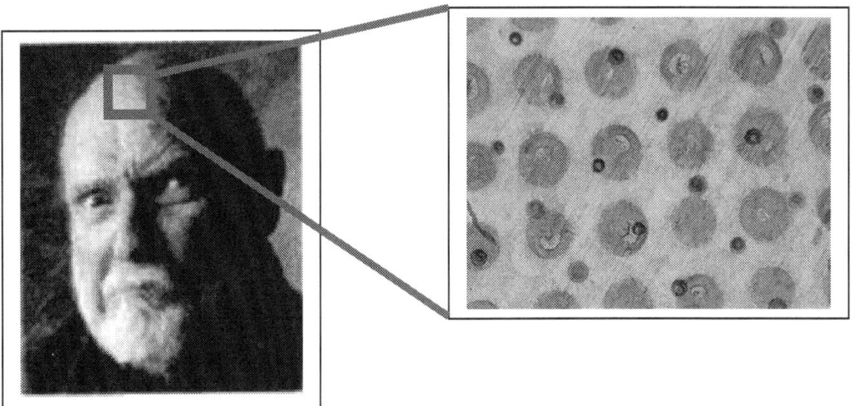

Fig. 10. A flexo printed bold face and its magnification.

The figure below shows the influence of the cell volume of a flexo printed conductive ink.

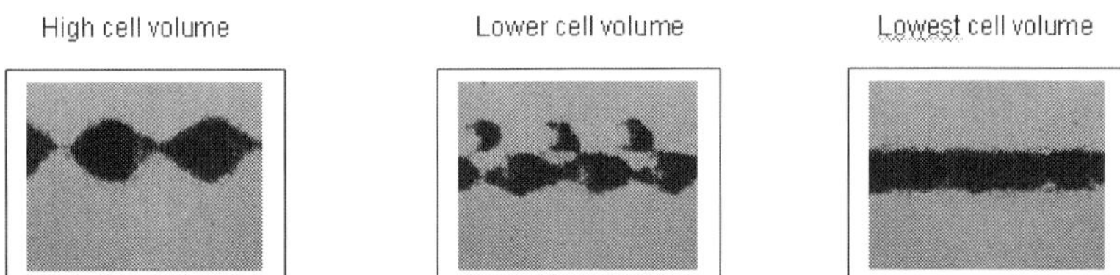

Fig. 11. Influence cell volume

Drying

Another critical step in the RFID antenna print process is the drying of the ink, as it determines the final properties of the antenna. When drying is not sufficient, the conductivity will be too low and some of the mechanical properties may fail.

Drying can be achieved with conveyer ovens, box ovens, hot air, infrared heaters or contact driers. Of great importance are the printing speed and the applied layer thickness, as they have a significant influence on the amount of heat required to completely dry the ink.

5. Typical properties and RF-performance

The properties of RFID inks can be divided into wet and dry properties and, of course, the RF-performance. The key parameter is the conductivity or resistance of the dried ink. After the ink is dried, conductivity is normally the first characteristic that is examined. Other important properties are the adhesion/cohesion and flexibility of the inks.

To determine the conductivity or the resistance of the dried ink, the resistance over a track of known dimensions is measured. To compensate for differences in layer thickness, the value is normalized to a thickness of 25 μm. Figure 12 shows how the resistance value is determined.

$$Rp(\Omega/sq./\,\mu m) = \frac{Rt \times t \times W}{25 \times L} \quad \text{or} \quad Rt = \frac{Rp \times L \times c}{25 \times W}$$

Rp = product resistance (ohm/sq./25μm)

Rt = track resistance (ohm)

t   = dry coating thickness (μm)

W = track width (mm)

L   = track length (mm)

Fig. 12. Resistance calculation

Because of the roughness of certain printed inks, it is sometimes difficult to determine the exact layer thickness. When this is the case, the resistance is determined in Ohm/square.

Another important property is the adhesion of the ink to various substrates and there are several ways to measure adhesion. One of the most common methods is ASTM-3359B, as illustrated in figure 12. Using a special device, squares are cut in the ink following which a special tape is put on the ink and removed. The removed quantity of ink determines the measure for its adhesion, rated as in figure 13.

Rating	Appearance of the crosscut after adhesion test
5B	no removal
4B	
3B	
2B	
1B	
0B	More than 65% removal

❖ Rating

➤ 5B: no removal

➤ 4B: <5 % removed

➤ 3B: 5-15 % removed

➤ 2B: 15-35 % removed

➤ 1B: 35-65 % removed

➤ 0B: >65 % removed

Fig. 13. Adhesion test ASTM-3359B

RFID tags can be used in a variety of environments and, to ensure the antenna maintains its performance, the resistance should remain consistent regardless of varying environmental conditions. Using a chamber which is 85°C and has a relative humidity of 85%, harsh conditions can be simulated. The resistance of a defined track of dried ink is measured and put in the 85/85 chamber. After a set period of time, the percentage of change in resistance is measured to determine the environmental stability of the ink.

Table 2 illustrates s some typical properties of inks used for RFID applications.

Type of ink		Solvent	Solvent	Water	Solvent
RFID Segment		HF	UHF	UHF	Plating*
Application		Screen	Flexo	Flexo	Gravure
Wet properties	unit				
Solids	%	68	60	83	72
Density	kg/m³	2390	1920	3200	2340
Viscosity	mPa.s	17500	2000	4000	4400
Storage Temp	°C	<25°C	<35°C	<35°C	<35°C
Shelf life	Months	12	12	12	12
Dry properties					
Sheet resistance	Ohm/sq/mil	< 0.010	<0.015	<0.035	<0.010
Adhesion		5B	5B	5B	5B
Cohesion		No failure	No failure	No failure	No failure
Stability 85/85		No change	No change	No change	No change
Reading distance	m	<0.10	~ 3	~ 3	NA
*The use of a plating ink is a new and cost effective technique to make					
antennas. The next paragraph will briefly discuss this topic					

Table 2. Some typical wet and dry ink properties

Ultimately, the success of an RFID antenna is determined by its reading performance which, of course, depends on more than only the ink.

Table 3 provides some reading distances of a UHF ink. The printing, drying and die bonding, were conducted   under the same circumstances. In this instance,   although there is a wide spread in conductivity caused by the substrate, there is no clear influence on the reading performance. This indicates that conductivity is important, but not the only factor that influences the reading performance.

		speed	anilox	anilox	Shorted R	read distance
Ink	Substrate	m/min	ml/m2	L/cm	(Ohm)	m
17 μm Cu	PET				0.4	3.1
UHF ink	M-Cote	37	18.5	55	31.1	2.6
UHF ink	Fasson PP	37	18.5	55	25.9	2.5
UHF ink	Transfer matt	37	18.5	55	21	2.6
UHF ink	Cast-Cote	37	18.5	55	17.6	2.6
UHF ink	Algro Finess 2000	35	18.5	55	15.7	2.7

Table 3. Reading distance of a dried UHF ink, printed on different substrates and compared to 17 μm copper.

6. The plating approach

A relatively new technology is the production of an HF-antenna by using copper plating technology. Figure 14 illustrates the plating process steps.

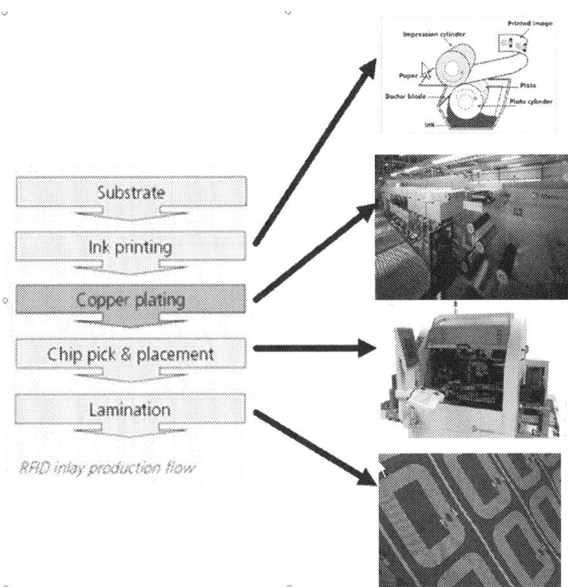

Fig. 14. Different steps of the plating process

A very thin plating or receptor ink is applied first. This can be achieved with different printing techniques, but rotogravure is probably the most cost-effective and efficient method. After drying/curing the ink, copper is electrochemically deposited onto the plating ink.

To be able to plate the ink with copper, certain conductivity is required. Since the printed layer is very thin -- only 2-4 μm -- it is critical that the plating ink is rather conductive. The ink should dry quickly and have high peel strength to ensure good antenna to substrate adhesion.

7. Conductive adhesives

Although adhesives are different from inks, the basic elements are similar. Conductive adhesives usually perform very well in combination with inks, therefore they are briefly discussed here.

After the antenna is printed, the die or die-strap (figure 15) is attached. For die strap attach, Isotropic Conductive Pastes (ICPs), which are conductive in the X, Y, and Z direction, are used. For direct die attach, Anisotropic Conductive Pastes (ACPs) can be used and, for this application, only conductivity in the Z-direction in needed. In some cases, the bare die contains spikes which will connect with the antenna material. A non-conductive adhesive paste (NCP) can be applied to keep the die in place.

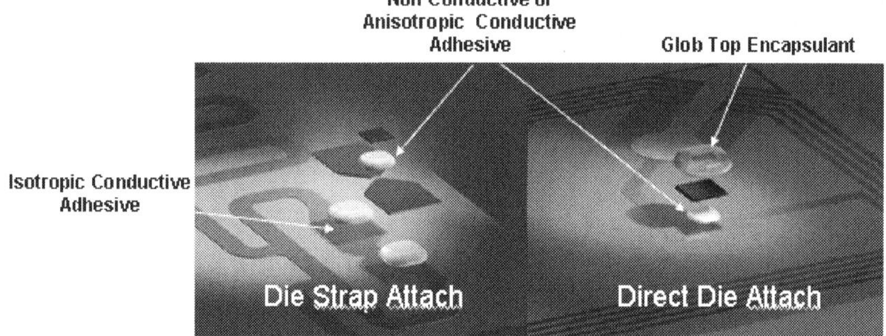

Fig. 15. Die strap and direct die attach

Common application methods of die attachment are jet dispensing and screen printing. The curing of the adhesives should be quick, and process speed is dictated by the printing method used.

Some other key properties are shown in table 4.

Property	Test Method	Positive Control	Negative Control
Cure <10s at <130°C	hot plate	Cure <10 s	Cure >10s
Initial contact resistance	DC resistance measured on shorted vd Cu strap bonded to stamped Al trace on PET	Initial contact resistance <1 Ohm across strap	Initial contact resistance >1 Ohm across strap
Contact resistance stability after 2 weeks 85'C/85% RH	DC resistance measured on shorted vd Cu strap bonded to stamped Al trace on PET	Contact resistance <1 Ohm across strap after conditioning	Contact resistance >1 Ohm across strap after conditioning
Jettability	Asymtek jet dispenser - dots/hour	>40,000 dots/hour with no skipped dots	<10,000 dots/hour with skipped dots
Good mechanical strength	Bend test on ¼" diameter mandrel; 20 bends face up, 20 face down	<15% change in contact resistance from initial	>15% change in contact resistance from initial
Good work life	Viscosity change at room temperatue	<25% change from initial	>25% change from initial

Table 4. General RFID adhesive properties

Because antennae can be made of silver, aluminum and copper, there are also different adhesives available.

8. New developments

Future developments in RFID ink technology are focused on cost efficiency, the environment and improving conductivity.

To address the needs of the low-cost RFID market, manufacturing inks using less silver would be extremely beneficial. Many years ago the silver price was rather steady, but in recent years, silver prices have had large fluctuations, which have had a tremendous impact on ink pricing. Reducing the silver content would help stabilize pricing.

Environmental responsibility also plays an important role in materials development. Reducing the halogen content as well as the amount of solvent used are efforts that are underway to ensure the environmental friendliness of modern RFID materials.

However, many of these environmentally friendly systems generally exhibit lower conductivity values than conventional systems. New and non traditional pigments are needed to accommodate these issues.

Silver-coated copper – a method by which copper base pigments are covered with a very thin layer of silver -- is currently one of the most studied materials. Although silver-coated coppers are not new to the market, those previously developed have not been suitable for printing inks.

Recently, however, more suppliers of silver-coated copper have entered the market, resulting in rapid improvements to the properties of these materials.

Water-based systems are available for UHF RFID and environmentally appealing because they do not contain harmful solvents, yet provide good conductivity at room temperature.

Several experiments have been done with water-based systems on a wide variety of substrates, and results suggest that these systems work well with numerous substrates. One drawback to water-based inks is that they do tend to dry quickly under certain conditions. If the ink dries on the flexo plate, it may prevent subsequent printing. However, good viscosity control and an enclosed doctor blade system may eliminate these issues.

With current materials, there remains a gap between the conductivity values of water-based inks and conventional, solvent-based inks. Water-based systems reach conductivities in the 20 mOhm/sq/mil range, whereas solvent-based systems can reach conductivities of less than 10 mOhm/sq/mil. New developments are aiming to narrow this gap, while also achieving more cost-effective pigments.

UV-curable systems provide many benefits, not the least of which is their user familiarity, friendliness for RFID applications. Flexo printers are very amenable to this technology, as the ink will not dry on the flexo plate and can be left exposed to air for some time without any adverse effects. . Because there are little to no solvents in these systems,   coverage is very high and environmental emissions are limited. Table 5 gives illustrates results of an experimental UV-curable UHF ink as compared to a water-based UHF ink; both deliver the same RF-performance.

UV-curable conductive inks have high silver loading to create conductivity. Because of this, a relatively high amount of UV-energy is needed to fully cure the material. This can be mitigated, however, by an additional heat exposure, either through hot air or infrared, will overcome some of this and promote curing to yield better conductivity development.

Like water-based conductive technologies, the conductivity of UV-curable inks has not yet reached the level of conventional solvent-based systems. Efforts are ongoing to increase these conductivity levels, while also improving cost-effectiveness of these systems.

Ink	Substrate	Shorted R (Ohm)	Heat treat (Ohm)	Read distance m
17 µm Cu	PET	0.4		3.1
water	Algro Finess 2000	25.8	15.7	2.7
water	Algro Finess 2000	16.2	9.7	2.7
UV	Algro Finess 2000	9.5	4.2	2.7
UV	Algro Finess 2000	19.2	10.5	2.6

Table 5. Read distance water-based UHF ink versus UV-curable UHF ink

9. Conclusion

Conductive inks are very relevant for certain applications within the RFID market, with performance being consistent with that of aluminum or copper antennas. Printing and drying of the ink are key to manufacturing a functional antenna. RFID can be very cost-effective for existing print houses, as no additional investment is required for market entry. Printing is an additive process, which means less waste, and screens and flexo plates are relatively inexpensive, which enables frequent switching to other designs.
However, there are drawbacks to using inks, not the least of which is cost. Development of lower-cost inks is underway, and successful formulation would certainly encourage more widespread use in the RFID market.

10. References

Syed, A.; Demarest, K. & Deavours, D.D.(2007). Effects of Antenna Material on the Performance of UHF RFID Tags. IEEE RFID, March 26 – 28, Grapevine, TX, p 57-62 Performance of UHF RFID Tags

Nikitin, P.V.; Lam, S. & Rao, K.V.S. (2005). Low Cost Silver Ink RFID Tag Antennas. IEEE Antennas and Propagation Symposium, Washington.

Hoornstra J.; Weeber A. W.; de Moor H. H. C.& Sinke W. C. (1997). *The importance of paste rheology in improving fine line, thick film screen printing of front side metallization, 14th EPSEC*, Barcelona, 823-826

Geiger R. J.& Henderson C. A. (2001) *Radtech report*, 15-19, July/August 2001

Inkjet Printing and Alternative Sintering of Narrow Conductive Tracks on Flexible Substrates for Plastic Electronic Applications

Jolke Perelaer and Ulrich S. Schubert
Friedrich-Schiller-University Jena
Germany

1. Introduction

Although inkjet printers are widely used for graphical applications, it was only within the last decades that inkjet printing has grown to a mature patterning technique. As a consequence, it has gained specific attention in scientific research because of its high precision and its additive nature: only the necessary amount of functional material is dispensed (Tekin *et al.*, 2008). Furthermore, the absence of physical contact between print head and substrate allows many potential applications, such as inkjet printing of labels onto rough curved surfaces, or surfaces that are sensitive to pressure. Inkjet printing is utilised to dose many different kinds of materials, such as conductive polymers and nanoparticles (Osch *et al.*, 2008; Perelaer *et al.*, 2008ab), sol-gel materials (Berg *et al.*, 2007), cells (Derby, 2008), structural polymers (Gans & Schubert, 2004), ceramics (Reis *et al.*, 2005) and even molten metals (Attinger *et al.*, 2000).

Inkjet printing, and in particular drop-on-demand inkjet printing, can compete with lithography techniques, since it places material on demand and in a direct way, which reduces the number of processing steps and the amount of material required, thereby also reducing time, space and waste consumption within production. Furthermore, inkjet printing can also be combined with roll-to-roll production (Forrest, 2004). Typical applications can be seen in the field of plastic electronic devices, which are microelectronic devices that are prepared on flexible polymer substrates, including radio frequency identification tags or electrodes for thin-film transistor circuits.

Similarly, over the few last years, there has been a growing interest in the inkjet printing of conductive materials. One of the compound types that has been frequently used is the poly(3,4-ethylenedioxythiophene):poly(4-styrenesulfonate) (PEDOT:PSS), due to its relatively low costs. However, this polymer does not have a high conductivity (Yoshioka & Jabbour, 2006). Besides conductive polymers, inks that contain metals have been used to create microstructures on polymer substrates (Smith *et al.*, 2006; Kim *et al.*, 2007). It has been shown that inkjet printing of conductive materials is a relatively cheap alternative for the fabrication of electronic devices when compared to other micro- and nanopatterning techniques (Menard *et al.*, 2007), such as photo-lithography (Liu *et al.*, 2005) or laser patterning (Cuk *et al.*, 2000).

Although metals like copper (Hong & Wagner, 2000) and gold (Molesa *et al.*, 2003) have been used for inkjet printing applications, direct inkjet printing of conductive silver tracks onto flexible substrates has gained interest due to silver having the lowest resistivity value and the relatively simple synthesis of silver nanoparticles (Schmid, 2004). Therefore, it has been used for many applications, such as interconnections for a circuitry on a printed circuit board (Szczech *et al.*, 2002), disposable displays and radio frequency identification (RFID) tags (Huang *et al.*, 2004; Potyrailo *et al.*, 2009), organic thin-film transistors (Kim *et al.*, 2007; Gamerith *et al.*, 2007), and electrochromic devices (Shim *et al.*, 2008).

This chapter will describe how inkjet printing techniques can be used for the fabrication of conductive tracks on a polymer substrate. The selective sintering of inkjet printed silver nanoparticles is described by using microwave radiation. This not only sinters the particles into a conductive feature, but it also reduces the sintering time significantly from hours to minutes or even seconds. Furthermore, techniques to improve the printing resolution will be discussed and the fabrication of conductive tracks of 40 μm wide will described.

Before going in detail on inkjet printing of advanced nanoparticle inks, we first review the history of inkjet printing.

2. Historical overview of inkjet printing

The origin of inkjet printing goes back to the eighteenth century when Jean-Antoine Nollet published his experiments on the effect of static electricity on a stream of droplets in 1749 (Nollet & Watson, 1749). Almost a century later, in 1833, Felix Savart discovered the basics for the technique used in modern inkjet printers: an acoustic energy can break up a laminar flow-jet into a train of droplets (Savart, 1833). It was, however, only in 1858 that the first practical inkjet device was invented by William Thomson, later known as Lord Kelvin (Thomson, 1867). This machine was called the *Siphon recorder* and was used for automatic recordings of telegraph messages.

The Belgian physicist Joseph Plateau and the English physicist Lord Rayleigh studied the break-up of liquid streams and are, therefore, seen as the founders of modern inkjet printing technology. The break-up of a liquid jet takes place because the surface energy of a liquid sphere is smaller than that of a cylinder, while having the same volume – see Figure 1 (Goedde & Yuen, 1970).

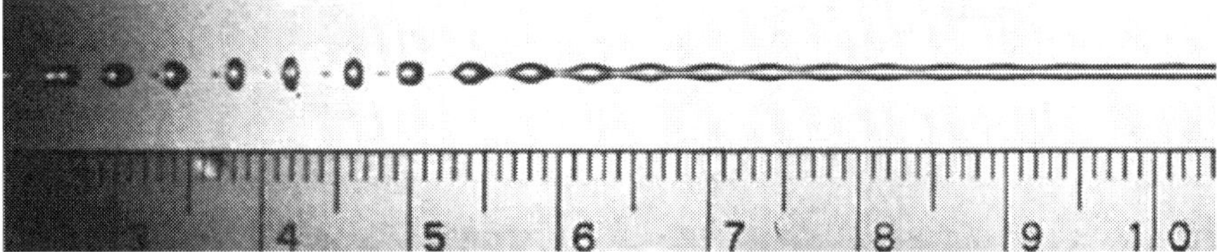

Fig. 1. Break-up of a laminar flow-jet into a train of droplets, because of Rayleigh-Plateau instability (cm scale). Reprinted from (Goedde & Yuen, 1970).

When applying an acoustic energy, the frequency of the mechanical vibrations is approximately equal to the spontaneous drop-formation rate. Subsequently, the drop-formation process is synchronised by the forced mechanical vibration and therefore produces ink drops of uniform mass. Lord Rayleigh calculated a characteristic wavelength λ for a fluid stream and jet orifice diameter d given by (Rayleigh, 1878):

$$\lambda = 4.443d \tag{1}$$

The numerical value was later slightly corrected to 4.508 (Bogy, 1979). However, it took another 50 years before the first design of a continuous inkjet printer, based on Rayleigh's findings, was filed as a patent by Rune Elmqvist (Elmqvist, 1951). He developed the first inkjet electrocardiogram printer that was marketed under the name *Mingograf* by Elema-Schönander in Sweden and *Oscillomink* by Siemens in Germany (Kamphoefner, 1972).

In the beginning of the 1960s, two continuous inkjet (CIJ) systems were developed simultaneously, with a difference only in function of the electrical driving signals (Keeling, 1981). The first system was developed by Richard Sweet at Stanford University. He made a high frequency oscillograph, where droplets were formed at a rate of 100 kHz and controlled with respect to their direction by the electrical signal (Sweet, 1965). Later, in 1968, the A. B. Dick Company elaborated upon Sweet's invention to produce a device that was used for character printing and named it the *Videojet 9600*: this was the first commercial continuous inkjet printer. In parallel at the Lund Institute of Technology in Sweden, Hertz *et al.* had developed a similar system where an electrical signal was used to disperse the droplets into a mist, which enables frequencies up to 500 kHz (Hertz & Simonsson, 1969). However, since their technique used a narrower nozzle diameter, 10 μm *versus* 50 μm, the chance of nozzle clogging was greater (Heinzl & Hertz, 1985).

Instead of firing droplets in a continuous method, it is also possible to produce droplets when required, hence an impulse jet, or better known as drop-on-demand (DoD). In the late 1940s, Clarence Hansell invented the DoD device, at the Radio Corporation of America (Hansell, 1950). Figure 2 shows the schematics of his invention, which was never developed into a commercial product at that time. It took until 1971 when the Casio Company released the model *500 Typuter*, which was an electrostatic pull DoD device.

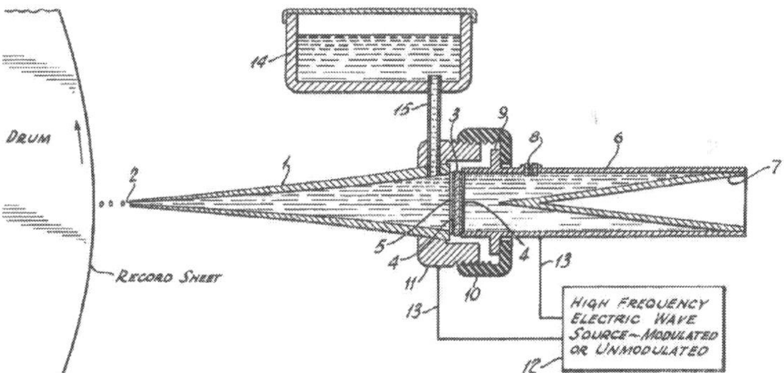

Fig. 2. Schematic drawing of the first drop-on-demand piezoelectric device. Reprinted from (Hansell, 1950).

Despite the fact that the basis of thermal inkjet (TIJ) DoD devices in the form of the sudden stream printer had already been developed in 1965 at the Sperry Rand Company (Naiman, 1965), this idea was picked up much later by the Canon company, when in 1979 they filed the patent for the first thermal inkjet printhead (Endo *et al.*, 1979). Simultaneously, Hewlett-Packard independently developed a similar technology that was first filed in 1981 (Vaught *et al.*, 1984). Thermal inkjet printers are actuated by a water vapour bubble, hence their name bubble jet. The bubble is created by a thermal transducer that heats the ink above its boiling point and, thereby, causes a local expansion of the ink, resulting in droplet formation. The location of the thermal transducer can be either at the top of the reservoir – as used by HP – or at its side, which is the technique Canon uses.

At the beginning of the 1970s the piezoelectric inkjet (PIJ) DoD system was developed (Carnahan & Hou, 1970). At the Philips laboratories in Hamburg printers operating on the DoD principle were the subject of investigation for several years (Döring, 1982). In 1981 the *P2131* printhead was developed for the Philips *P2000T* microcomputer, which had a *Z80* microprocessor running at 2.5 MHz. Later the inkjet activities of Philips in Hamburg were continued under the spin-off company Microdrop (nowadays Microdrop Technologies, www.microdrop.com). The first piezoelectric DoD printer on the market was the serial character printer Siemens *PT80* in 1977.

Four different modes for droplet generation by means of a piezoelectric device were developed in the 1970s, which are summarised in Figure 3, and further explained below (Brünahl & Grishin, 2002).

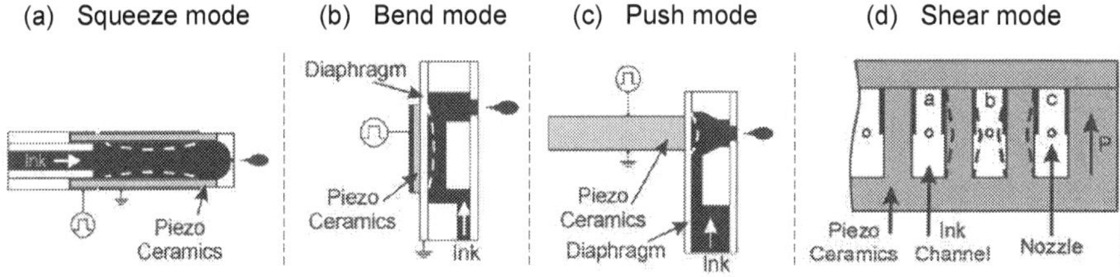

Fig. 3. Different piezoelectric drop-on-demand technologies. Reprinted from (Brünahl, 2002).

Firstly, the squeeze method, invented by Steven Zoltan (Zoltan, 1972), uses a hollow tube of piezoelectric material, that squeezes the ink chamber upon an applied voltage (Figure 3a). Secondly, the bend-mode (Figure 3b) uses the bending of a wall of the ink chamber as method for droplet ejection and was discovered simultaneously by Stemme of the Chalmers University in Sweden (Stemme, 1972) and Kyser of the Silonics company in the USA (Kyser & Sears, 1976). The third mode is the pushing method by Howkins (Figure 3c), where a piezoelectric element pushes against an ink chamber wall to expel droplets (Howkins, 1984). Finally, the shear-mode (Figure 3d) was found by Fishbeck, where the electric field is designed to be perpendicular to the polarization of the piezo-ceramics (Fishbeck & Wright, 1986).

Besides the continuous and drop-on-demand inkjet technique, a third type of inkjet printing is known, which is based on the electrostatic generation of ink droplets (Winston, 1962). The system is weakly pressurised, causing the formation of a convex meniscus of a conductive ink. An electrostatic force, which exceeds the meniscus' surface tension, is applied between the ink hemisphere and the flat electrode by setting a voltage. Depending on the nature of the electrical potential the system can either be a continuous or drop-on-demand inkjet: the pulse duration determines whether the ejected ink is a continuous stream or a stream of droplets. As a summary of the different inkjet printing technologies, Figure 4 schematically represents a classification thereof.

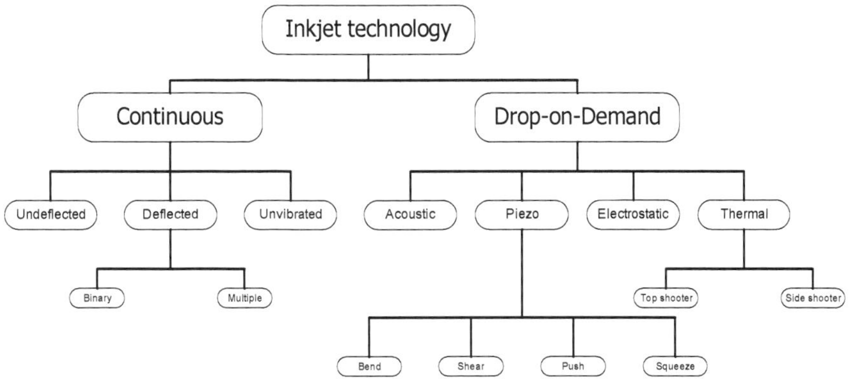

Fig. 4. Classification of inkjet printing technologies, adapted from (Le, 1998).

Although inkjet printing offers a simple and direct method of electronic controlled writing with many advantages, including high speed production, silent, non-impact and fully electronic operation, inkjet printers failed to be commercially successful in their beginning: print quality as well as reliability and costs were hard to combine in a single printing technique. Whereas CIJ provides high throughput, it also requires high costs to gain good quality. Nowadays this technique is used in lower quality and high speed graphical applications such as textile printing and labelling. On the other hand, PIJ usually provides good quality but lacks high printing velocities: although this can be compensated for by using multi nozzle systems, but this increases the production costs as well. TIJ changed the image of inkjet printing dramatically. Not only could thermal transducers be manufactured in much smaller sizes, since they require a simple resistor instead of a piezoelectric element, but also at lower costs. Therefore, thermal inkjet printers dominate the colour printing market nowadays (Kipphan, 2004).

In scientific research piezoelectric DoD inkjet systems are mainly used because of their ability to dispense a wide variety of solvents, whereas thermal DoD printers are more compatible with aqueous solutions (Gans *et al.*, 2004). Furthermore, the rapid and localised heating of the ink within TIJ induces thermal stress on the ink. Nevertheless, research has been conducted using TIJ printers, for example to form conductive patterns, either by printing the water soluble conjugated polymer PEDOT:PSS (Yoshioka & Jabbour, 2006), or by printing aqueous solutions of conductive multi-walled carbon nanotubes (Kordás, 2006).

3. Methods for sintering nanoparticle inks

Conductive materials that are suitable for inkjet printing can be either solution-based or particle based. The former one is usually based on a metallo-organic decomposition (MOD) ink, in particular silver neodecanoate dissolved in an aromatic solvent (Dearden *et al.*, 2005; Smith *et al.*, 2006). These MOD inks have been used for inkjet printing since the late 1980s (Vest *et al.*, 1983). In order to obtain metal features, a conversion of organometallic silver inks is required, which usually takes place at relatively low temperatures below 200 °C (Wu *et al.*, 2007), although temperatures below 150 °C have been reported as well (Smith *et al.*, 2006, Perelaer *et al.*, 2009a). The typical metal loading of organometallic inks is 10 to 20 wt%.

In contrast to metal containing inks based on complexes, inks consisting of a dispersion of nanoparticles have been investigated as well, with the ability to have a silver loading >20 wt% being one of the reasons. Such a dispersion contains metallic nanoparticles with a diameter between 1 and 100 nm. It was found that gold nanoparticles with a diameter below 100 nm reveal a significant reduction in their melting temperature (Buffat & Borel, 1976), as depicted in Figure 5a from their bulk melting temperature of 1064 °C to well below 300 °C when the diameter is below 5 nm. Ten years later, Allen and co-workers showed that this reduction of the melting temperature is also valid for other metals, including tin, lead and bismuth (Allen *et al.*, 1986). In a graph of the melting temperature against the reciprocal of the particle radius the data exhibit near-linear relationships, as depicted in Figure 5b. It was also found that plates instead of spheres do not show a reduced melting temperature. This suggests that the size dependence of melting particles is related to the internal hydrostatic pressure caused by the surface stress and by the large surface curvature of the particles, but not by the planar surfaces of platelets.

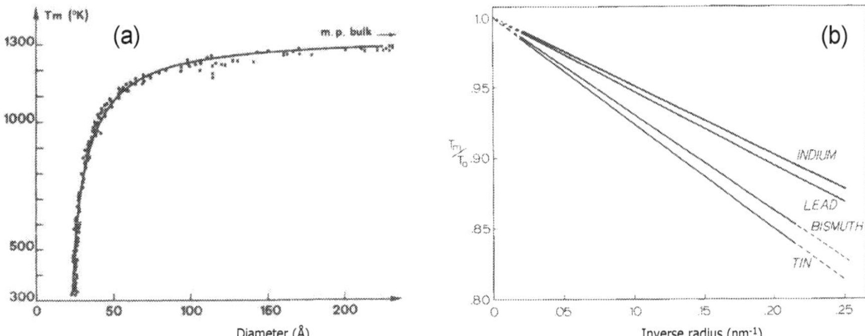

Fig. 5. Influence of the gold (a) and lead, bismuth, tin and indium (b) particle diameter on their melting temperature. Reprinted from (Buffat & Borel, 1976; Allen *et al.*, 1986), respectively.

Given the reduced melting temperature of nanoparticles, these particles represent ideal candidates for dispersion in a liquid medium and, subsequently, for inkjet printing. However, when two or more particles are in contact, merging of nanoparticles into larger clusters can take place due to the large surface curvature of the individual nanoparticles. This process is called sintering and takes place with small particles within the medium and at room temperature. Therefore, the nanoparticles have to be protected by a shell to prevent agglomeration in solution and to obtain a stable colloidal dispersion, as schematically depicted in Figure 6 (Lee *et al.*, 2006).

In non-polar solvents usually long alkyl chains with a polar head, like thiols, amines or carboxylic acids, are used to stabilise the nanoparticles (Perelaer *et al.*, 2008a). Steric stabilisation of these particles in non-polar solvents substantially screens van der Waals attractions and introduces steep steric repulsion between the particles at contact, which avoids agglomeration (Bönnemann & Richards, 2001). In addition, organic binders are often added to the ink to assure not only mechanical integrity and adhesion to the substrate, but also to promote the printability of the ink.

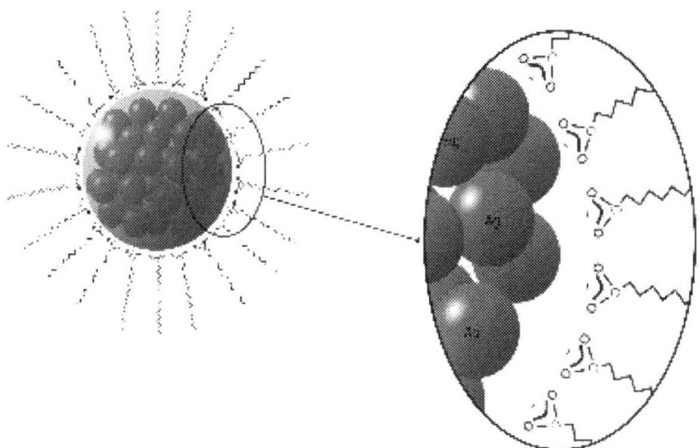

Fig. 6. Schematic illustration of a silver nanoparticle with carboxylic acids as capping agent.

After silver containing inks have been inkjet printed, and solvent evaporation has occurred, another processing step is necessary to form conductive features since the organic shell inhibits close contact of the nanoparticles. Although evaporation of the solvent forces the particles close together, conductivity only arises when metallic contact between the particles is present and a continuous percolating network is formed throughout the printed feature. An organic layer between the silver particles as thin as a few nanometers is sufficient to prevent electrons moving from one particle to the other (Lovinger, 1979). The adsorbed dispersant stays on the surface of the particles and, typically, is removed by an increase in temperature.

Mostly, particulate features have been rendered conductive by applying heat. This thermal sintering method usually requires temperatures above 200 °C (Chou *et al.*, 2005). Other techniques that have been used to for conductive features include LASER sintering (Ko *et al.*, 2007), exposure to UV radiation (Radivojevic *et al.*, 2006), high temperature plasma sintering (Groza *et al.*, 1992) and pulse electric current sintering (Xie *et al.*, 2003). However, most of these techniques are not suitable for polymer substrate materials due to the large overall thermal energy impact. In particular, when using common polymer substrates, like polycarbonate (PC) and polyethylene terephthalate (PET), that have their glass transition temperature (Tg) well below the temperature required for sintering. In fact, only the expensive high-performance polymers, like polytetrafluoroethylene, polyetheretherketone and polyimide (PI) can be used at high temperatures, which represents a serious drawback for implementation in a large area production of plastic electronics and is not favourable in terms of costs.

In the field of sintering two properties are very important: firstly, the lowest temperature at which printed features become conductive, which is mainly determined by the organic additives in the ink (Liang *et al.*, 2004). Secondly, obtaining the lowest possible resistance of the printed features at the lowest possible temperature. To achieve a low resistance, sintering of the particles is required to transform the initially very small contact areas to thicker necks and, eventually, to a dense layer. High conductivities, hence low resistance, can then be obtained through the formation of large necks, which decrease constriction resistance and eventually form a metallic crystal structure with a low number of grain boundaries.

In the low temperature regime, the driving forces for sintering are mainly surface energy reduction due to the particles large surface-to-volume ratio, a process known as *Ostwald ripening* (Ostwald, 1896). This process triggers surface and grain boundary diffusion rather than bulk diffusion within the coalesced particles, as schematically depicted in Figure 7. Grain boundary diffusion allows for neck formation and neck radii increase, which is diminished by the energy required for grain boundary creation (Greer *et al.*, 2007). Therefore, the process will stall eventually, leaving a porous structure behind, which leads to lower conductivity values when compared to the bulk material.

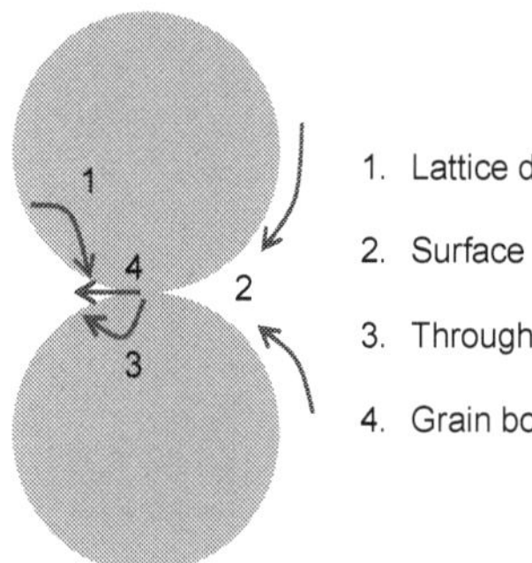

1. Lattice diffusion (no densification)

2. Surface diffusion (no densification)

3. Through-lattice diffusion (densification)

4. Grain boundary diffusion (densification)

Fig. 7. A schematic representation of various atomic diffusion paths between two contacting particles. Paths 1 and 2 do not produce any shrinkage whilst paths 3 and 4 enable the sphere centres to approach one another, resulting in densification. Reprinted from (Greer *et al.*, 2007).

At high temperatures, however, lattice diffusion leads to closure of pores and densification. However, long sintering times are necessary for creating dense conductive features in a thermal process and obstruct the feasibility for an efficient industrial production processes. In order to reduce production costs, alternative techniques that sinter silver nanoparticles in a selective manner without harming the underlying polymer substrate need to be found.
The properties of thermal sintering will be discussed in the next paragraphs, after which a technique that uses microwave radiation will be described as possible candidate for a selective sintering process.

3.1 Thermal sintering of inkjet printed silver lines
A major concern with printed electronics involves not only the control of the morphology of the tracks (Smith *et al.*, 2006; Berg *et al.*, 2007), but also the stability and adhesion of the obtained conductive tracks, although this has scarcely been investigated (Kim *et al.*, 2006). However, the main focus in plastic electronics lies in the low curing temperature of the conductive ink. For particle-based inks, the curing temperature is defined as the temperature where particles loose their organic shell and start showing conductance by direct physical contact. Whereas sintering (which is often mistakenly used instead of curing temperature) takes place at a higher temperature when all the organic material has been burnt off and necks begin to form between particles. The lowest temperature at which printed features become conductive is mainly determined by the organic additives in the ink (Liang *et al.*, 2004). Often high temperatures – typically up to 300 °C – are required to burn off the organic additives and to stimulate the sintering process to realise a more densely

packed silver layer and a lower resistivity (Smith *et al.*, 2006; Yoshioka & Jabbour, 2006). It is therefore of utmost importance if further progress is to be made to identify an optimum between time, temperature and the obtained conductivity.

In order to reveal first structure-property relationships and to later develop the new ink, the sintering behaviour of inkjet printed silver tracks based on commercial inks was studied.

The critical curing temperature is defined in this case as the temperature at which the sample becomes conductive, *i.e.* having a resistance lower than 40 MΩ which is the upper measuring limit of the used multi-meter. Single lines with a length of 1 cm of the specific ink were inkjet printed onto boron-silicate glass and subsequently heated to 650 °C in an oven at a heating rate of 10 °C min^{-1}. During heating the resistance was measured online in a semi-continuous way, by measuring every 2 seconds. Using this dynamic scan approach, differences between the various inks can be determined.

Typical resistance results for the Cabot and Nippon inks are shown in Figure 8a and Figure 9a, respectively. The resistance of the lines for both inks decreases rapidly when heated above the critical curing temperature. The critical curing temperature for the Cabot silver ink is 194 °C, which is lower than the Nippon ink, 269 °C. According to the particle size measurements, 52.4 ± 11.0 nm for the Cabot ink and 10.8 ± 6.7 nm for the Nippon ink (see Figure 8b and Figure 9b), it was expected that the smaller particles would sinter at the lower temperature because of their higher sintering activity (Buffat & Borel, 1976; Allen *et al.*,

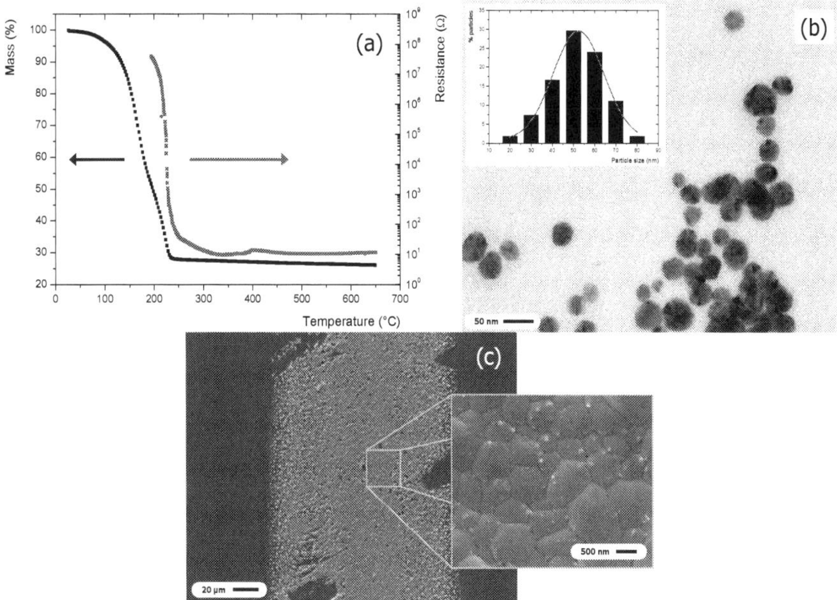

Fig. 8. Resistance over a single inkjet printed line with a length of 1 cm as function of temperature and thermogravimetric analysis (TGA) of Cabot silver ink (a). Transmission electron microscopy (TEM) image and particle size distribution of Cabot silver nanoparticles (b). Scanning electron microscopy (SEM) image of sintered Cabot silver nanoparticles at a temperature of 650 °C (c). Reprinted from (Perelaer et al., 2008a).

1986). This indicates that the organic additives in the ink strongly affect the critical curing temperature. Unfortunately, the nature of the organic additives in these commercially available inks is not disclosed.

To elaborate on this the mass decrease upon heating by means of thermogravimetric analysis (TGA) was also investigated. It should be mentioned that all inks have been dried prior to measuring by heating to 50 °C for 20 minutes, which removed volatile solvents. The TGA curve for Cabot silver ink shows a decrease of 72 wt%, which is not only the organic binder that is around each nanoparticle but also the non-volatile solvent ethylene glycol which is present in the ink (Figure 8a). The critical curing temperature corresponds to a temperature at which the initial sharp weight loss slows down. The first step in the removal of the organic materials has ended at this temperature. The steep decrease in resistance relates to the temperature range in which the last part of the organics is burnt off. Apparently, all organics have to be removed before the sintering of the Ag particles can proceed in a fast way. This is indicative of an additive that is strongly adsorbed on the surface of the silver particles.

The lines printed with the Nippon ink reveal a critical curing temperature and a fast decrease in resistance when only about 15% of the organic additives are removed (Figure 9a). Obviously, these particles can make metallic contact long before all the organics are gone. In addition, sintering proceeds very fast due to the small particle size. At the temperature where the organics have been completely burnt off, only a small additional decrease in resistance occurs. In this ink, only a minor part of the organic additives interferes with the sintering process but it does shift the critical curing temperature to a high value. For both inks, however, the resistance value levels off at a certain temperature. At this temperature all organics are burnt off and, apparently, the sintering process has ended and a silver layer with a final density and morphology has formed.

Figure 8c shows a scanning electron microscopy (SEM) image of a Cabot silver track that has been heated to 650 °C. As can be seen the particles have sintered to a dense continuous line.

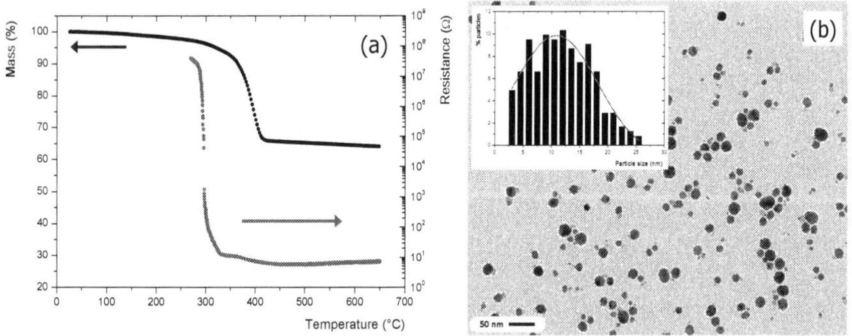

Fig. 9. Resistance over a single inkjet printed line with a length of 1 cm as function of temperature and thermogravimetric analysis (TGA) of Nippon silver ink (a). Transmission electron microscopy (TEM) image and particle size distribution of Nippon silver nanoparticles (b). Reprinted from (Perelaer et al., 2008a).

The electrical resistivity ρ of the inkjet printed lines was calculated after heating to 650 °C, using

$$\rho = R \cdot A / \ell \tag{2}$$

with the lines resistance R, its length λ, and its cross sectional area A, and compared to the value of bulk silver (1.59×10^{-8} Ω m) (Fuller *et al.*, 2002). The resistivity was calculated to be 3.10×10^{-8} Ω m (51%) and 3.06×10^{-8} Ω m (52%) for Cabot and Nippon, respectively. The values in brackets indicate the percentage of conductivity ($1/\rho$) of bulk silver.

In summary, typical sintering temperatures of above 200 °C are required, which limits the usage of many potentially interesting substrate materials, such as common polymer foils or paper. Moreover, the long sintering time of 60 minutes or more that is generally required according to the ink supplier to create conductive features, also obstruct industrial implementation, e.g. roll-2-roll applications.

One selective technique for nanoparticle sintering that has been described in literature is based on an Argon ion LASER beam that follows the as-printed feature and selectively sinters the central region. Features with a line width smaller than 10 µm have been created with this technique (Ko *et al.*, 2007). However, the large overall thermal energy impact together with the low writing speed of 0.2 mm s^{-1} of the translational stage are limiting factors (Chung *et al.*, 2004). In fact, with this particularly technique low writing speeds are required for good electrical behaviour since the resistance increases for faster write speeds (Smith *et al.*, 2006). Thus, other techniques have to be used in order to facilitate fast and selective heating of the printed structures only. Microwave heating fulfils these requirements (Nüchter *et al.*, 2004).

3.2 Selective sintering of silver nanoparticle by using microwave radiation

Microwave heating is widely used for sintering of dielectric materials, conductive materials, and in synthetic chemistry (Wiesbrock *et al.*, 2004). It offers the advantage of uniform, fast and volumetric heating.

The dielectric response to a field is given by the complex permittivity

$$\varepsilon_r = \varepsilon' + i\varepsilon'' = \varepsilon' + i\frac{\sigma}{\omega \cdot \varepsilon_0} \tag{3}$$

where ε' accounts for energy storage, ε'' for energy loss of the incident electromagnetic wave or so-called dissipation, i the imaginary unit, σ the conductivity and ω the angular frequency. The ratio of the imaginary to the real part of the permittivity defines the capability of the material to dissipate power compared to energy storage and is generally know as the loss tangent:

$$\tan\delta = \frac{\varepsilon''}{\varepsilon'} \tag{4}$$

Depending on their loss characteristic, and thus their conductivity, materials can be opaque, transparent or an absorber. For bulk metals, being good electronic conductors, no internal electrical field is generated and the induced electrical charge remains at the surface of the sample (Agrawal, 2006). Consequently, metals reflect microwaves; while bulk metals do not absorb until they have been heated to about 500 °C, powders with particle sizes within the micrometer-region are rather good absorbers (Cheng, 1989). It is believed that the conductive particle interaction with microwave radiation, *i.e.* inductive coupling, is mainly based on Maxwell-Wagner polarisation, which results from the accumulation of charge at the materials interfaces, electric conduction, and eddy currents. However, the main reasons for successful heating of metallic particles through microwave radiation are not yet fully understood.

The penetration depth d is defined as the distance into the material at which the incident power is reduced to $1/e$ (36.8%) of the surface value and is given by

$$d = \frac{c\varepsilon_0}{2\pi f \varepsilon''} = \frac{1}{\sqrt{\pi f \mu \sigma}} \tag{5}$$

with c being the speed of light and f the frequency of the microwave radiation. Typically, highly conductive materials (*e.g.* metals) have a very small penetration depth. For example, the penetration depth of microwaves with a frequency of 2.45 GHz for metal powders of silver and copper is 1.3 and 1.6 µm, respectively. In contrast to the relatively strong microwave absorption by the conductive particles, the polarisation of dipoles in thermoplastic polymers below the Tg is limited, which makes the polymer foil's skin depth almost infinite, hence transparent, to microwave radiation.

Microwave sintering can only be successful if the dimension of the object perpendicular to the plane of incidence is of the order of the penetration depth. The average height of a single inkjet printed track of silver nanoparticles was measured to be 4.1 µm. The calculated penetration depth of the microwave irradiation into silver at a frequency of 2.45 GHz using equation (5) is only 1.3 µm. Therefore, it is to be expected that microwave heating will not be uniform throughout the complete line. However, since silver is a good thermal conductor in comparison to the polymer substrate, the silver tracks will be heated uniformly by thermal conductance.

Unsintered non-conductive silver lines were treated in a microwave reactor operating in constant power mode (300 W). The sintering times are significantly shortened in the microwave, from 60 minutes or more down to 240 seconds, as shown in Figure 10a. Within the reactor vessel the temperature reaches 200 °C, which is near the sintering temperature of 220 °C for conventional thermal sintering. Longer sintering times did not increase the conductivity, but sometimes resulted in deformation or decomposition of the substrate at the edges of the silver lines and the substrate.

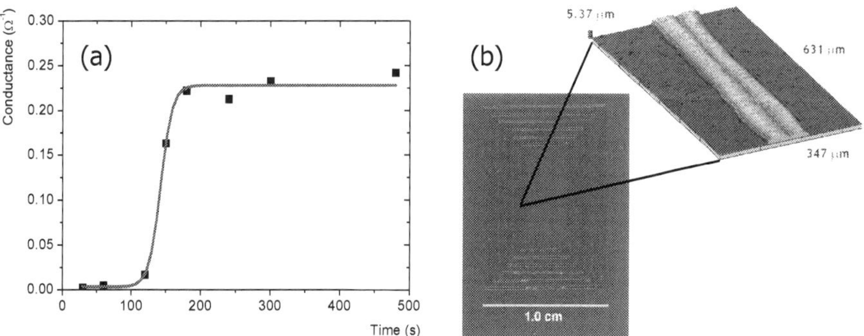

Fig. 10. Conductance as function of time for the microwave sintering of silver tracks (a) printed onto a polyimide substrate (b). Reprinted from (Perelaer et al., 2006).

A typical resistivity value is 3.0×10^{-7} Ω m, which is approximately 20× the bulk silver value. With thermal sintering in an oven (220 °C, 60 minutes) similar values are obtained, which is in agreement to what is reported by other authors (Cheng *et al.*, 2005).

It was recently discovered that conductive antenna structures are more susceptible for absorption of microwaves than the printed feature by itself (Perelaer *et al.*, 2009b). Therefore, conductive antenna structures have been applied onto the polymer foil and were found to improve the sintering process, and thereby the obtained conductivity, significantly. These antennae were used both to measure the resistance of the single ink line and to capture the electromagnetic waves, which was possible since the electrodes were composed of particles that are able to absorb microwaves, as schematically depicted in Figure 11.

A single silver ink line was inkjet printed over the metallic probes and shortly cured in an oven for 1 to 5 minutes at a temperature of 110 °C. This relatively short time was chosen to stimulate solvent evaporation, but to minimize thermal curing. After this treatment, the single line had a relatively high resistance in the order of 10^2 to 10^4 Ω. The sample was subsequently exposed to microwave radiation for at least 1 second, while applying the lowest set-power of 1 W. This resulted in a pronounced decrease of the resistance of which the exact outcome depends on the initial resistance.

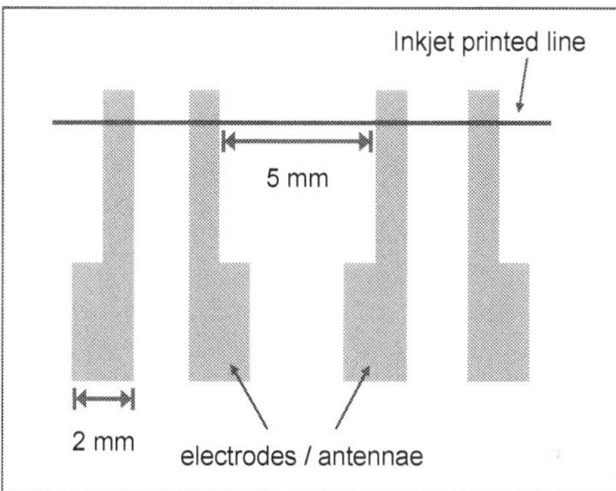

Fig. 11. Schematic representation of the printed template (a), with four silver electrodes/antennae in gray and a single silver line inkjet printed on top of the antennae in black. The total length of the line is 1.6 cm. Reprinted from (Perelaer et al., 2009b).

The *antenna effect*, which reflects the capability of absorbing microwaves into the material, was studied systematically by altering the surface area of the electrodes of the template. When increasing the size of the electrodes a rapid decrease of the resistance after microwave exposure was revealed, as is shown in Figure 12 for pre-dried samples. This may be explained by the improved absorption of the microwaves due to an increased surface area of the electrodes.

The antenna effect, however, is larger when the initial line resistance is small (Figure 12a), which is likely due to enhanced heat conduction from the electrodes to the ink line. For ink lines with an initially large resistance (Figure 12b), the energy transfer is still very effective, although the total antenna area has less impact on the final resistance. The data obtained in the absence of antennae (A = 0 mm²) clearly demonstrate that the energy absorption by the printed line is negligibly small at these short times.

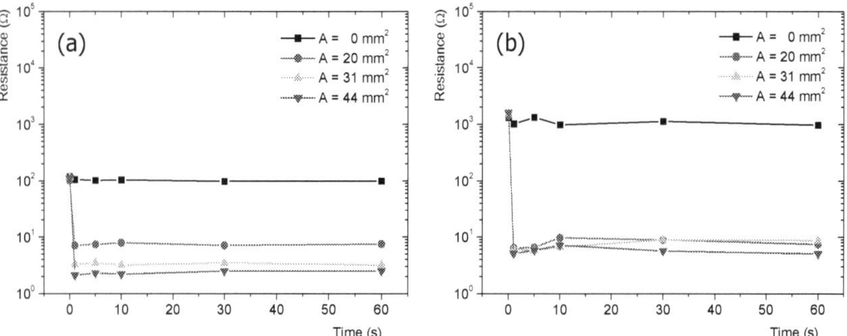

Fig. 12. Influence of the total surface of the four electrodes on the template for an initial line resistance of 100 Ω (b) and 1 kΩ (c) on microwave flash exposure for 1 to 60 seconds. Reprinted from (Perelaer et al., 2009b).

The absorption of microwave radiation may be improved by the presence of antennae due to a subsequently smaller impedance mismatch between air and sample. The intrinsic impedance Z of a circuit relative to free space is given by:

$$Z = Z_0 \sqrt{\varepsilon^*}$$ (6)

where Z_0 is the impedance of free space (Z_0 = 377 Ω) and ε^* is the complex permittivity of the circuit relative to free space (Zuckermann & Miltz, 1994). The complex permittivity is related to the dielectric constant and the loss factor, according to equation (3), where the real part ε' is the ability of the material to store energy and where the imaginary part ε'' accounts for the losses via energy dissipation. The reflection, *i.e.* impedance mismatch Z/Z_0, scales with the square root of the complex permittivity ε^* and thus, in our experiments, with the resistance of the electrodes, which depends on their total surface area.

The electrical resistivity ρ of an inkjet printed line was subsequently calculated from equation 2. The conductivity $(1/\rho)$ for the maximum surface area (44 mm²) of the antennae was found to be 34% when compared to the bulk silver value. A conductivity of 10% was revealed for the antennae with a surface area of 20 mm². This value, however, is significantly larger than the 5% that was reported in the previous section.

This process of flash sintering in the presence of conductive antenna structures may be implemented in a roll-to-roll production for sintering of inkjet printed silver tracks. The antennae do not need to make contact with the unsintered features, which makes recycling of the antennae possible; it was found that sintering took place even with a gap up to 0.5 cm between the antennae and the printed line. Increasing the distance reduces, however, the final conductivity. Thus, the antenna structures need to be close enough to the unsintered features, since the relay of electromagnetic waves is limited.

4. Improved resolution of direct inkjet printed conductive silver tracks on untreated polymer substrates

Another important aspect of conductive tracks, besides their conductivity properties, is the tracks dimensions, in particular the width of the track, hence the resolution of the printed feature. Typical dimensions of inkjet printed features depend on the nozzle diameter and are usually not below 100 µm. The most obvious way to minimise the feature size, *i.e.* line width, is by reducing the nozzle diameter (Le, 1998). However, this introduces a narrow window with respect to the applicable surface tension and viscosity of the inks, and thereby, the range of inks that can be printed. Furthermore, when printing dispersions the particles should be sufficiently smaller than the nozzle diameter; otherwise nozzle clogging occurs. When using piezo-electric based DoD inkjet printers, smaller droplets can also be produced by modifying the waveform (Chen & Basarana, 2002).

Other techniques to minimise feature sizes include an increased substrate's temperature, which will stimulate solvent evaporation and leave smaller droplets on the substrate due to a limited spreading (Perelaer *et al.*, 2006), and fluid-assisted dewetting effects (Dockendorf *et al.*, 2006). The latter and relatively new method was described as follows: a line of gold nanoparticle dispersion in toluene was deposited on a glass substrate, after which a water droplet was dispensed over the line. Subsequently, the pattern shrank due to a transport of the toluene from the gold dispersion into the water region, triggered by controlled heat addition from the substrate, whose temperature was set to 95 °C. Toluene and water are practically immiscible at room temperature, but at this temperature toluene mixes very well with water. During the dewetting phase, the three-phase-contact line is pulled by the uncompensated Young's force. Furthermore, the authors explain the dewetting dynamics by the action of thermocapillarity enhanced by the convection microflow generated in the water layer.

Much research has been done on predefined (surface energy) patterns on a substrate that forces material to remain in a preferred area on the surface (Menard *et al.*, 2007; Hendriks *et al.*, 2008; Sirringhaus *et al.*, 2000). These techniques rely on the use of expensive masks and conventional photolithography, which subsequently increases production costs. To reduce these costs a method to produce narrow conductive silver tracks without pre-patterning or modifying the surface energy of the substrate is required. Preferably, the substrate's surface energy should not be too low, because printing then introduces bulges into the printed features (Duineveld, 2003), for example with polytetrafluorethylene (PTFE) foils, as can be seen on the left-hand side in Figure 13. Commonly used polymer substrates, like polyethylene terephthalate (PET) or polyimide (PI), have a relatively high surface energy, shown on the right-hand side in Figure 13. Although printing on these substrates leads to continuous and straight lines, broad lines are obtained over the whole printed feature, due to the relatively good wetting of the solvent with the substrate. Moreover, unwanted drying effects, such as the coffee ring effect may appear (Deegan *et al.*, 2000; Soltman & Subramanian, 2008). Clearly, an optimum between surface energy and solvent is necessary. Polyarylate polymer foils fulfil this need, since they have a surface energy between the value of PTFE and PI.

The silver ink from Cabot (Cabot Printing Electronics and Displays, Albuquerque, USA) has been inkjet printed using a cartridge able to dispense droplets with a volume as small as 1 pL. Besides a decrease in nozzle size, and thus a higher print resolution, further decrease in line diameter was realised by heating the sample holder of the printer to its maximum temperature (60 °C), which stimulates the evaporation of the solvent and prevents broadening of the lines (Perelaer *et al.*, 2006).

Figure 14a shows the dependency of line width on dot spacing; a decrease in line width is observed when the dot spacing is increased. Obviously, with increased dot spacing less material is deposited per unit length, resulting in smaller structures. Partially continuous lines were formed, when a dot spacing larger than 25 µm was used and further increase led

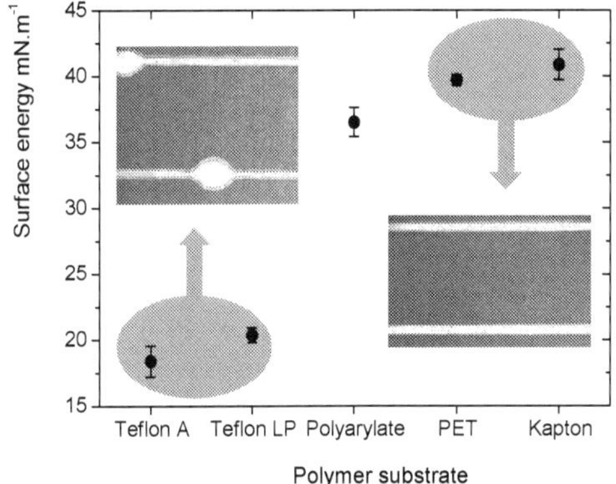

Fig. 13. Surface energy of five commercially available polymer substrates and an impression of the printed lines on these surfaces. Reprinted from (Osch et al., 2008).

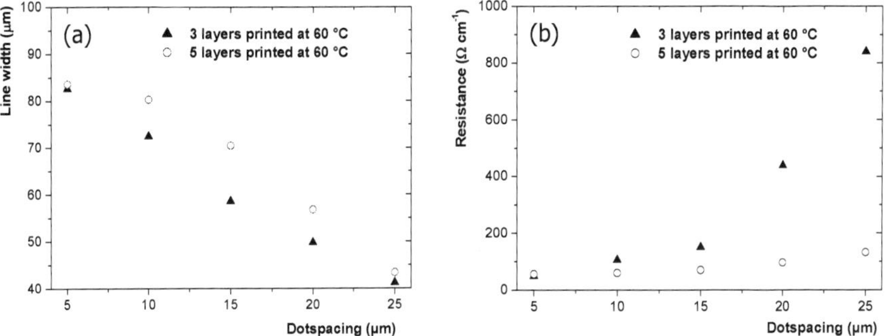

Fig. 14. Line width (a) and resistance (b) as function of dot spacing for three and five layers subsequently printed on top of each other. Reprinted from (Osch et al., 2008).

to individual droplets. Therefore, using a dot spacing of 25 µm resulted in the smallest line width of 40 µm, which corresponds to a resolution of approximately 600 dots per inch.[1]

In the same way, the resistance will increase when the dot spacing is increased, as depicted in Figure 14b. However, the number of layers printed on top of each other strongly influences the dependency. The resistance of lines consisting of 3 layers strongly increases with dot spacing, whereas the resistance shows a much more gradual increase with dot spacing when 5 layers are printed on top. This can be explained by the formation of more parallel percolating pathways when more material is deposited per unit length.

Typical dimensions of printed silver tracks onto polyarylate films are shown in Figure 15. Inkjet printed features on polyarylate foil with a line width of 85 µm and 40 µm without any defects such as bulges or coffee drop effects were obtained using a dot spacing of 5 µm (a) and 25 µm (b), respectively. The silver lines were sintered after drying in an oven at 200 °C for 1 hour. Subsequently, the resistance was measured by using the 4-point method. The

---

[1] 600 dots per inch (DPI) correspond to a single dot diameter of 2.54/600 cm = 42.3 µm.

conductivity $(1/\rho)$ was 23% of the bulk silver value for tracks printed with a dot spacing of 5 µm and 13% when using a dot spacing of 25 µm.

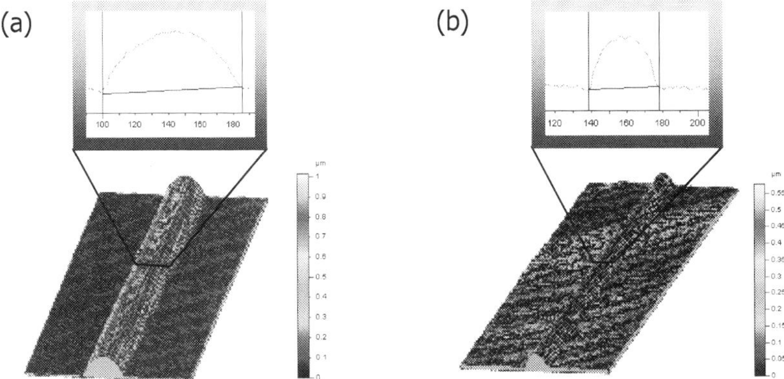

Fig. 15. Cross-sectional image and 3D image of inkjet printed silver tracks on polyarylate films using a dot spacing of 5 µm (a) and 25 µm (b). The substrate was heated to 60 °C and five layers were printed on top of each other. Reprinted from (Osch et al., 2008).

5. Conclusions and outlook

In order to develop better and alternative sintering methods, a basic understanding of thermal sintering and how particles sinter has been described. The conductivity development of commercially available silver inks has been discussed. Hereto, the resistance was measured on-line during the heating of the silver tracks from room temperature to 650 °C. The standard method, however, for sintering is by applying heat, typically above 200 °C, which is not compatible with the common polymer foils that are being considered as substrates for plastic electronics applications.

By using microwave radiation instead of conventional radiation-conduction-convection heating, the sintering time of silver nanoparticles was shortened by a factor of 20, *i.e.* instead of 60 minutes three minutes were sufficient for sintering. The polymer substrate is virtually transparent to microwave radiation, whereas the conductive silver nanoparticles absorb the microwaves strongly and sinter.

Furthermore, the presence of conductive antennae promotes nanoparticle sintering in pre-cured ink lines to an extent that depends on the total area of the antennae. For cured nanoparticle inks that are connected to antennae, sintering times of 1 second are sufficient to obtain pronounced nanoparticle sintering. The antenna effect is greater if the ink line already exhibits conductivity. It is believed that this is due to the decreased mismatch between the impedance of air and sample. Using metal antennae, it was revealed that 1 second is sufficient to obtain pronounced sintering by microwave heating. The degree of sintering for these short exposure periods, however, strongly depends on the initial resistance of the pre-cured ink lines. After microwave flash sintering, the tracks revealed conductivity values of 10 to 34% compared to the bulk silver value, which is significantly higher compared to conventional heating methods.

The procedure of printing and subsequently microwave flash sintering can be used, for example, in roll-to-roll (R2R) production applications, such as large area fabrication of RFID tags or solar cells. Its major advantage pertains to both the high process speed and the low processing temperature, which reduce processing costs, as common polymer foils like polyethylene naphthalate (PEN) can be used.

Very recently, another alternative sintering technique has been reported that uses a low pressure argon plasma exposure (Reinhold *et al.*, 2009). Typical resistivity values of 2.5 to 3× the bulk silver value were achieved. The process shows an evolution starting from a sintered top layer into bulk material, which determines the resistivity of the sintered material. Through-sintering does not occur with greater thicknesses than the penetration depth of the plasma species.

Several techniques have been discussed to improve the resolution of inkjet printed features. In order to reach the smallest line width for printed silver tracks, it is necessary to tune the ink surface tension to the substrates energy, hence its wettability. Furthermore, dot spacing, surface temperature and in-flight droplet diameter strongly affect the resolution as well and need to be optimised.

In general, it can be concluded that inkjet printing is capable of preparing high-resolution conductive features on polymer substrates. Together with inkjet printing, (ink) materials can be saved, since the ink is only dispensed on demand. It is, however, necessary to tune the polymer substrate as well the conductive inks properties. Alternative and selective sintering methods open new routes to produce conductive features on common polymer foils that have a relatively low glass transition temperature. This combination may be then employed in roll-to-roll printed plastic microelectronic devices.

6. References

Agrawal, D. (2006). Microwave sintering of ceramics, composites and metallic materials, and melting of glasses. *Trans. Ind. Ceram. Soc.*, 65, 129-144.

Allen, G. L.; Bayles, R. A.; Gile, W. W. & Jesser, W. A. (1986). Small particle melting of pure metals. *Thin Solid Films*, 144, 297-308.

Attinger, D.; Zhao, Z. & Poulikakos, D. (2000). An experimental study of molten microdroplet surface deposition and solidification: transient behavior and wetting angle dynamics. *J. Heat Transfer*, 122, 544-556.

Berg, van den, A. M. J.; Laat, de, A. W. M.; Smith, P. J.; Perelaer, J. & Schubert, U. S. (2007a). Geometric control of inkjet printed features using a gelating polymer. *J. Mater. Chem.*, 17, 677-683.

Bogy, D. B. (1979). Drop formation in a circular liquid jet. *Ann. Rev. Fluid. Mech.*, 11, 207-228.

Bönnemann, H. & Richards, R. M. (2001). Nanoscopic metal particles – synthetic methods and potential applications. *Eur. J. Inorg. Chem.*, 2455-2480.

Brünahl, J. & Grishin, A. M. (2002). Piezoelectric shear mode drop-on demand inkjet actuator. *Sens. Actuators A*, 101, 371-382.

Buffat, P. & Borel, J.-P. (1976). Size effect on the melting temperature of gold particle. *Phys. Rev. A*, 13, 2287-2298.

Carnahan, R. D. & Hou, S. L. (1977). Inkjet technology. *IEEE Trans. Ind. Appl.*, IA-13, 95-105.

Chen, A. U. & Basaran, O. A. (2002). A new method for significantly reducing drop radius without reducing nozzle radius in drop-on-demand drop production. *Phys. Fluids*, 14, L1-L4.

Cheng, D. K. (1989). *Field and Wave Electromagnetics*, Addison-Wesley Co. Inc., ISBN 0-201-12819-5, Reading.

Cheng, K.; Yang, M.-H.; Chiu, W. W. W.; Huang, C.-Y.; Chang, J.; Ying, T.-F. & Yang, Y. (2005). Ink-jet printing, self-assembled polyelectrolytes, and electroless plating: low cost fabrication of circuits on a flexible substrate at room temperature. *Macromol. Rapid Commun.*, 26, 247-264.

Chou, K.-S.; Huang, K.-C. & Lee, H.-H. (2005). Fabrication and sintering effect on the morphologies and conductivity of nano-Ag particle films by the spin coating method. *Nanotechnology*, 16, 779-784.

Chung, J.; Bieri, N. R.; Ko, S.; Grigoropoulos, C. P. & Poulikakos, D. (2004). In-tandem deposition and sintering of printed gold nanoparticle inks induced by continuous gaussian laser irradiation. *Appl. Phys. A*, 79, 1259-1261.

Cuk, T.; Troian, S. M.; Hong C. M. & Wagner, S. (2000). Using convective flow splitting for the direct printing of fine copper lines. *Appl. Phys. Lett.*, 77, 2063-2065.

Dearden, A. L.; Smith, P. J.; Shin, D.-Y.; Reis, N.; Derby, B. & O'Brien, P. (2005). A low curing temperature silver ink for use in ink-jet printing and subsequent production of conductive tracks. *Macromol. Rapid Commun.*, 26, 315-318.

Deegan, R. D.; Bakajin, O.; Dupont, T. F.; Huber, G.; Nagel, S. R. & Witten, T. A. (2000). Contact line deposits in an evaporating drop. *Phys. Rev. E*, 62, 756-765.

Derby, B. (2008). Bio printing: inkjet printing proteins and hybrid cell-containing materials and structures. *J. Mater. Chem.*, 18, 5717-1521.

Dockendorf, C. P. R.; Choi, T.-Y.; Poulikakos, D. & Stemmer, A. (2006). Size reduction of nanoparticle ink patterns by fluid-assisted dewetting. *Appl. Phys. Lett.*, 88, 131903 (3 pp).

Döring, M. (1982). Inkjet printing. *Philips Tech. Rev.*, 40, 192-198.

Duineveld, P. C. (2003). The stability of ink-jet printed lines of liquid with zero receding contact angle on a homogeneous substrate. *J. Fluid Mech.*, 477, 175-200.

Elmqvist, R. (1951). Measuring instrument of the recording type. US Patent 2,566,443.

Endo, I.; Sato, Y.; Saito, S.; Nakagiri, T. & Ohno, S. (1979). Liquid jet recording process and apparatus therefor. GB Patent 2,007,162.

Fishbeck, K. H. & Wright, A. T. (1986). Shear mode transducer for drop-on-demand liquid ejector. US Patent 4,584,590.

Forrest, S. R. (2004). The path to ubiquitous and low-cost organic electronic appliances on plastic. *Nature*, 428, 911-918.

Fuller, S. B.; Wilhelm E. J. & Jacobson, J. M. (2002). Inkjet printed nanoparticle microelectromechanical systems. *J. Microelectromech. Syst.*, 11, 54-60.

Gamerith, S.; Klug, A.; Schreiber, H.; Scherf, U.; Moderegger E. & List, E. J. W (2007). Direct ink-jet printing of Ag-Cu nanoparticle and Ag-precursor based electrodes for OFET applications. *Adv. Funct. Mater.*, 17, 3111-3118.

Gans, de, B.-J; Duineveld, P. C. & Schubert, U. S. (2004). Inkjet printing of polymers: state of the art and future developments. *Adv. Mater.*, 16, 203-213.

Gans, de, B.-J. &. S. Schubert, U. S. (2004). Inkjet printing of well-defined polymer dots and arrays. *Langmuir*, 20, 7789-7793.

Goedde, E. F. & Yuen, M. C. (1970). Experiments on liquid jet instability. *J. Fluid Mech.*, 40, 495-511.

Greer, J. R. & Street, R. A. (2007). Thermal cure effects on electrical performance of nanoparticle silver inks. *Acta Mater.*, 55, 6345–6349.

Groza, J. R.; Risbud, S. H. & Yamazaki, K. (1992). Plasma activated sintering of additive-free AlN powders to near-theoretical density in 5 minutes. *J. Mater. Res.*, 7, 2643-2645.

Hansell, C. W. (1950). Jet sprayer actuated by supersonic waves. US Patent 2,512,743.

Heinzl, J. & Hertz, C. H. (1985). Ink-jet printing. *Adv. Electron. Electron Phys.*, 65, 91-171.

Hendriks, C. E.; Smith, P. J.; Perelaer, J.; Berg, van den, A. M. J. & Schubert, U. S. (2008). "Invisible" silver tracks produced by combining hot-embossing and inkjet printing. *Adv. Funct. Mater.*, 18, 1031-1038.

Hertz, C. H. & Simonsson, S.-I. (1969). Intensity modulation of ink-jet oscillographs. *Med. & Biol. Eng.*, 7, 337-340.

Hong, C. M. & Wagner, S. (2000). Inkjet printed copper source/drain metallization for amorphous silicon thin-film transistors. *IEEE Electron Device Letters*, 21, 384-386.

Howkins, S. D. (1984). Inkjet method and apparatus. US Patent 4,459,601.

Huang, D.; Liao, F.; Molesa, S.; Redinger, D. & Subramanian, V. (2004). An ink-jet-deposited passive component process for RFID. *IEEE Trans. Electron Devices*, 51, 1978-1983.

Kamphoefner, F. J. (1972). Inkjet printing. *IEEE Trans. Electron Devices*, 19, 584-593.

Keeling, M. R. (1981). Inkjet printing. *Phys. Technol.*, 12, 196-203.

Kim, D.; Jeong, S.; Park, B. K. & Moon, J. (2006). Direct writing of silver conductive patterns: imporvement of film morphology and conductance by controlling solvent compositions. *Appl. Phys. Lett.*, 89, 264101 (3 pp).

Kim, D.; Jeong, S.; Lee, S.; Kyun Park, B. & Moon, J. (2007). Organic thin film transistor using silver electrodes by the inkjet printing technology. *Thin Solid Films*, 515, 7692-7696.

Kipphan, H. (2004). *Handbook of print media: Technologies and manufacturing processes*, Springer, ISBN 3-540-67326-1, Berlin.

Ko, S. H.; Pan, H.; Grigoropoulos, C. P.; Luscombe, C. K.; Fréchet, J. M. J. & Poulikakos, D. (2007). All-inkjet-printed flexible electronics fabrication on a polymer substrate by low-temperature high-resolution selective laser sintering of metal nanoparticles. *Nanotechnology*, 18, 345202 (8 pp).

Kordás, K.; Mustonen, T.; Tóth, G.; Jantunen, H.; Lajunen, M.; Soldano, C.; Talapatra, S.; Kar, S.; Vajtai, R. & Ajayan, P. M. (2006). Inkjet printing of electrically conductive patterns of carbon nanotubes. *Small*, 2, 1021-1025.

Kyser, E. L. & Sears, S. B. (1976). Method and apparatus for recording with writing fluids and drop projection means therefore. US Patent 3,946,398.

Lee, K. J.; Jun, B. H.; Kim, T. H. & Joung, J. (2006). Direct synthesis and inkjetting of silver nanocrystals toward printed electronics. *Nanotechnology*, 17, 2424-2428.

Le, H. P. (1998). Progress and trends in ink-jet printing technology. *J. Imaging Sci. Technol.*, 42, 49-62.

Liang, L. H.; Shen, C. M.; Du, S. X.; Liu, W. M.; Xie, X. C. & Gao, H. J. (2004). Increase in thermal stability induced by organic coatings on nanoparticles. *Phys. Rev B.*, 70, 205419 (5 pp).

Liu, J. G.; Chen, C. H.; Zheng, J. S. & Huang, J. Y. (2005). CO_2 laser direct writing of silver lines on epoxy resin from solid film. *Appl. Surf. Sci.*, 245, 155-161.

Lovinger, A. J. (1979). Development of electrical conduction in silver-filled epoxy adhesives. *J. Adhesion*, 10, 1-15.

Menard, E.; Meit, M. A.; Sun, Y.; Park, J.-U.; Jay-Lee Shir, D.; Nam, Y.-S.; Jeon, S. & Rogers, J. A. (2007). Micro- and nanopatterning techniques for organic electronic and optoelectronic systems. *Chem. Rev.*, 107, 1117-1160.

Molesa, S.; Redinger, D. R.; Huang, D. C. & Subramanian, V. (2003). High-quality inkjet-printed multilevel interconnects and inductive components on plastic for ultra-low-cost RFID applications. *Mat. Res. Soc. Symp. Proc.*, 769, H8.3.1-H8.3.6.

Naiman, M. (1965). Sudden stream printer. US Patent 3,179,042.

Nollet, J.-A. & Watson, W. (1749). Recherches sur les causes particulieres des phénoménes électriques, et sur les effects nuisibles ou avantageux qu'on peut en attendre. *Philos. Trans.*, 46, 368-397.

Nüchter, M.; Ondruschka, B.; Bonrath, W. & Gum, A. (2004). Microwave assisted synthesis - a critical technology overview. *Green Chem.*, 6, 128-141.

Osch, van, T. H. J.; Perelaer, J.; Laat, de, A. W. M. & Schubert, U. S. (2008). Inkjet printing of narrow conductive tracks on untreated polymeric substrates. *Adv. Mater.*, 20, 343-345.

Ostwald, W. (1896). *Lehrbruck der Allgemeinen Chemie*, Vol. 2, Part 1, Leipzig.

Perelaer, J.; Gans, de, B.-J. & Schubert, U. S. (2006). Ink-jet printing and microwave sintering of conductive silver tracks. *Adv. Mater.*, 18, 2101-2104.

Perelaer, J.; Laat, de, A. W. M.; Hendriks, C. E. & Schubert, U. S. (2008a). Inkjet-printed silver tracks: low temperature curing and thermal stability investigation. *J. Mater. Chem.*, 18, 3209-3215.

Perelaer, J.; Smith, P. J.; Hendriks, C. E.; Berg, van den, A. M. J. & Schubert, U. S. (2008b). The preferential deposition of silica micro-particles at the boundary of inkjet printed droplets. *Soft Matter*, 4, 1072-1078.

Perelaer, J.; Hendriks, C. E.; Laat, de, A. W. M. & Schubert, U. S. (2009a). One-step inkjet printing of conductive silver tracks on polymer substrates. *Nanotechnology*, 20, 165303 (5 pp).

Perelaer, J.; Klokkenburg, M.; Hendriks, C. E. & Schubert, U. S. (2009b). Microwave flash sintering of inkjet printed silver tracks on polymer substrates. *Adv. Mater.*, 21, 4830-4834.

Potyrailo, R. A.; Surman, C. & Morris, W. G. (2009). Combinatorial screening of polymeric sensing materials using RFID sensors: combined effects of plasticizers and temperature. *J. Comb. Chem.*, 11, 598-603.

Reinhold, I.; Hendriks, C. E.; Eckardt, R.; Kranenburg, J. M.; Perelaer, J.; Baumann, R. R. & Schubert, U. S. (2009). Argon plasma sintering of inkjet printed silver tracks on polymer substrates. *J. Mater. Chem.*, 19, 3384-3388.

Reis, N.; Ainsley, C. & Derby, B. (2005). Ink-jet delivery of particle suspensions by piezoelectric droplet ejectors. *J. Appl. Phys.*, 97, 094903 (6 pp).

Radivojevic, Z.; Andersson, K.; Hashizume, K.; Heino, M.; Mantysalo, M.; Mansikkamaki, P.; Matsuba, Y. & Terada, N. (2006). Optimised curing of silver ink jet based printed traces. *Proceedings of 12th Intl. Workshop on Thermal investigations of ICs*, pp. 133-138, ISBN 2-916187-04-9, Nice, September 2006, TIMA Editions, Nice.

Rayleigh, L. (1878). On the instability of jets. *Proc. London Math. Soc.*, 10, 4-13.

Savart, F. (1833). Mémoires sur la constitution des veines liquides lancées par des orifices circulaires en mince paroi. *Ann. Chim. Phys.*, 53, 337-386.

Schmid, G. (2004). *Nanoparticles: from theory to applications,* Wiley, ISBN 3-527-30507-6, Weinheim.

Szczech, J. B. Megaridis, C. M. Gamota D. R. & Zhang, J. (2002). Fine-line conductor manufacturing using drop-on-demand PZT printing technology. *IEEE Trans. Electron. Pack.,* 25, 26-33.

Shim, G. H.; Han, M. G.; Sharp-Norton, J. C.; Creager, S. E. & Foulger, S. H. (2008). *J. Mater. Chem.,* 18, 594-601.

Sirringhaus, H.; Kawase, T.; Friend, R. H.; Shimoda, T.; Inbasekaran, M.; Wu, W. & Woo, E. P. (2000). High-resolution inkjet printing of all-polymer transistor circuits. *Science,* 290, 2123-2126.

Smith, P. J.; Shin, D.-Y.; Stringer, J. E.; Reis, N. & Derby, B. (2006). Direct inkjet printing and low temperature conversion of conductive silver patterns. *J. Mater. Sci.,* 41, 4153-4158.

Soltman, D. & Subramanian, V. (2008). Inkjet-printed line morphologies and temperature control of the coffee ring effect. *Langmuir,* 24, 2224-2231.

Stemme, N. G. E. (1972). Arrangement of writing mechanics for writing on paper with a colored liquid. US Patent 3,747,120.

Sweet, R. G. (1965). High frequency recording with electrostatically deflected ink jets. *Rev. Sci. Instr.,* 36, 131-136.

Tekin, E.; Smith, P. J. & Schubert, U. S. (2008). Inkjet printing as a deposition and patterning tool for polymers and inorganic particles. *Soft Matter,* 4, 703-713.

Thomson, W. (1867). Improvements in telegraphic receiving and recording instruments. UK Patent 2,147.

Vaught, J. L.; Cloutier, F. L.; Donald, D. K.; Meyer, J. D.; Tacklind, C. A. & Taub, H. H. (1984). Thermal ink-jet printer. US Patent 4,490,728.

Vest, R. W.; Tweedell, E. P. & Buchanan, R. C. (1983). Inkjet printing of hybrid circuits. *Intl. J. Hybrid Microelectron.,* 6, 261-267.

Wiesbrock, F.; Hoogenboom, R. & Schubert, U. S. (2004). Microwave-assisted polymer synthesis: state-of-the-art and future prospectives. *Macromol. Rapid Commun.,* 25, 1739-1764.

Winston, C. R. (1962). Method and aparatus for transfering ink. US Patent 3,060,429.

Wu, Y.; Li, Y. & Ong, B. S (2007). A simple and effective approach to a printable silver conductor for printed electronics. *J. Am. Chem. Soc.,* 129, 1862-1863.

Xie, G.; Ohashi, O.; Yamaguchi, N. & Wang, A. (2003). Effect of surface oxide films on the properties of pulse electric-current sintered metal powders. *Metall. Mater. Trans. A,* 34A, 2655-2661.

Yoshioka, Y. & Jabbour, G. E. (2006). Desktop inkjet printer as a tool to print conducting polymers. *Synth. Met.,* 156, 779-783.

Zoltan, S. I. (1972). Pulsed droplet ejecting system. US Patent 3,683,212.

Zuckerman, H. & Miltz, J. (1994). Changes in thin-layer susceptors during microwave. *Packag. Technol. Sci.,* 7, 21-26.

17

Tag4M, a Wi-Fi RFID Active Tag Optimized for Sensor Measurements

Silviu Folea[1] and Marius Ghercioiu[2]
[1]*Technical University of Cluj-Napoca, Department of Automation,*
[2]*Tag4M,*
[1]*Romania,*
[2]*USA*

1. Introduction

Tag4M is a Wi-Fi RFID active tag with the functionality of a multifunctional Input/Output measurement device. The tag offers a combination of Wi-Fi radio and measurement capabilities for sensors and actuators that generate output as voltage, current, or digital signal. Tag4M is very suitable for prototyping of wireless sensor measurements and also for teaching wireless measurement using the existing Wi-Fi infrastructure.

In many applications cables need to be removed from measurement setups and replaced with wireless devices that are connected to sensors and send data wirelessly to the network and to computers. Wireless measurement devices that replace cabling need to be small and cheap and reliable in order to be a valid replacement for cabling. Mobile type measurement applications like monitoring of rotating machinery or moving objects also benefit from wireless measurement devices. Inside the class of wireless measurement devices there are those running on batteries. These devices are built around low or very low power microcontrollers, have the capability of going to sleep for long periods of time, and implement some kind of radio and associated communication protocol that are designed to save battery power.

Wireless USB, ZigBee, Bluetooth and ultra low-power Wi-Fi are the most common radio platforms used in wireless measurement and communication. Basic performance benchmarks for comparison of these technologies, things like application domains, typical range, network connectivity, network topology and key attributes are available in the reference (Sidhu et al., 2007).

Wireless USB devices, like the wireless mouse for example, are mostly used as computer peripherals. Bluetooth devices are more power hungry therefore this wireless technology is used in PDAs and computers that can be (re)charged overnight. The strength of Bluetooth lies in its ability to allow interoperability and replacement of cables.

ZigBee and ultra low power Wi-Fi are the two wireless technologies best suited for sensor measurement. The one major difference between ZigBee and ultra-low power Wi-Fi is that ZigBee nodes use the ZigBee protocol and not any native Internet protocol like TCP/IP or UDP, and therefore ZigBee nodes need a dedicated Access Point that translates ZigBee into TCP/IP in order for the data to be sent over the network. ZigBee networks can support a larger number of devices and in most cases, longer range between devices than Bluetooth for example. ZigBee is cheaper and has lower power consumption but its transfer rate is quite small if larger amount of information has to be sent (Labiod et al., 2007).

By comparison, Wi-Fi wireless LAN adapters are much more powerful and capable of reaching data transmission rates approaching 54Mbps. Wi-Fi products also have strong security protocols (WEP/WPA), which make them a better network solution. If key attributes for Wi-Fi are wider bandwidth and flexibility, for ZigBee are cost and power.

Inside the wide spectrum of existing Wi-Fi solutions, Tag4M chose a Wi-Fi radio that is ultra low power. The ultra-low power radio makes the tag suitable for sensing applications where battery power management is critical. The batteries must deliver a current peak up to 0.5A, but the pulse duration is very short, of about 1-2ms, due to high transmission rate. The ultra-low power Wi-Fi radio was chosen because of its small form factor, and capability to "talk" native Internet language TCP/IP and UDP. Tag4M does not need a specialized Access Point to reach the network. An off-the-shelf Access Point that is configured to "see" Wi-Fi tags will be able to route data from tags to the network and further to data client computers.

Tag4M exposes I/O terminal blocks, very similar to a data acquisition device. The tag user can build wireless sensor solutions for a wide range of applications by attaching Sensor wires to tag terminal blocks. Optimized tags with a lower cost to build can be built for custom applications. Wi-Fi networked sensors send measurements to web pages which in effect become "Web Instruments". Web Instruments of all kind will be built and posted on the Internet to allow users of sensors to bring measurements into computers.

2. Tag4M – hardware description

2.1 Tag hardware components

Tag4M is a Wi-Fi 802.11 b/g tag solution for sensor measurements. The board is built upon G2 Microsystems' 2.4GHz G2M5477 Wi-Fi radio module.

The G2M5477 is an embedded system incorporating a sensor interface, a 32-bit CPU, memory, operating system, complete Wi-Fi networking solution, TCP/IP network stack, crypto accelerator, power management system and real time clock (G2 Microsystems, 2008 a). WEP and WPA with a 4Mbit/s throughput sustained TCP/IP are the tag security suites. The G2C547 chip is the second generation of ultra low-power Wi-Fi SoCs from G2 Microsystems. Technical solutions based on the first ultra-low power chip, G2C501, were presented in (Ghercioiu, 2007) and (Folea & Ghercioiu, 2008). G2M5477 is the smallest, lowest power 802.11b/g module available.

The module supports adhoc and enterprise networking modes. Additionally, the radio module contains a 14-bit Analog-to-Digital Converter (ADC) that is used for the analog sensor interfaces. The tag offers four digital lines for general purpose I/O. The tag provides direct connections to an onboard thermistor for temperature reading, and to its own battery for voltage reading. The tag can be powered from a CR123A 3.0V battery if inserted in the tag battery holder, or from an external 3.3V DC adapter power supply. The Tag4M module is programmed and controlled with web, C++ and LabVIEW interfaces. Once the Tag4M module is powered it will scan to find an access point, associate, authenticate, and connect over any Wi-Fi network. Tag dimensions are 4.7cm x 7.0cm (Tag4M, 2009). A placement diagram of the tag system is presented in Figure 1.

A general bloc diagram of the tag system is presented in Figure 2. The tag offers the following analog interface:

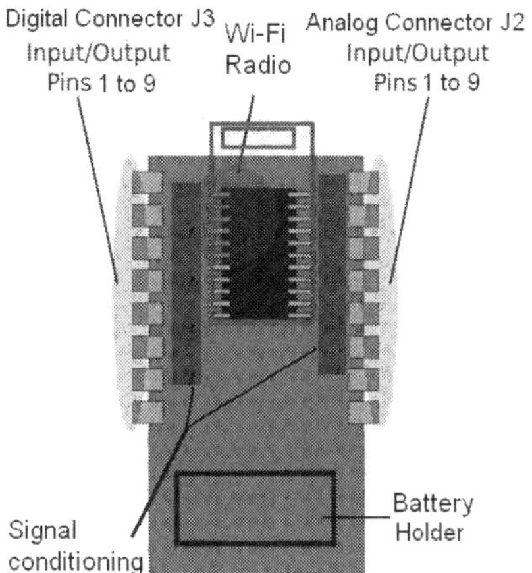

Fig. 1. Tag4M Placement Diagram

- one channel for 0-10V voltage input range,
- one channel for 4-20mA current input range,
- three channels for 0-400mV input range, channels with current generators capability between 0.2uA and 200uA to be connected to sensor extensions,
- one on-board temperature sensor implemented with a 10k 1% thermistor, and
- battery voltage measurement capability.

Conversion time for the onboard 14-bit analog-to-digital converter is between 35ms and 35us, with 1% gain error and 0.01% linear error accuracy.

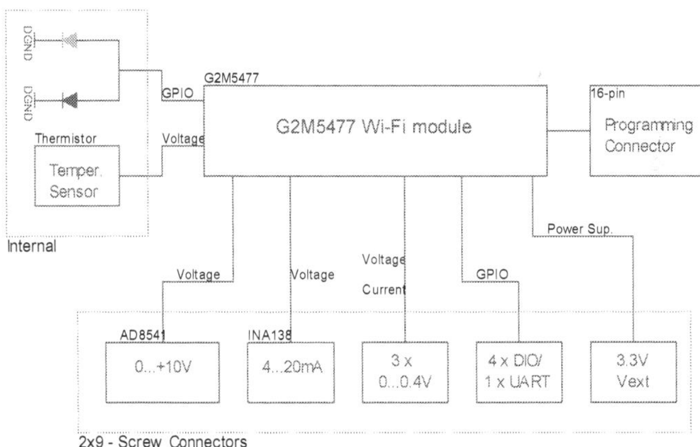

Fig. 2. Tag4M Bloc Diagram

Figure 3 displays a real size 1:1 scale picture of the Tag4M device. The tag has two I/O connectors marked J2 and J3 on the board. J3, the connector on left side is for digital signals and power supply while J2, the connector on right side is for the analog signal lines, 0-10V,

0-400mV, 4-20mA, and their reference Analog Ground (AGND). The battery holder is located on the lower part of the tag Printed Circuit Board (PCB).

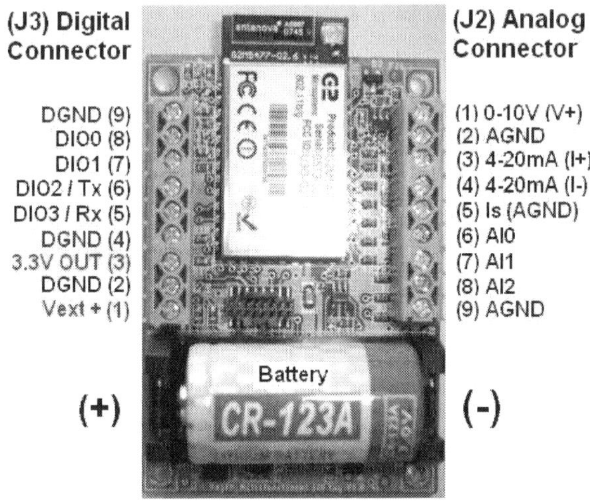

Fig. 3. Tag4M I/O Connector Blocks

Besides Wi-Fi radio, two I/O connector blocks, and conditioning circuitry, the tag implements protection circuitry against reversed battery mounting into tag battery holder, as shown in Figure 4.

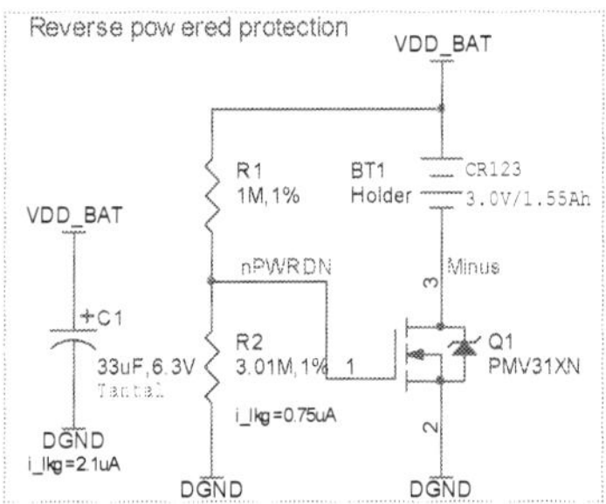

Fig. 4. Reverse Power Supply Protection

The reversed protection circuit is implemented using a MOSFET transistor in position Q1 and two resistors R1 and R2. Reversed protection is needed in case the battery is plugged-in reversed to protect the tag from malfunctioning. Capacitor C1's role is to reduce the peak current from battery.

2.2 Sensor attachments

There will be a separate discussion regarding sensor attachments to the tag in Chapter 4. This section presents the measurement strategy in terms of channel selection when using the tag. If the signal is in the range of 0-10V, it is recommended to wire it to the 0-10V (V+) and AGND I/O connector terminals. 0-10V measurement is done with a 14-bit precision ADC with accuracy in the +/- 5mV range. If the signal is in the range of 0-400mV, is recommended to wire it to one of the channels AI0, AI1, or AI2 referenced in AGND to increase the resolution at 14-bit. Channels AI0, 1, 2 have current drive capability which can be enabled in software and it should be used when reading thermistors or other sensors that need current excitation. Channel 4-20mA is assigned to current sensors. DIO lines 0, 1, 2, 3 are general input/output digital lines TTL compatible.

2.3 Power consumption

Tag power consumption is very much determined by the task the tag runs at any given moment in time. Generally speaking, the tag is running in power cycles, each cycle contains a period of sleep and a period of wake time. Inside a cycle, during the periods of sleep and wake, the tag executes distinct tasks in terms of amount of powered required and time to execute. Figure 5 bellow shows the current consumption and length of time for each of these tasks, during one cycle, when the tag is powered by a 3.0V, 1.55Ah Lithium battery. Total cycle length of time is between 150 and 500msec.

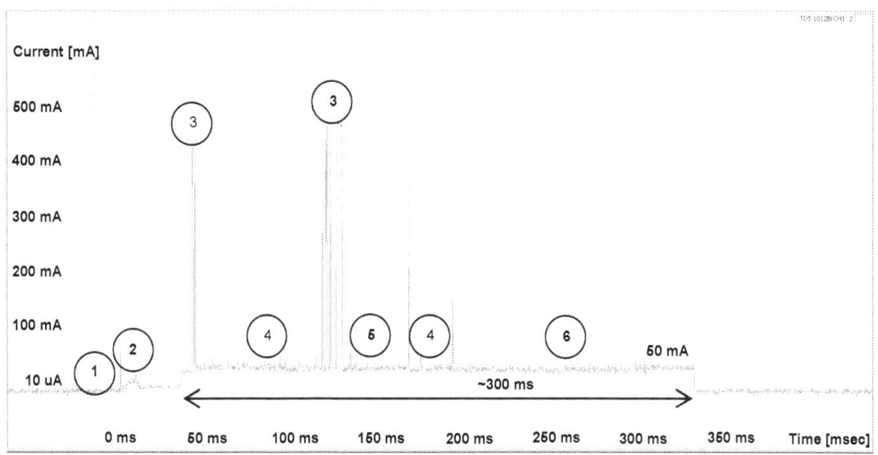

Fig. 5. Current consumption and Function Periods

The tag Wi-Fi radio module was designed for three distinct power domains in order to achieve flexible power management and lower power consumption. These power domains are (G2 Microsystems, 2008 c):

Sleep Mode: when the tag is in sleep mode the tag executes the minimum amount of tasks and by that consumes the minimum amount of power to stay alive.

In sleep mode power consumption is in the microwatts range, while the CPU and almost all tag components are not available. The following limited number of functions will run in this mode:

- decrement timers and detect expiry;
- detect the state change of switch sensors;
- monitor the sampled comparator and detect when external parameters pass preset thresholds;

-   respond to battery brownout (low voltage) etc.

Doze Mode: when the tag is in doze mode the 1.3V rail will be powered and it will remain powered in order to facilitate the tag to respond very quickly to interrupt source (45ns). Tag CPU will not be clocked in doze mode.

Awake Mode: when the tag is in awake mode the 3.3V rail will be powered. The functions that will run when the tag is in the tag awake mode are:

-   loading and executing programs from flash memory,
-   communication based on Wi-Fi radio,
-   measurements using the sensor interface,
-   utilization of GPIO, SPI, SDIO and UART interfaces, reading and writing flash memory or NVM (Non Volatile Memory).

Transition from sleep to awake is triggered by an awake event, which is a subset of some specific interrupts.

The tag battery is capable of providing the peak power required for all tasks, including data transmission which is the most demanding in terms of power consumption. Obviously the battery cannot supply energy to make the tag operate in its high-power state indefinitely. If the execution of long tasks that are power hungry is required in an application, then the designer of this application needs to break long tasks into smaller tasks, and insert low-power states in between them. Let us look into the tasks that are executed during a full cycle. The tag is in sleep mode at the beginning of each cycle. Sleep is the very first task executed in each cycle (Ghercioiu et al., 2007)

Sleep period (Figure 5, #1). Sleep is the least power consuming state of a cycle. Sleep is also the longest tag state in terms of time (as in seconds, minutes, etc) occupied per cycle and also in terms of total tag life time. The tag consumes in average 4-10uA or about 100uW while it performs its sleep function. If a tag is not in sleep mode (and not in doze mode either as we do not activate this mode in firmware) then the tag is in awake mode. A tag wake-up event has five distinct periods as marked and seen in Figure 5.

Boot-up period (Figure 5, #2). The tag does two things during boot-up: it reads a pre-loaded application from flash and it starts execution. It takes an average of 20-30ms to execute these two tasks. Boot-up time may be reduced if using a fixed image that does not need to be searched for in tag flash. The tag will use 10-15mA of current during boot-up.

Transmit Data period (Figure 5, #3). The tag transmits data during transmit periods. You do want your transmit period to be the shortest period of time because it is the most power hungry of all periods in the life cycle. One transmit period can take anywhere between 1 to 10msec, and tag power consumption during this period may go up to 500mA while transmitting 802.11b frames.

CPU processing period (Figure 5, #4). The tag radio has a microcontroller which does all processing that is needed for processing of collected data and transmission via wireless. The CPU processing period is the period of time in the cycle occupied by the microcontroller doing the processing. This period can take anywhere from 100 to 150msec, depending on the amount and length of tasks that need to be processed by the microcontroller. The tag consumes 50mA in this state.

Measurement period (Figure 5, #5). This is the period of time occupied by the microcontroller for collecting data (or data acquisition). The measurement period is short, in the range of 12msec during which the Tag consumes on average 50mA. Measurement period precise timing and power consumption is presented in figure 6.

Receive Data period (Figure 5, #6). This is the period of time following a transmission when the tag is waiting for a command that may be coming from the Access Point. There may be no command, in fact most of the times there is none. The receive period was defined for cases when it is necessary to configure and reconfigure the tag for a different set of measurements during the same application. Tag configuration is done by sending commands to the tag immediately following a data transmission. Tag receive period may take between 0 and 500msec. The length of the receive period is software configurable.

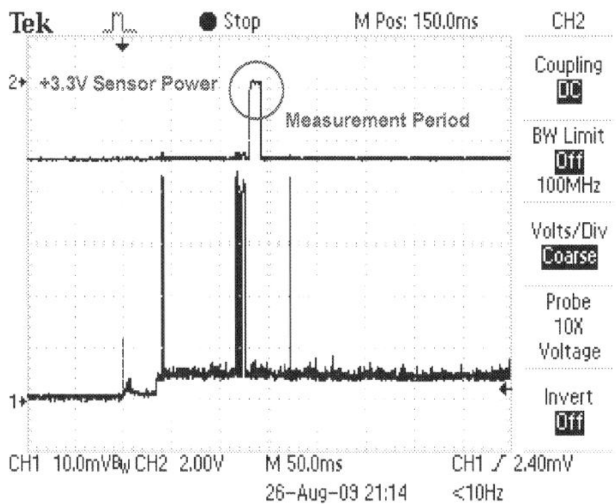

Fig. 6. Measurement Period

The average current consumption during an entire wake-up period is below 70mA. A Tag battery Life-Time Prediction Table was created based on the above power consumption period model and a CR123A lithium battery with a total charge capacity of 1550mAh:

Length of Sleep Period	Battery Life Time
1 s	52 hours
10 s	13 days
100 s	4 months
500 s	2 years

Table 1. Battery Life Time

The tag battery life time depends on two things: length of the sleep and wake-up times as shown in Table 1, but also on the functionality choices that are made in firmware, things like:
- If wireless security is turned ON/OFF,
- If the mechanism used to obtain network access is DHCP with fix IP address or a dynamic one,
- If the protocol used to deliver reports is UDP.

3. Tag description - software

The Tag4M software architecture contains two layers of software:
- Tag Firmware - is downloaded, it resides and runs from tag EEPROM and ROM memories,

- Tag Application Software - runs on a computer and interacts with the tag firmware with the purpose of controlling the tag application. Examples of application software packages used in tag applications are: Web page, LabVIEW, or a C++.

The tag application can run in two modes: Web Mode and Local Mode.

In **Web Mode,** tags send data to an Access Point (AP) that is located in the tag vicinity, and it is connected to the Internet. Data will be routed into the network by the AP and it will travel to the computer hosting the Web Instrument, wherever this is located. Other computers connected to the network can access the web page from their browser.

In **Local Mode,** tags send data to an Access Point (AP) that is located in the tag vicinity, and this Access Point may or may not be connected to the Internet. Data will be broadcasted by the AP to address 255.255.255.255 local wireless network and it will be captured at local PC Port 50007 by the computer that is hosting the local application software built tag instrument.

3.1 Tag firmware

The Tag4M firmware provides the infrastructure required by a tag application. Tag firmware contains an embedded operating system (eCos), a TCP/IP stack (LWIP), start-up code, an application loader, power saving features, and a device driver. This functionality resides in, and is run, from a ROM location on the tag, so it does not need to be loaded at start-up. The device driver contains definitions of tag opcodes (operation code), which are numerical codes that describe tag settings and tag measurement functions, and are loaded from an EEPROM location at start-up.

The tag firmware can perform the following functions:
- set sleep time,
- get sleep time,
- set receive period,
- get receive period,
- get total tag sleep period in msec,
- get total tag wake up period in msec,
- get tag battery voltage,
- get temperature value from on-board tag thermistor,
- get voltage reading from the 0-10V line with respect to AGND,
- get voltage reading from AI0, AI1, AI2 lines with respect to AGND,
- get current reading from 4-20mA I+, I- lines,
- get DIO0, 1, 2, 3 digital lines state,
- get tag RSSI value (Received Signal Strength Indication),
- get tag MAC address (Media Access Control),
- get tag IP address (Internet Protocol).

An opcode (operation code) is a pair of numbers that describes a certain function. The first number in the pair is the function code, while the second number is the value of that particular function. Let us look at opcode (1, 3000) as an example. The first parameter (1) is the code for tag sleep time, which in this example is set to 3000msec. The list of all supported opcodes in tag firmware is presented in Table 2.

To activate an input the application sends opcode 131, followed by a pair of numbers indicating the function to activate and current value. The set number is 12-bit format, with 1 for an active input is presented in Table 3.

(1, 3000)	Sleep time = 3000ms
(2, 120)	Tag receive period = 120ms
(3, 303)	Total power down period = 303msec
(4, 28)	Total power up period = 28msec
(13, xxx)	check value (last message acked)
(132, 25686)	Tag thermistor temperature value = 25.686 Deg. C (25686/1000)
(2180, 18809)	Current measured at Tag 4-20mA (I+) and (Is) is 0.188mA (18809/100000)
(2436, 300829)	Voltage measured at Tag 0-10V(V+) with respect to AGND channel is 3.008V (300829/100000)
(388, 28000)	Voltage measured at AI0(6) with respect to AGND(9) channel is 280.0mV
(644, 32000)	Voltage measured at AI1(7) with respect to AGND(9) channel is 320.0mV
(900, 2900)	Voltage measured at AI2(8) with respect to AGND(9) channel is 29.0mV
(1156, N)	State of Digital Line DIO0
(1412, N)	State of Digital Line DIO1
(1668, N)	State of Digital Line DIO2
(1924, N)	State of Digital Line DIO3
(2948, -39)	Tag RSSI value = -39dBm
(2692, 2844)	Tag Battery Voltage value = 2.844V (2844/1000)

Table 2. Opcodes in Tag Firmware

Temp.	A0	A1	A2	DIO0	DIO1	DIO2	DIO3	4-20mA	0-10V	Vbatt	RSSI

Table 3. The Function to Activate

3.2 Web instrument

A Web Instrument as shown in Figure 7 is a web page hosted by a computer which allows the user of tags to configure tags for measurement, read the data and display the information in a user friendly interface that is easy to use.

Web Instruments are true virtual instruments because the physical part of the instrument may be located very far away from the instrument panel. The web instrument leverages the network as infrastructure technology.

A Web Instrument is very configurable, each tag that is displayed can be individually configured. By clicking on the tag MAC address, the Web Instrument opens a configuration window that allows the tag to be configured for the set of measurements that need to be made and duration of sleep time. The tag can be calibrated by filling in the calibration coefficients under the Calibration tab. Finally, by clicking on the Settings tab the IP address of the host computer, which is the computer running the web page application can be changed. The configuration window is presented in figure 8.

MAC	TEMP	0-10V	4-20mA	AI 0	AI 1	AI 2	DIO 0	DIO 1	DIO 2	DIO 3	VBATT	RSSI	SLEEP
0012B8002001	25.0 C	3.015 V	0.086 mA	301.26 mV	301.28 mV	301.00 mV	0	0	0	0	2.94V	-77dBm	1s
0012B8000585	24.7 C	0.193 V	0.157 mA	259.25 mV	255.85 mV	254.56 mV	0	0	0	1	3.11V	-46dBm	0.1s
0012B8000647	23.3 C	0.144 V	-	254.23 mV	255.67 mV	254.66 mV	-	0	0	-	2.81V	-62dBm	10s
0012B800078F	23.3 C	0.215 V	-	256.07 mV	-	-	0	0	0	0	2.68V	-65dBm	10s
0012B8000798	23.3 C	-	0.000 mA	-	255.31 mV	253.39 mV	0	-	-	0	2.28V	-63dBm	10s
0012B80007FF	23.4 C	0.163 V	-	255.32 mV	257.29 mV	-	-	0	-	0	2.86V	-60dBm	10s

Fig. 7. Web Instrument Example

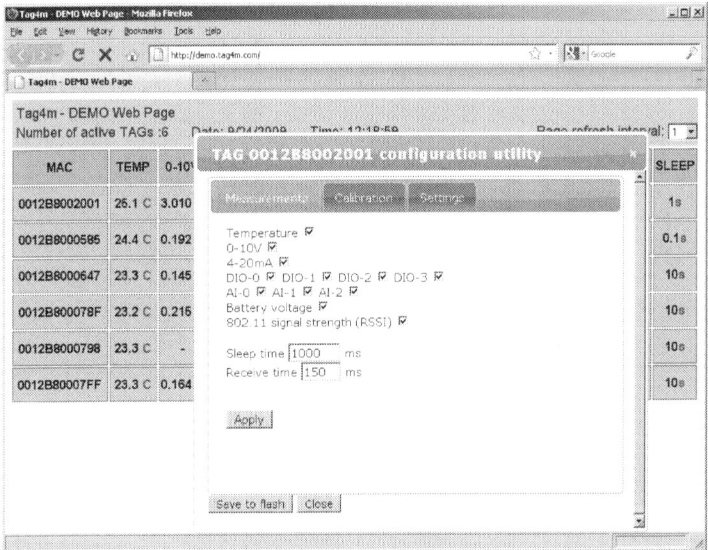

Fig. 8. Configuration Panel presented by the Web Instrument Example

3.3 LabVIEW driver

If a tag measurement application is running in Local Mode, chances are high that LabVIEW is used as PC application environment. Tag4M tags have a LabVIEW Driver which is a set of ready-made instruments that allow tag users to talk to their tags from inside LabVIEW running on a local PC. The LabVIEW driver contains the following VI's:
-    *WiFiTag_Read_MAC.VI*: this VI should run first to determine the tag MAC address and also its allocated IP address. The panel VI is presented in figure 9.

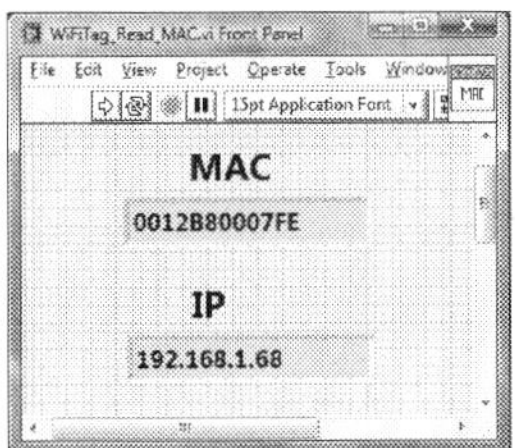

Fig. 9. WiFiTag_ReadMAC.VI

- *WiFiTag_Read_Temperature.VI*: this VI returns the temperature reading from a tag's thermistor. The panel VI is presented in figure 10.

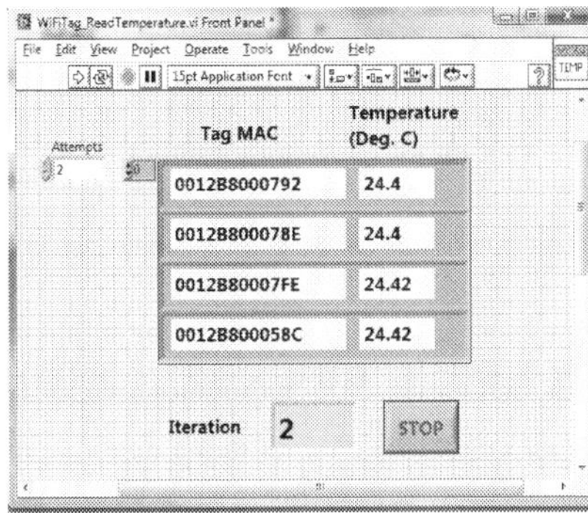

Fig. 10. WiFiTag_ReadTemperature.VI

- *WiFiTag_Read_RSSI.VI*: this VI returns tag RSSI value with regard to the AP. You can calculate the distance in meters between the tag and the associated AP based on the RSSI reading (Folea & Ghercioiu, 2008).
- *WiFiTag_Read_BatteryVoltage.VI*: this VI returns the tag battery voltage.
- *WiFiTag_Read_0-10V.VI*: this VI returns voltage value red from tag channel 0-10V. The panel VI is presented in figure 11.

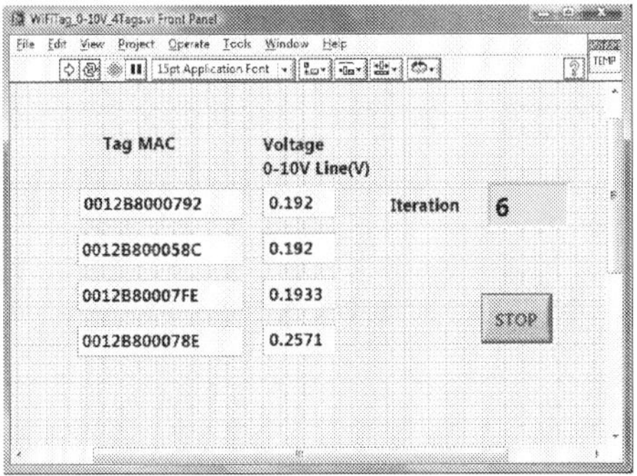

Fig. 11. WiFiTag_0-10V_4Tags VI

- *WiFiTag_Read_4-20mA.VI*: this VI returns current red from tag channel 4-20mA.
- *WiFiTag_Read_AI012.VI*: this VI returns (low) voltage red from tag channels AI0, AI1, and AI2.
- *WiFiTag_Read_DIO.VI*: this VI returns the binary values of tag digital lines DIO0, 1, 2 and 3.
- *WiFiTag_Change_SleepTime.VI*: this VI sets sleep time and receive time in tag firmware. This VI is a write (to tag) type VI.

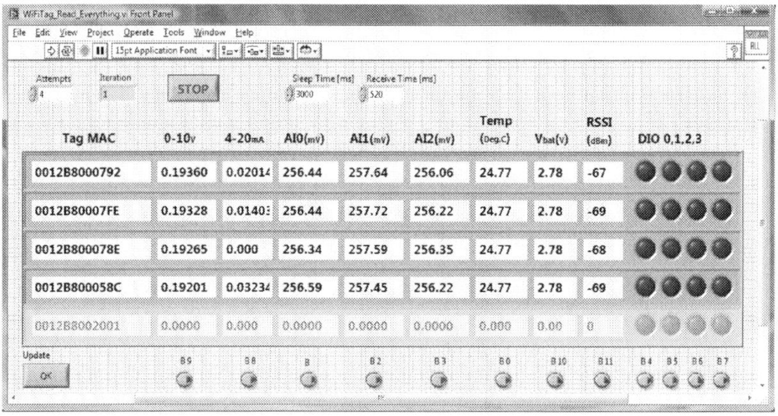

Fig. 12. LabVIEW Instrument Example

The Run WiFiTag_Read_Everything VI which includes all the previously described tasks under one panel is presented in figure 12.

The tag LabVIEW Driver offers a Calibration VI which allows the user to calculate calibration constants and write them to either tag NVR memory for testing or tag EEPROM for permanent storage.

3.4 C++ tag instrument

A C++ tag instrument was developed for users that do not have the LabVIEW programming environment installed in their computer. This instrument when run it opens its panel and displays the tags that are powered with measurements and sleep time, very similar to the web instrument. The C++ instrument example is presented in figure 13.

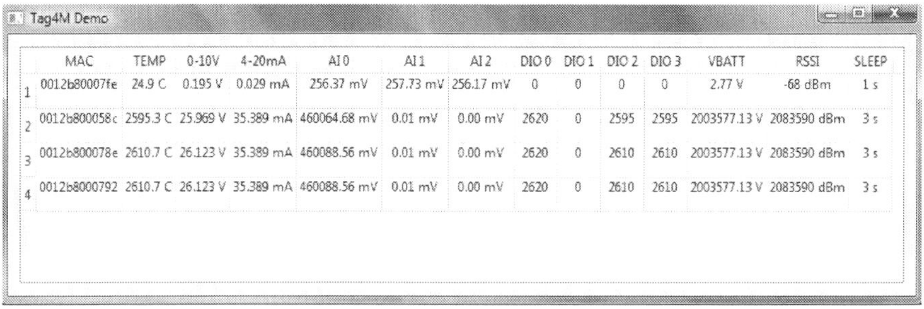

	MAC	TEMP	0-10V	4-20mA	AI 0	AI 1	AI 2	DIO 0	DIO 1	DIO 2	DIO 3	VBATT	RSSI	SLEEP
1	0012b80007fe	24.9 C	0.195 V	0.029 mA	256.37 mV	257.73 mV	256.17 mV	0	0	0	0	2.77 V	-68 dBm	1 s
2	0012b800058c	2595.3 C	25.969 V	35.389 mA	460064.68 mV	0.01 mV	0.00 mV	2620	0	2595	2595	2003577.13 V	2083590 dBm	3 s
3	0012b800078e	2610.7 C	26.123 V	35.389 mA	460088.56 mV	0.01 mV	0.00 mV	2620	0	2610	2610	2003577.13 V	2083590 dBm	3 s
4	0012b8000792	2610.7 C	26.123 V	35.389 mA	460088.56 mV	0.01 mV	0.00 mV	2620	0	2610	2610	2003577.13 V	2083590 dBm	3 s

Fig. 13. C++ Instrument Example

3.5 Calibration program

The LabVIEW Calibration VI was developed as a tool that can be used from inside the LabVIEW environment to re-calibrate the tag. This VI calculates and writes to tag EEPROM calibration coefficients – offset and slope - for all measurements and their associated tag channels.

The application panel for calibration VI is presented in figure 14.

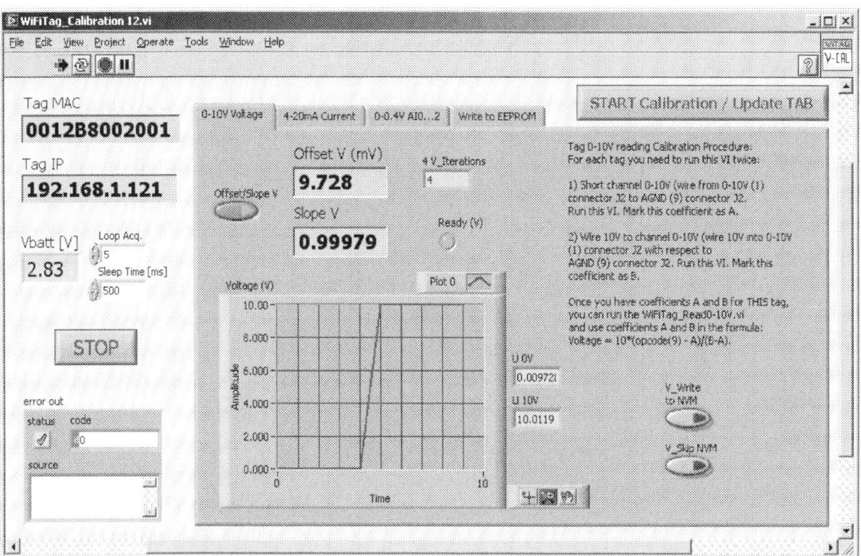

Fig. 14. LabVIEW Calibration Program

The LabVIEW diagram for calibration VI is presented in figure 15 and is an example of LabVIEW graphical programming used.

4. Applications

Wi-Fi tags are used in applications where there is a need for a small wireless tag that can do measurements and can send data to the network (Morariu et al., 2009). The following is a list of such applications:

- Patience monitoring: tremors, heart pulse, temperature, arterial tension.
- Environment: temperature, humidity, light, sound, infrared, barometer, beta and gamma radiation.
- Defence-related sensor networks: battlefield surveillance, treaty monitoring, transportation monitoring.
- Inventory Control: know where your products are and their status.
- Product quality monitoring:
  - Temperature, humidity, impact and vibration, weight and strain.
  - Failure analysis and diagnostic information.

Most of these applications require that a certain sensor or set of sensors is attached to tag measurement lines. For some applications a small extension or attachment board can be

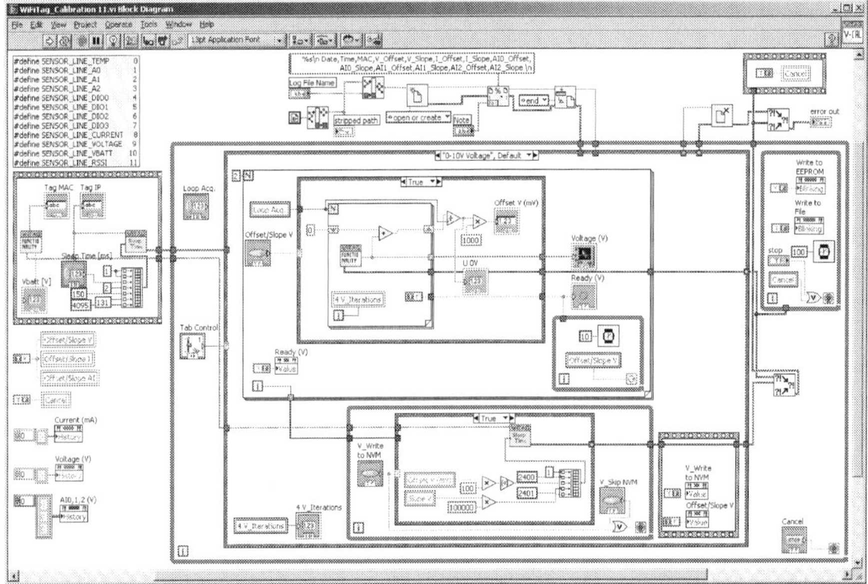

Fig. 15. LabVIEW Calibration Program - diagram

built that contains the sensor, sensor conditioning and power circuitry all on a small PCB. This section presents some of the sensor extension boards that were designed with the idea of showing tag users some examples of what can be done.

4.1 RS232 interface to PC

The RS232 tag extension is used to connect a PC to the tag via RS-232 cable.

The RS232 extension board presented in Figure 16 contains a MAX3221 line driver/receiver and a DB9 connector. This extension module can be used to read data from the tag into a PC with the purpose of debugging the application software. A very similar module containing a 16-pin connector, Reset and Wake_Up buttons is used for downloading firmware into the tag.

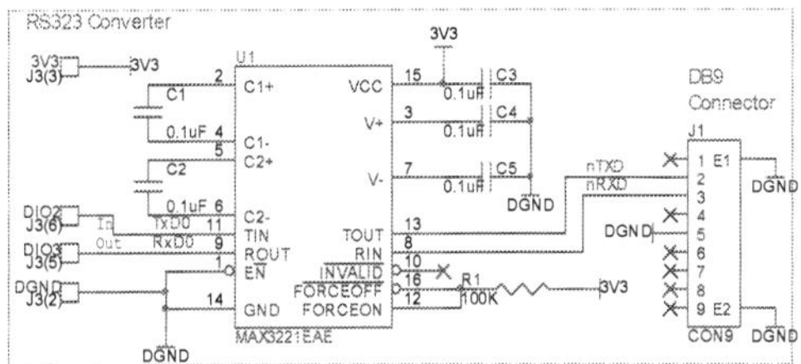

Fig. 16. RS232 Interface to PC

4.2 External power supply

The External Power Supply tag extension presented in Figure 17 is a special connector that allows external power to be connected to the tag.

The external power supply option is needed for power hungry applications like those with a high data acquisition rate and very short sleep time, or with a continuous output that is generated by the tag and needs to be maintained.

The external power voltage range is between +1.8…+3.3V with +3.7V as absolute maximum value and 1A current capability. The tag battery should not be connected if external power is used.

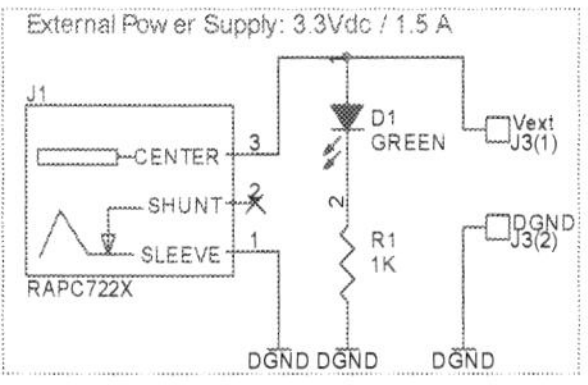

Fig. 17. The Tag4M External Power Supply

4.3 Temperature and humidity sensor

The Temperature and Humidity Sensor extension presented in Figure 18 is used to measure ambient temperature and humidity in the tag vicinity.

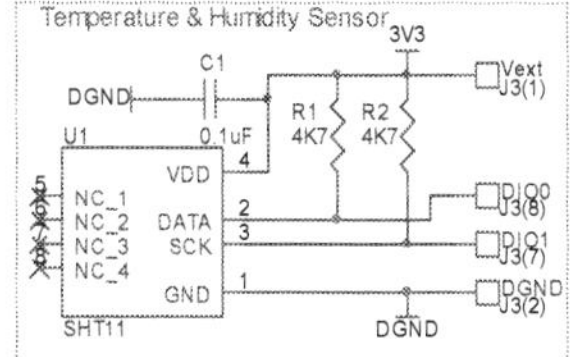
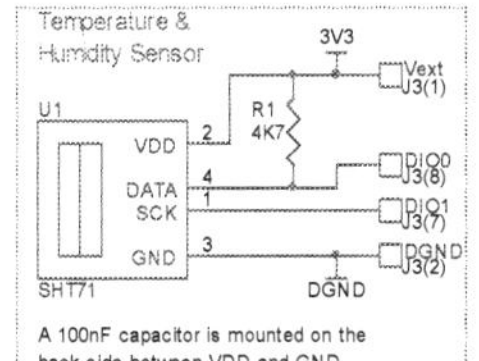
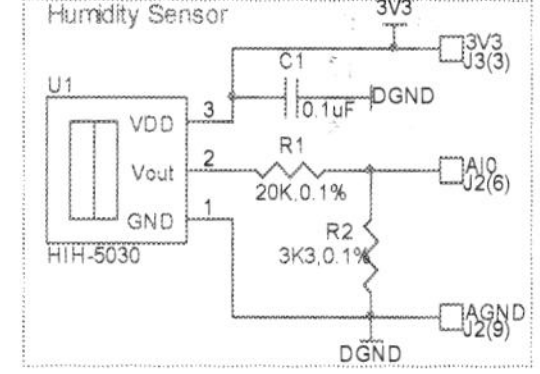

Fig. 18. Temperature and Humidity (SHT11/71) and Humidity (HIH-5030) Sensors

The three different sensors for this type of application were investigated: SHT11 and SHT71 for temperature and humidity and HIH-5030 for humidity only.

SHT11 and SHT71 are digital sensors with an I2C bus for communication. These sensors are very accurate but also more expensive that a thermocouple or RTD (Resistance Temperature Detector). The I2C interface between the sensor and the tag is relatively slow, you can make a reading every 3 seconds (Sensirion, 2005). The HIH-5030 is an analog sensor (voltage) which means it is faster. Generally speaking the tag performs faster readings when sensor output is current or voltage (Honeywell, 2009).

For RTD sensors like the PT100 or 1…10kohm NTC thermistors the internal current generator on lines AI0, 1, 2 can be activated for excitation (G2 Microsystems, 2008 b).

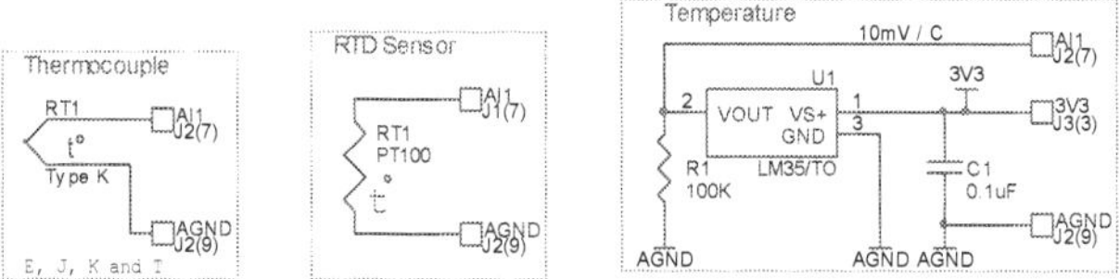

Fig. 19. Temperature: Thermocouple (Type K), RTD (PT100) and LM35 Sensors

The LM35 temperature sensor, Figure 19 right, allows for temperature measurement in the 0-100°C range, generating an output voltage of 10mV/°C, which is proportional with the temperature value. This circuit does not need further calibration and it is quite precise at ±1/4°C at 20°C (National Semiconductor, 2000).

4.4 Light and pressure sensor

The Light Sensor extension presented in Figure 20 is used to measure ambient light in the tag vicinity. The sensor used to measure light intensity is LX1972, a low cost silicon light sensor with spectral response that closely emulates the human eye. Sensor circuitry produces peak spectral response at 520nm, with IR response less than ±5%, of the peak response, above 900nm. Usable ambient light conditions range is from 1 to more than 5000 Lux (Microsemi, 2005).

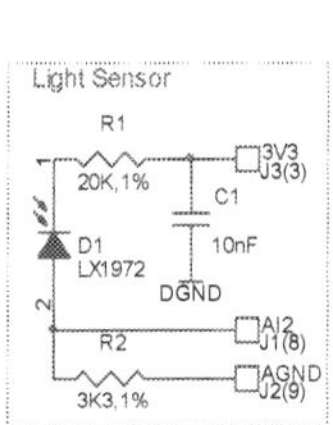
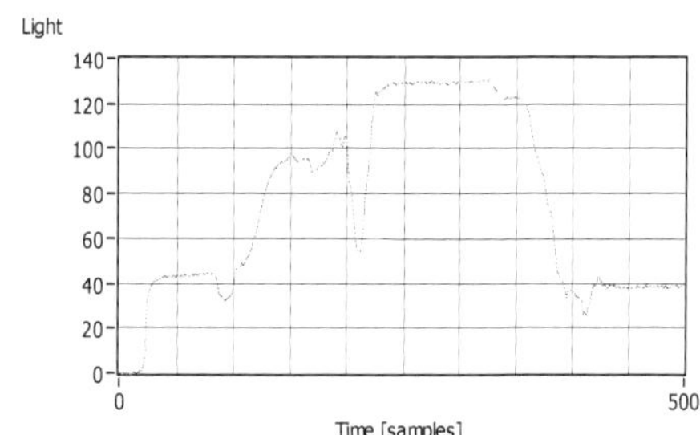

Fig. 20. Light Sensor and Signal

The Pressure Sensor extension is used to measure atmospheric pressure in weather applications. The board mounts the MPX5100 pressure sensor and signal conditioning circuitry as shown in Figure 21 (Freescale Semiconductor, 2009). Here is an example of measured atmospheric pressure for a period of 2 hours.

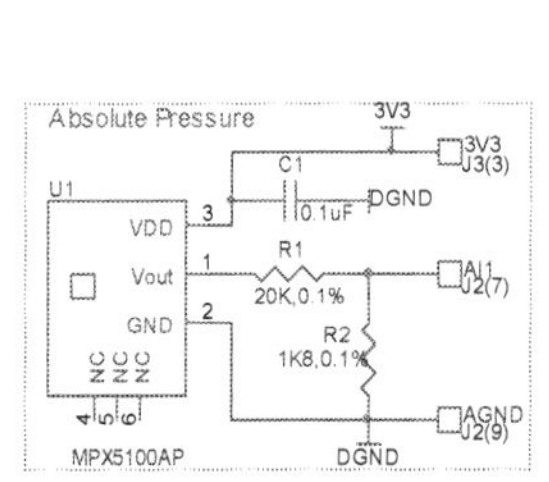
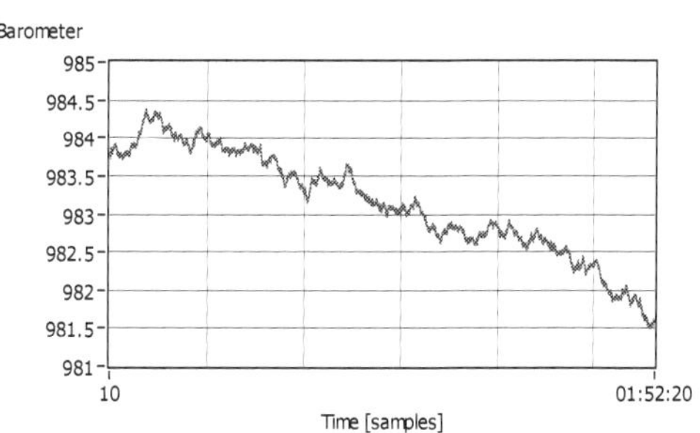

Fig. 21. Pressure Sensor and Signal

4.5 Volt free input and output

The Volt Free Input extension presented in Figure 22 is used to detect if contacts are closed or open. This extension board does not need external components or power supply to complete the circuit with exception of a contact, open or close connection.

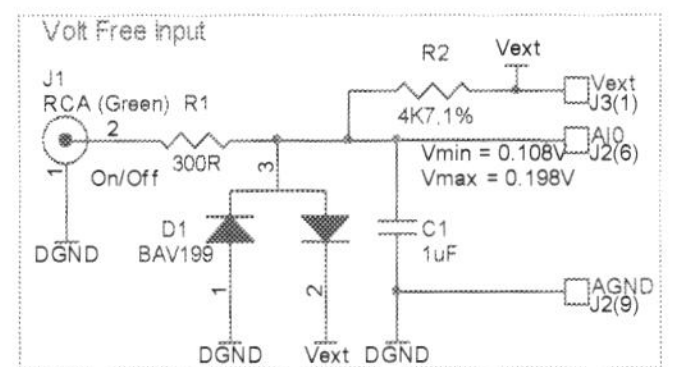
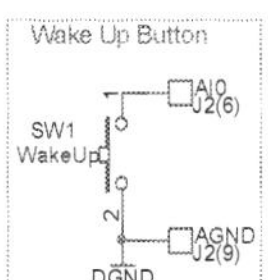

Fig. 22. Volt Free Input and Wake Up Button

The wake up button implementation is a cheaper version of the volt free input board, which does not offer protection against signal spikes.

The Volt Free Output extension presented in Figure 23 is used to control a light source, AC/DC motor, etc. The extension board circuitry is based on a solid state relay AQV202A that works in both AC and DC at 60V, 0.4A, and has optical isolation (Panasonic, 2005).

The external latch, TC7WH74 IC, is necessary to keep the output unchanged while the tag is in sleep mode which sets signal lines CLK and VFO to floating state (Toshiba, 2001).

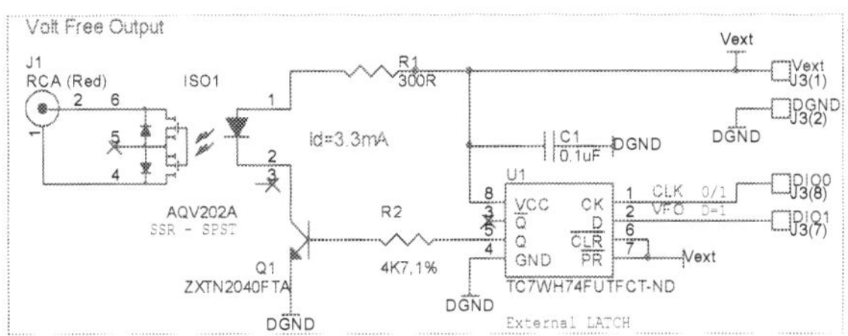

Fig. 23. Volt Free Output

The Volt Free Output extension is only recommended for applications where an external power supply is used.

4.6 Voltage input

The 0-10V Voltage Input extension presented in Figure 24 is connected to low-voltage signal lines AI0, 1, 2 in order to extend their measurement range to 0-10V. This extension increases the number of 0-10V channels to 4 for one tag.

The analogue voltage input range on this extension board is easy to change to almost any range. The circuitry implements over voltage protection too.

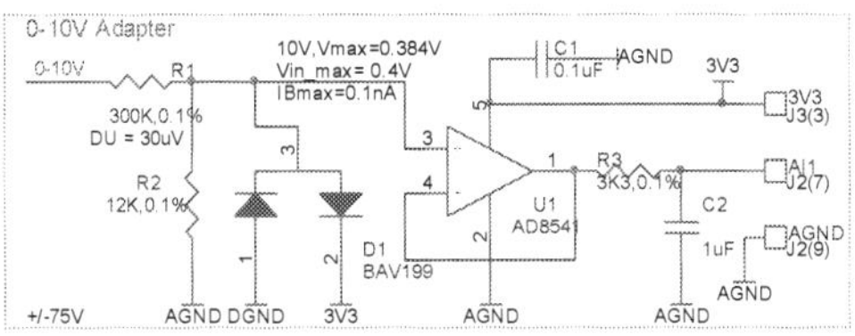

Fig. 24. 0-10V Voltage Input Circuit

The voltage conversion formula from 0-10V in 0-0.4V is (1):

$$U_{AI} = V_{[0-10\,V]} \cdot \frac{R_2}{R_1 + R_2} [V] \qquad (1)$$

Where R_1=300kΩ, 0.1% and R_2=12kΩ, 0.1% are the input divider resistors.

This circuitry implements a divider by 26 which is implemented using two precision resistors and with the AD8541 IC, a low input current (4pA) rail-to-rail amplifier (Analog Devices, 2008).

4.7 Current input

The 4-20mA Current Input extension presented in Figure 25 is used to convert current to voltage in order to allow current signals from sensors to be red by using tag voltage channels.

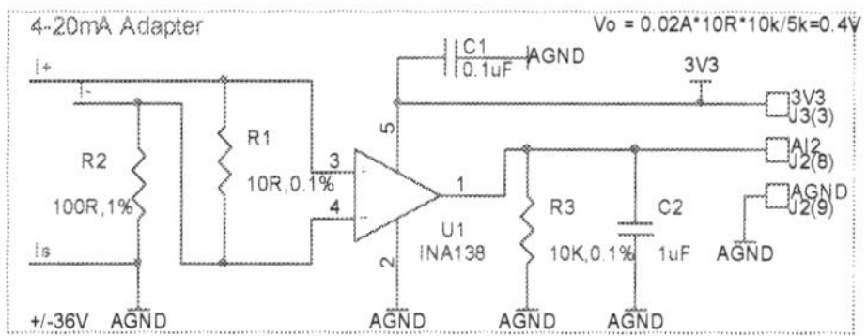

Fig. 25. 4-20 mA Current Input Circuit

The current-to-voltage conversion formula from 4-10mA to 0-0.4V is (2):

$$U_{AI} = I_{[4-20\,mA]} \cdot R_1 \cdot 2 [V] \qquad (2)$$

Where R_1=10Ω, 0.1% is the measurement resistor which needs to be very precise.

The current to voltage converter is implemented using the INA138 IC which has a wide input common-mode voltage range, low quiescent current and tiny SOT23 packaging enable used in a variety of applications (Burr-Brown, 2005).

4.8 Tilt Sensor

The Tilt Sensor extension presented in Figure 26 is used to detect tilt.

This extension is built using the SQ-SEN-200 series sensor. The tilt sensor acts like a normally closed switch which chatters open and closed as it is tilted or vibrated. Unlike other rolling-ball sensors, the SQ-SEN-200 is truly an omnidirectional movement sensor. It will function regardless of how it is mounted or aligned (SignalQuest, 2009).

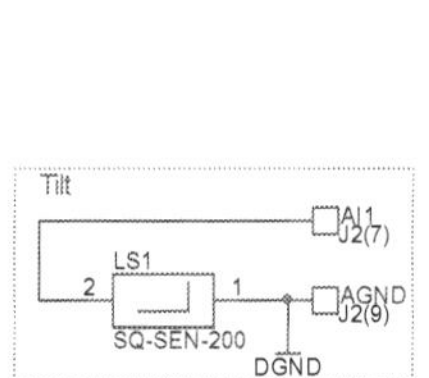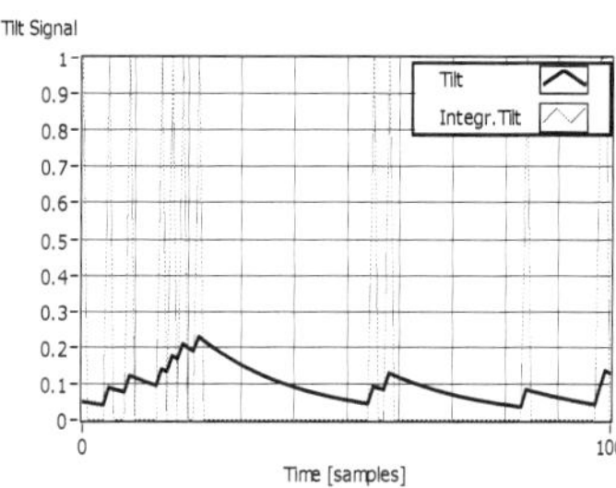

Fig. 26. Tilt Sensor

Each edge coming from the switch signal is software integrated to obtain the function shown in Figure 26, right (SignalQuest, 2006).

4.9 3-Axis acceleration sensor

The 3-Axis Acceleration Sensor extension presented in Figure 27 is used to detect position and to a lesser extent to calculate acceleration.

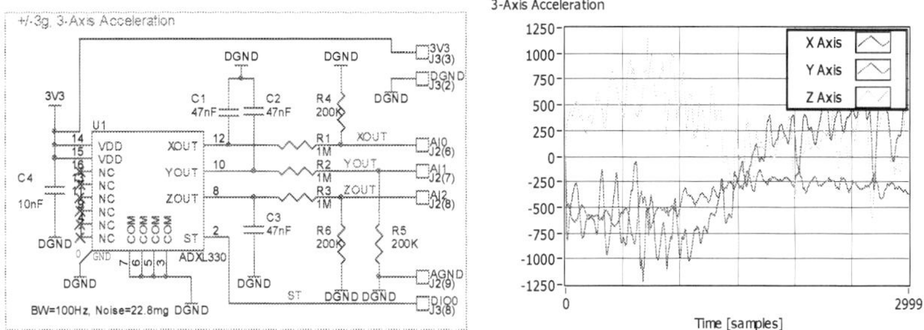

Fig. 27. 3-Axis Acceleration Sensor, Schematic and Output Signal

The ADXL330 (Figure 27) was used in experiments. It is an acceleration sensor of iMEMS type, from Analog Devices. This sensor allows measurement of static or dynamic acceleration on three axes, in the range of ±3g. External components are used to hardware set output signal period in the range of 2 to 1000ms, frequency being limited to the range of 0.5Hz to 1.6kHz. Typical sensor noise level is 280µg/√Hz rms which allows for a precision of less than 5mg level. The extension can also be used for vibrations measurement (Analog Devices, 2009).

Possible applications of the Wi-Fi tag with the 3-Axis Acceleration Sensor extension are the stud of different types of tremors, in a health telematic network. An experimental, wireless tremor telemonitoring system could be implemented. The system is composed of an optional number of portable devices integrating three-axis acceleration mini-sensors which are connected to Wi-Fi tags with transmission capabilities. The main advantages of the design system consist of the possibilities to monitor simultaneously many body parts of one or multiple subjects on local or more extended areas both for scheduled assessments and in an everyday life environment (Bilodeau et al., 2007).

4.10 LEDs, buzzer output and 8-Bit expander

The LED and Buzzer output extensions presented in Figure 28 are used to visualize DIO lines states and to create a buzzer sound. The LED extension board contains LEDs connected

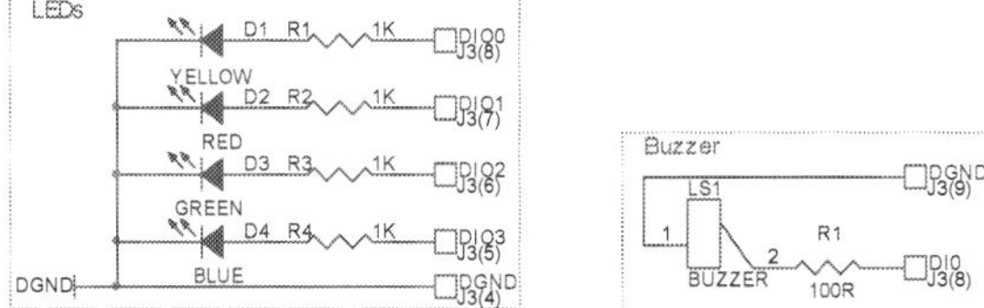

Fig. 28. LEDs and buzzer output

to tag digital outputs lines via current limitation resistors. An auto-oscillated buzzer is used on the Buzzer extension.

The Programmable Expander board presented in Figure 29 can be used in applications where a large number of digital input/output lines are used.

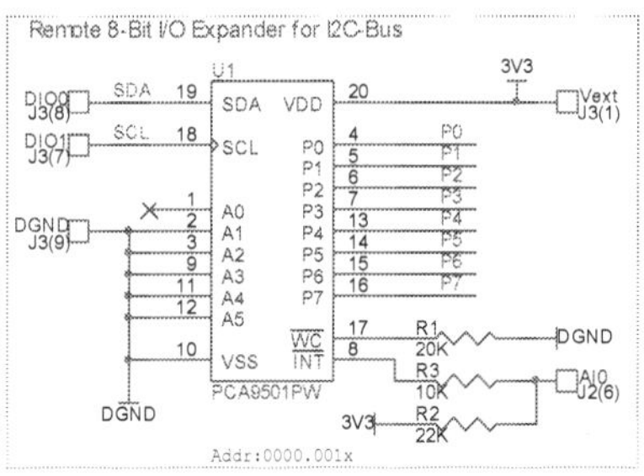

Fig. 29. Remote 8-Bit IO Expander for I²C Bus

The I/O expandable eight quasi-bidirectional data pins can be independently assigned as inputs or outputs to monitor board level status or activate indicator devices such as LEDs. The data for each input or output is kept in the corresponding input or output register (NXP, 2009).

The PCA9501 active LOW open-drain interrupt output is activated when any input state differs from its corresponding input port register state. It is used to indicate that at a certain time an input state has changed and the device needs to be interrogated.

The PCA9501 has six address pins with internal pull-up resistors allowing up to 64 devices to share the common two-wire I²C-bus software protocol serial data bus.

4.11 Microphone and PIR

The Microphone extension board presented in Figure 30 is used to capture sound. Security applications may use sound intensity for alarm condition detection. In Figure 30 right, is presented the first experiment that use sound capture, where there is a longer period of sound.

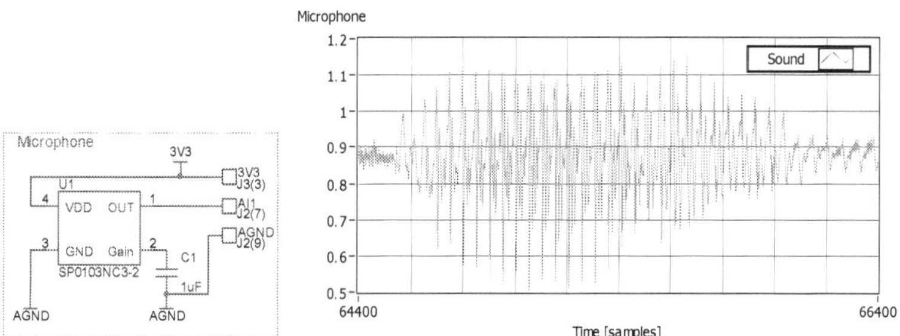

Fig. 30. Microphone, Schematic and Signal

In Figure 31 is presented the second experiment that use sound capture, where the sound period is very short, as in one clap of the hands.

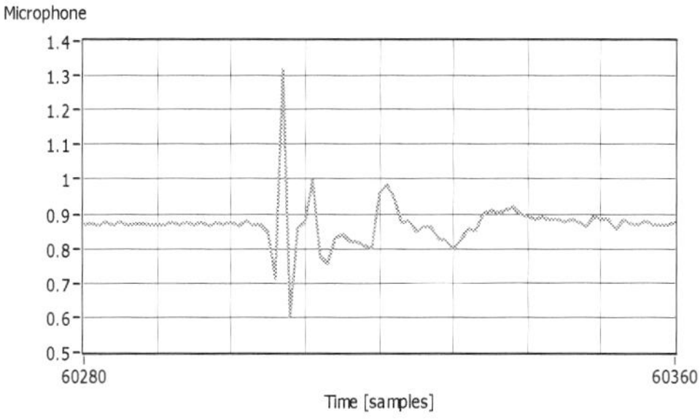

Fig. 31. Microphone Signal

The PIR extension board presented in Figure 32 is used to capture movement.

The extension board mounts the AMN41122 sensor. This sensor detects changes in infrared radiation caused by a person's movement (or an object's movement) which has a different temperature from the surroundings (Panasonic, 2006).

The sensor output is digital, and its current consumption is very low in the range (46…60uA). The sensor outputs current up to 100uA.

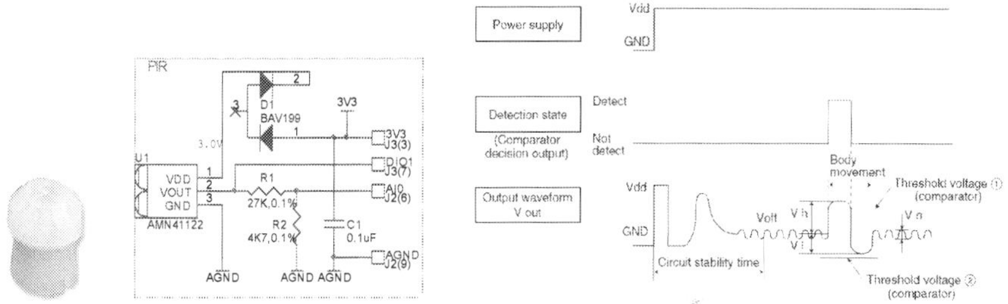

Fig. 32. PIR detector, Schematic and Signal Output

4.12 USB to serial interface

The USB to Serial extension board presented in Figure 33 is used to connect to the tag using the USB interface. The USB interface is capable of powering the tag with up to 500mA.

The FT232R was used. It is a USB to serial UART interface device which simplifies USB to serial designs and reduces external component count by fully integrating an external EEPROM, USB termination resistors and an integrated clock circuit which requires no external crystal, into the device (FTDI, 2009).

Tag applications with continuous high sample analog input rate require more power and therefore are well suited for using the USB to RS232 extension board.

5. Conclusions

This presentation introduces a Wi-Fi tag named Tag4M. The novelty of the design is Wi-Fi with ultra-low power, in a very small package that is running on a battery for years and offers a platform for sensor measurements using the existing (Internet) network

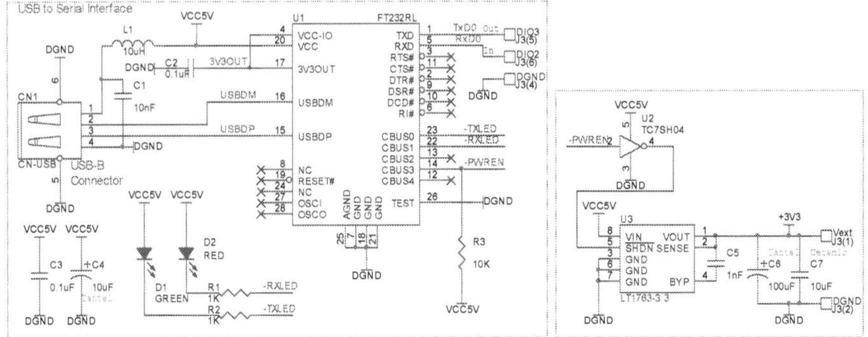

Fig. 33. USB to Serial Interface and +3.3 V, 500 mA LDO

infrastructure. The tag firmware contains communication and measurement commands implemented in the radio microcontroller. The tag has application software support for WEB pages, LabVIEW and C++ instruments.

Sensors can be directly attached to the tag I/O connector or to tag extension boards that may contain conditioning and power circuitry. Several of these dedicated tag extension boards were presented with the idea of showing what can be done in terms of sensor connectivity to the tag in applications for temperature, light, movement, sound, vibration, etc. The tag has its own limitations mostly related to battery power consumption. Data acquisition rate has to be below 30kSps, and tag not being well suited for video applications or very fast and deterministic real time control applications.

For future research is targeted developing portable instruments based on this platform and also integrating the tag in devices used by industrial applications.

6. References

Analog Devices, (2008). *General-Purpose CMOS Rail-to-Rail Amplifiers*, Technical Report, pp. 1-3, www.analog.com

Analog Devices (2009). *Small, Low Power, 3-Axis ±3g iMEMS Accelerometer*, pp. 1-16, D05677-0-6/07(A), pp. 1-4, www.analog.com

Bilodeau, M., Bisson, E., DeGrace, D., Despres, I., & Johnson, M. (2007). *Muscle activation characteristics associated with differences in physiological tremor amplitude between the dominant and non-dominant hand*. Journal of Electromyography and Kinesiology, vol.: 19. Elsevier Publisher. pp. 131–138.

Burr-Brown, (2005). *High-Side Measurement Current Shunt Monitor*, Sbos122c – December 1999 – Revised November 2005, pp. 1-10

Folea, S. & Ghercioiu, M. (2008). *Ultra-Low Power Wi-Fi Tag for Wireless Sensing*, Automation, Quality and Testing, Robotics, AQTR 2008, IEEE International Conference on Volume 3, Issue , 22-25 May 2008, pp. 247 – 252, ISBN: 978-1-4244-2576-1, Cluj-Napoca, Romania

Freescale Semiconductor, (2009). *Integrated Silicon Pressure Sensor On-Chip Signal Conditioned, Temperature Compensated, and Calibrated*, Technical Data, pp. 1-2, Rev 12, 3/2009, Freescale Semiconductor Literature Distribution Center, Denver, Colorado 80217

FTDI, (2009). *FT232R USB UART IC*, Datasheet Version 2.04, pp. 1-40, Future Technology Devices, International Ltd., Document No.: FT_000053

Ghercioiu, M.; Folea, S. & Monoses, I. (2007). *The WiTAG - a WiFi Sensor TAG*, The 2007 International Conference on Wireless Networks, pp. 376-380, ISBN 1-60132-039-6, Las Vegas, Nevada, USA

G2 Microsystems, (2008) a. *G2C547 SoC Data Sheet*, product brief, pp. 1-38, G2 Microsystems, Inc., Campbell, CA, USA

G2 Microsystems, (2008) b. *G2C547 Software, Example Application*, pp. 1-28, G2 Microsystems, Inc., Campbell, CA, USA

G2 Microsystems, (2008) c. *G2M5477 Wi-Fi Module Data Sheet*, product brief, pp. 1-10, G2 Microsystems, Inc., Campbell, CA, USA

Honeywell, (2009). *HIH-5030/5031 Series. Low Voltage Humidity Sensors*, product brief, pp. 1-8, 009050-1-EN IL50 GLO, Printed in USA, May 2009

Labiod de H.; Afifi H. & De Santis, C. (2007). *Wi-Fi, Bluetooth, ZigBee and WiMAX*, pp. 1-316, Springer, ISBN: 978-1402053962, Dordrecht, Netherlands

Microsemi, (2005). *Ambient Light Detector, LX1972*, pp. 1-9, Rev. 1.1b, 2005-10-31, pp. 1-2, www.microsemi.com

Morariu, A.; Folea, S. & Valean, H. (2009). *Reliable Agent Based Monitoring System*, ISCA 24th International Conference On Computers And Their Applications, CATA-2009, ISBN: 978-1-880843-70-3, pp. 99-104, April 8-10, 2009, New Orleans, Louisiana, USA.

National Semiconductor, (2000). *LM35 Precision Centigrade Temperature Sensors*, pp. 1-13, DS005516, 2000, www.national.com

NXP, (2009). *PCA9501 8-bit I2C-bus and SMBus I/O port with interrupt, 2-kbit EEPROM and 6 address pins*, Product data sheet, pp. 1-28, Rev. 04 – 10 February 2009, www.nxp.com

Panasonic, (2005). *Low on-resistance. Controls load voltage 40V to 400V. DC load type is available.*, Technical Report, pp. 1-5, Matsushita Electric Works, Ltd.

Panasonic, (2006). *Motion Sensor (Passive Infrared Type)*, Technical Report, pp. 1-8, Panasonic Electric Works Co., Ltd

Sidhu, B.; Singh, H. & Chhabra, A. (2007). *Emerging Wireless Standards - WiFi, ZigBee and WiMAX*, World Academy of Science, Engineering and Technology, volume 25, pp. 1345-1349, ISSN: 2070-3724, Sydney, Australia

SignalQuest, (2006). *SQ-SEN-200, Application Note Digital Filter and Motion Estimation Algorithm*, pp. 1-2, 10 Water St. Lebanon, NH 03766 USA

SignalQuest, (2009). *Datasheet Sq-Sen-200, Omnidirectional Tilt and Vibration Sensor*, pp. 1-6, 10 Water St. Lebanon, NH 03766 USA

Sensirion, (2005). *SHT1x / SHT7x Humidity & Temperature Sensor*, v2.04 May 2005, pp. 1-9, www.sensirion.com

Tag4M (2009). *Tag4M - The Multifunctional I/O Wi-Fi Tag*, Technical Report, Copyright 2007-2009, Austin, Texas, USA, tag4m.com

Toshiba, (2001). *TC7WH74 Series- D-Type Flip Flop With Preset And Clear*, Technical Report, pp. 1-7, www.semicon.toshiba.co.jp

18

Organic RFID Tags

Kris Myny[1,2,3], Soeren Steudel[1], Peter Vicca[1], Monique J. Beenhakkers[4],
Nick A.J.M. van Aerle[4], Gerwin H. Gelinck[5], Jan Genoe[1,2],
Wim Dehaene[1,3], and Paul Heremans[1,3]

[1]*IMEC vzw, Leuven,*
[2]*Katholieke Hogeschool Limburg, Diepenbeek,*
[3]*Katholieke Universiteit Leuven, Leuven,*
[4]*Polymer Vision, Eindhoven,*
[5]*TNO Science and Industry, Eindhoven,*
[1,2,3]*Belgium*
[4,5]*The Netherlands*

1. Introduction

In this chapter, we investigate the potential of organic RFID tags as a product label. The primary market for organic RFID tags could be barcode replacement, i.e. tags that generate a fixed code sequence when powered by an RF field. We infer this from the current state of the art of the technology: code generators that generate code sequences up to 128 bit are possible in organic electronics (Myny et al., 2009) and chips comprising 414 organic-based thin-film transistors (OTFTs) can today be integrated into fully functional organic RFID tags with HF communication frequency of 13.56 MHz (Myny et al., 2009). More complex RFID embodiments that comprise e.g. encryption, re-programmable code stored in a non-volatile memory, and true bi-directional communication with a reader are beyond the current state-of-the-art of organic electronics, but can be envisioned in a more distant future.

With 64 bit of data, that is read out in 10 to 20 ms, a realistic electronic tag for item-level identification can carry and read out the standard Electronic Product Code (EPC, http://www.epcglobalinc.org/). We have shown that these requirements to the complexity of the chip and the clock frequency can be obtained.

We further assume that product identification tags should preferably be passive tags, that do not include a battery, since the integration of a battery would considerably increase the cost of a tag. Passive tags are powered by the electromagnetic field of the reader, also called interrogator. To power passive organic RFID tags, high-quality organic rectifiers are needed and these need to be very carefully designed. We have been able to obtain sufficient DC voltage to power organic code generators by organic rectifiers, using reader fields that comply with the standards imposed by the safety rules concerning electromagnetic radiation.

Several research groups have published research results on organic RFID systems. In 2007, Cantatore et al. published a capacitively-coupled RFID system where a 64-bit code was read out at a base carrier frequency of 125 kHz (Cantatore et al., 2007). The 64-bit code generator was fully functional at a 30 V supply voltage. In that pioneering work, lower bit generators (up to 6 bit) could be read out using a base carrier frequency of 13.56 MHz by a capacitive antenna. Ullmann et al. demonstrated a 64-bit tag working at a bit rate exceeding 100 b/s, readout by inductive coupling at a base carrier frequency of 13.56 MHz (Ullmann et al., 2007). We review in this chapter recent advances in both the digital transponder chip and the analog front-end of organic RFID tags, and demonstrate that organic electronics can result in a tag with a realistic code size, bit rate and reading distance at a reasonable and allowed field strength (Myny et al., 2008; Myny et al., 2009). We also demonstrate significant increases in complexity of RFID transponder chips where the data have been Manchester encoded and a basic anti-collision protocol has been added that will allow the readout of multiple organic RFID tags in the field of a single reader at once (Myny et al., 2009).

The different sections in this chapter discuss the following building blocks of an organic RFID tag: the antenna coil, the HF-capacitor, the rectifier and the 64-bit transponder chip with integrated load modulator. The coil and HF-capacitor match the resonance frequency of 13.56 MHz, and absorb energy transmitted by the reader and power the organic rectifier with an AC voltage at 13.56 MHz. The rectifier generates the DC supply voltage for the organic transponder chip, which drives the modulation transistor between the on- and off-state with the code sequence.

2. Components of the organic RFID tag

The basic schematic of the organic RFID tag presented here is depicted in Fig. 1. The organic RFID tag consists of 4 different modules: the antenna coil, the HF-capacitor, the rectifier and the transponder chip with an integrated load modulator. The coil and the HF-capacitor form an LC tank resonating at the HF resonance frequency of 13.56 MHz, which energizes the organic rectifier with an AC voltage at 13.56 MHz. The rectifier generates the DC supply voltage for the 64-bit organic transponder chip, which drives the modulation transistor between the on- and off-state with a 64-bit code sequence. Load modulation can be obtained in two different modes, depending on the position of the load modulation transistor in the RFID circuit, shown in Fig. 1. AC load modulation, whereby the modulation transistor is placed in front of the rectifier, sets demanding requirements to the OTFT, since it has to be able to operate at HF frequency. This is not obvious, as a consequence of the limited charge carrier mobility of the OTFT, 0.1 – 1 cm²/Vs for pentacene as organic semiconductor. Therefore, load modulation at the output of the rectifier (DC load modulation) is preferred in organic RFID tags. In latter mode, the OTFT does not require to operate at HF frequency. The organic RFID tags in this chapter operate in DC load modulation mode. Nevertheless, organic RFID tags operating in AC load modulation mode have also been achieved (Myny et al., 2009).

2.1 Technology

In this part we describe the technology used to create high-performance organic RFID tags (Myny et al., 2009). As mentioned earlier, the tags are composed of four flexible foils, with the following components: an inductor coil, a capacitor, a rectifier and a transponder. The coil is made from etched copper on foil, and was manufactured by Hueck Folien GmbH.

The HF-capacitor consists of a metal-insulator-metal stack (MIM stack), processed on a 200-µm thick flexible polyethylene naphthalate (PEN) foil (Teonex Q65A, Dupont Teijin Films). The insulator material used for the capacitor is Parylene diX SR.

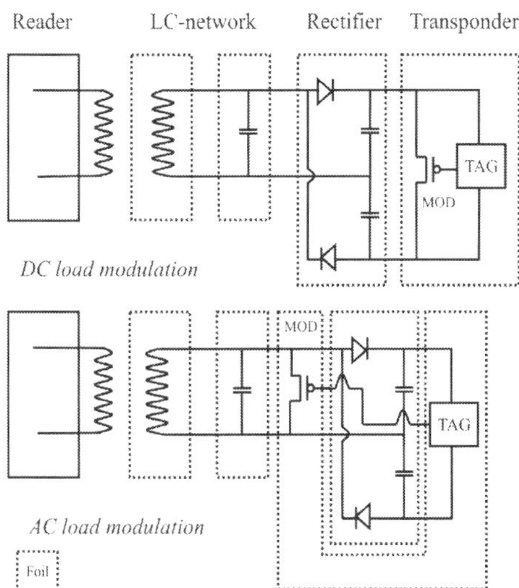

Fig. 1. Inductively-coupled organic RFID tags using DC (top pane) and AC (bottom panel) load modulation.

The rectifier comprises two vertical Schottky diodes, and two capacitors in a so-called double half-wave configuration. The schematic of the rectifier is shown in Fig. 9, and a photograph of the rectifier is depicted in Fig. 2. The substrate for manufacturing the rectifiers is a 200 µm thick, flexible 150 mm PEN foil, on which first a metal-insulator-metal (MIM) stack is processed for the capacitors in the circuit. The metal layers are 30 nm of gold (Au) and the insulator is Parylene diX SR, with a relative dielectric constant of ε_r of 3 and a thickness of 400 nm. Conventional photolithography is used to define the capacitors in the MIM stack. The top Au layer of the MIM stack is used as anode for the vertical diode. A 350 nm pentacene layer, the organic semiconductor, is evaporated through a shadowmask by HV-deposition. Last, an aluminum (Al) cathode is evaporated through a second shadowmask.

The organic 64-bit transponder chip is made on a 25 µm thin plastic substrate using organic bottom-gate thin-film transistors. The organic electronics technology that is used, was developed by Polymer Vision for commercialization in rollable active matrix displays and is described elsewhere (Huitema et al., 2003; Gelinck et al., 2004). The insulator layers and the semiconductor layer are organic materials processed from solution. The transistors, with a typical channel length of 5 µm, exhibit an average saturation mobility of 0.15 cm²/Vs. A micrograph picture of the 64-bit transponder chip and the 6″ wafer is depicted in Fig. 2.

2.2 The RFID measurement setup

The complete tag is realized by properly interconnecting the contacts of the four foils, which we achieved in an experimental set-up where we plug the individual foils into sockets as shown in Fig. 3. Alternatively, we have also achieved tags by lamination of the foils, whereby electrically conductive glue is used to interconnect the different contacts of the individual foils.

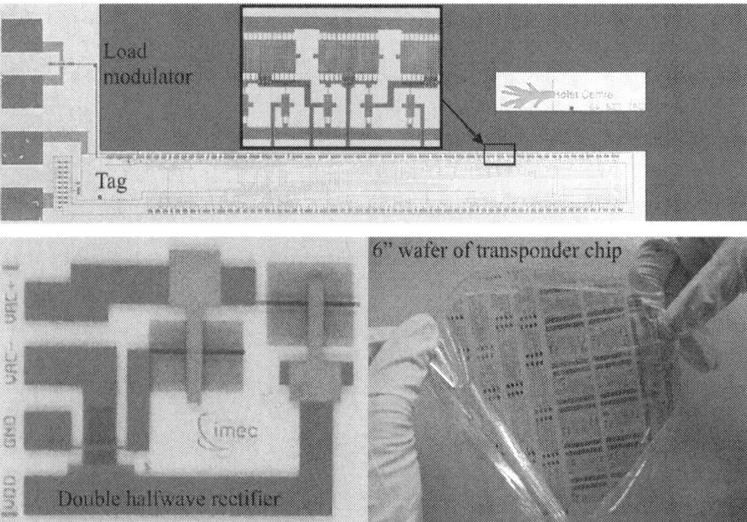

Fig. 2. Pictures of the transponder foil and the load modulator foil (top), the double half-wave rectifier foil (bottom left) and the 6" wafer full of transponder chips (bottom right).

The reader setup conforms to the ECMA-356 standard for "RF Interface Test Methods". It comprises a field generating antenna and two parallel sense coils (Fig. 3), which are matched to cancel the emitted field. By this method, only the signal sent by the RFID tag is read out at the reader side. The detected signal is then demodulated by a simple envelope detector (inset Fig. 12), being a diode followed by a capacitor and a resistor, and shown on an oscilloscope.

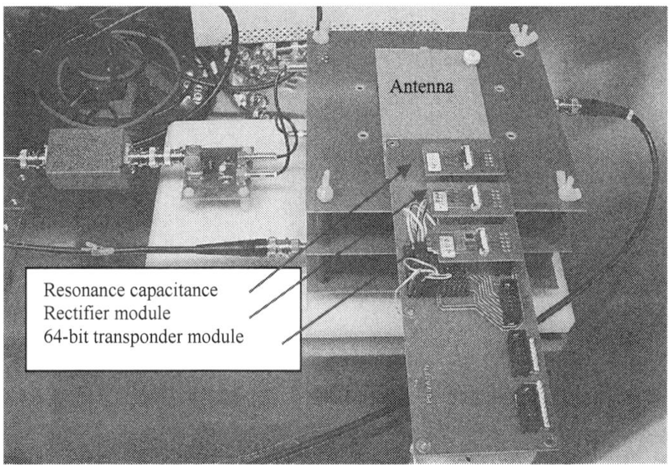

Fig. 3. Overview of the reader and RFID tag measurement setup. Foils are placed in sockets to ease manipulation.

3. Organic transponder chip

3.1 Design issues for the organic RFID transponder chip

In this part, we will discuss challenges in designing large, robust circuitry in present-day OTFT technology. A first challenge is created by the fact that in common OTFT technologies only unipolar, organic semiconductors are available. In our case, the design of the organic RFID transponder chip was limited to p-type only logic, since pentacene is a hole transport semiconductor. In addition, only a single threshold voltage is present in this technology. As a consequence of this, designs based on complementary logic or dual-threshold voltage logic (e.g. nMOS) cannot be implemented. Therefore, in the field of organic electronics, research towards complementary logic has gained considerable attention, potentially leading to an improvement of the robustness of organic circuitry. Very recently, the first 4-bit complementary organic RFID tag has been demonstrated (Blache et al., 2009).

A big challenge in designing complex organic circuitry is the intrinsic parameter spread, in particular on the threshold voltage and the mobility (De Vusser et al., 2006). To tackle this problem in the design, we used a Monte Carlo tool to distribute the transistor parameters randomly for each OTFT in the circuit and to vary these parameters randomly for every new simulation according to the nominal value and within the given spread values for each parameter. The outcome of multiple simulations provides an idea of the robustness of the design and allows to investigate design improvements for increased robustness. To allow such statistical design, the number of transistors in the organic RFID transponder chip has been kept very limited.

Possible gate design architectures in unipolar, single threshold voltage logic are depletion-mode logic or enhancement-mode logic (Cantatore et al., 2007). The gates in our organic RFID transponder chip are designed using the former, which is also known as zerovgs-logic. The choice for this type of logic was driven by the fact that the OTFTs used in this work show normally-on or depletion-mode behavior. The final design of the organic transponder chip has been made using only inverters and NAND-gates, both implemented in the zerovgs-logic as depicted in Fig. 4. The gain of such an inverter at the trip point, for supply voltages of 10 and 20 V, is 1.75 and 2.25, respectively. Nineteen-stage ring oscillators of inverters operate at a frequency of 627 Hz using 10 V supply voltage and at 692 Hz using 20 V supply voltage.

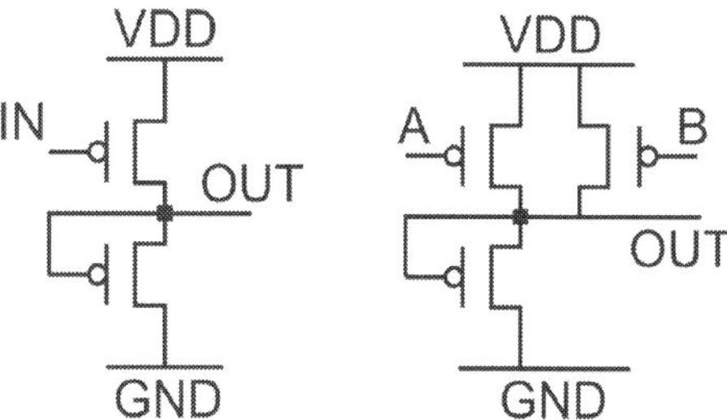

Fig. 4. p-type only inverter (left) and NAND-gate (right) using the zerovgs-logic.

3.2 Final design of the transponder chip

The schematic of the transponder foil is depicted in Fig. 5. A 19-stage ring oscillator generates the clock signal when powered. This clock signal is used to clock the output register, the 3-bit binary counter and the 8-bit line select. The 8-bit line select has an internal 3-bit binary counter and a 3-to-8 decoder. This block selects a row of 8 bits in the code. The 3-bit binary counter drives the 8:1 multiplexer, selecting a column of 8 bits in the code matrix. The data bit at the crossing of the active row and column, is transported via a 8:1 multiplexer to the output register, which sends this bit on the rising edge of the clock to the modulation transistor. The 3 bits of the 3-bit binary counter are also used in the 8-bit line select block for selecting a new row after all 8 bits in a row are transmitted.

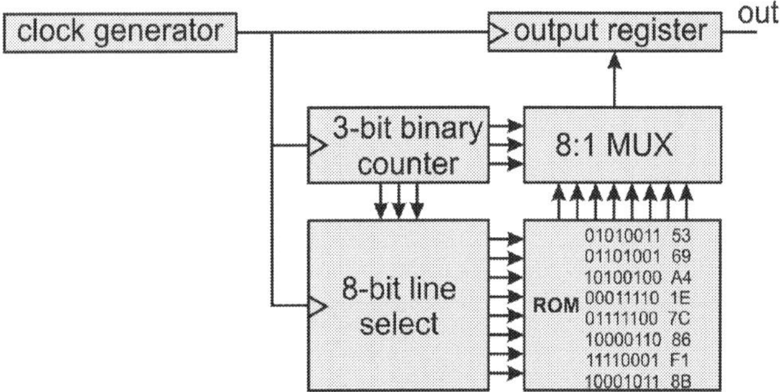

Fig. 5. Schematic overview of the digital logic part of the 64-bit transponder chip.

The 64-bit transponder foil comprises only 414 OTFTs. Figure 2 shows a micrograph image of the transponder foil. At 14 V supply voltage the 64-bit transponder foil generates the correct code at a data rate of 752 b/s, which is depicted in Fig. 6. Besides the 64-bit transponder foil, we also designed an 8-bit transponder foil. The main difference in the design is the complexity of the line select.

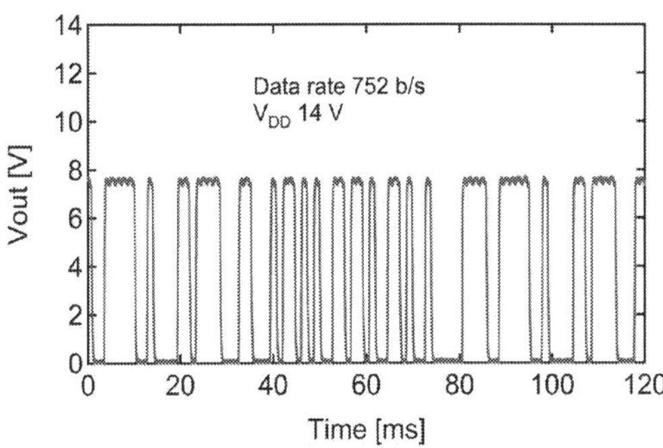

Fig. 6. Measured code of the 64-bit transponder chip at a supply voltage of 14 V.

4. Organic rectifier

4.1 Organic diode

The purpose of the rectifier in an RFID tag is to create a DC-voltage from the AC-voltage detected and generated by an antenna at the targeted base carrier frequency of 13.56 MHz. This frequency is selected because it is a standard in Si-based RFID tags, and will therefore enable partial compatibility with installed reader systems at 13.56 MHz. An important issue for organic RFID tags is the efficiency of the rectifier. A more efficient rectification will result in the required DC-voltage from lower AC-input voltage. This implies larger reading distances for the RFID tags (Myny et al., 2008; Myny et al., 2009).

A rectifier comprises diodes and capacitors. For organic diodes, two different topologies can be used, being a vertical Schottky diode (Steudel et al., 2005; Pal et al., 2008) and a transistor with its gate shorted to its drain. The transistor with shorted gate-drain node is often considered as the most favorable topology because its process flow is equal to that used for the transistors in the digital circuit of the RFID tag. In this work, however, we have chosen to use the vertical diode structure because of its better intrinsic performance at higher frequencies compared to transistors as diodes (Steudel et al., 2006).

The structure of a vertical, organic diode used to make the rectifier, is drawn in Fig. 7. As depicted, we fabricated hole-only organic diodes with a layer of pentacene (as organic semiconductor) sandwhiched between an Au- and Al-electrode on a 150 mm PEN foil. Because of their workfunctions, the Al-electrode blocks the injection of holes, whereas the Au-electrode permits the injection of holes.

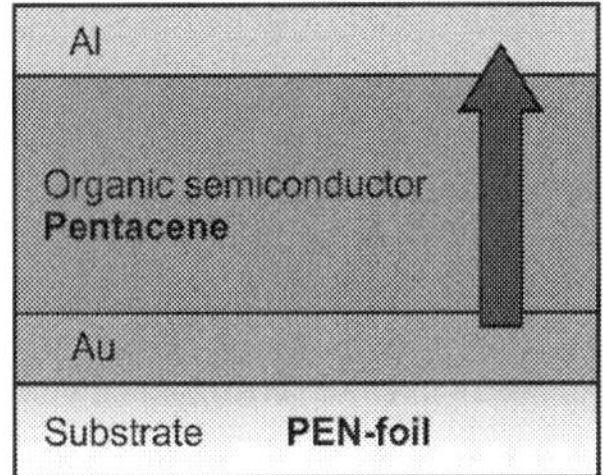

Fig. 7. Structure of a vertical, organic Schottky diode.

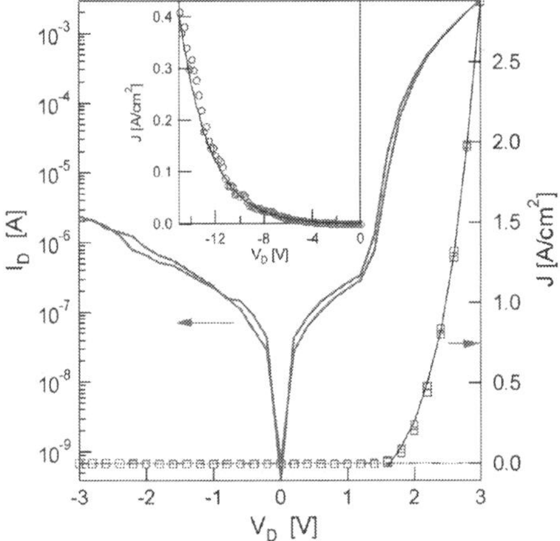

Fig. 8. $|I|$-V and J-V characteristics of an organic pentacene diode (area=500x200 μm^2) on linear (right axis) and logarithmic (left axis) scale. The inset shows the diode leakage current density between 0 and -15 V. The fits to the data of the current densities, both in forward and in reverse biases are represented by the solid black lines.

The electrical behavior of a single organic diode is examined by measurements performed in a nitrogen-filled glove box using an Agilent 4156C parameter analyzer. The current-voltage characteristics of a single pentacene diode are depicted in Fig. 8. The onset voltage of the diode is at about 1.2 V. However, 2 V is required to obtain a current density beyond 200 mA/cm². At 3 V forward bias, the current density is 2.88 A/cm². The leakage current of the diode shows an important increase when the reverse bias increases, limiting the maximum reverse voltage over the diode.

4.2 Organic double half-wave rectifier

A double half-wave rectifier comprises two diodes, each followed by a capacitor (Myny, 2008). Fig. 9a shows the schematic of this circuit. Both capacitors are 20 pF. The active area of the diodes is 500 μm x 200 μm.

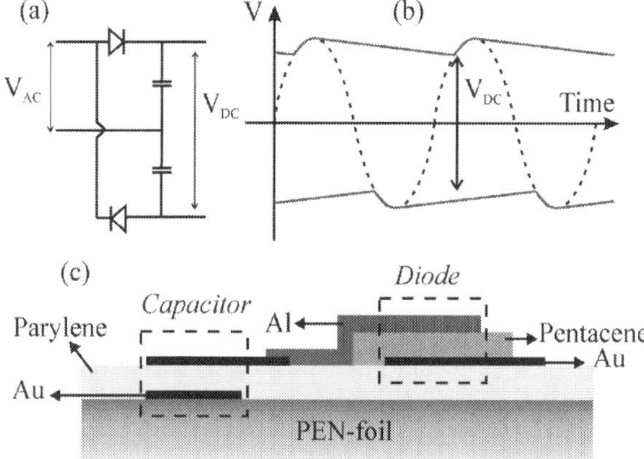

Fig. 9. (a) Schematic of a double half-wave rectifier, (b) schematic operation of a double half-wave rectifier, (c) vertical cross-section of the integrated capacitors and diodes on foil.

A double half-wave rectifier circuit consists of two single half-wave rectifiers connected between the same nodes, with diodes connected as shown in Fig. 9a. Both single half-wave rectifiers rectify the AC input voltage: one rectifies the upper cycles of the AC input voltage, the other single half-wave rectifier rectifies the lower cycles of the input voltage. This is schematically depicted in Fig. 9b. The power and the ground voltage for the digital logic of the RFID tag are taken between both rectified signals (Fig. 9a and b). Therefore, a double half-wave rectifier yields about double the rectified voltage compared to a single half-wave rectifier.

For AC measurements of the rectifiers, we constructed a small printed circuit board (PCB), comprising coax-connections for the AC input signal and the DC output signal. The AC signal is terminated with 50 Ω for limiting reflections at 13.56 MHz. The DC signal is sent to the oscilloscope (MSO6014A of Agilent Technologies) on which the DC voltage was read out, having an input impedance of 1 MΩ, which is the load of the rectifier. The rectifier foil is placed into a socket which is plugged in the PCB. We used a Si-based double half-wave rectifier circuit (assembly of 1N4148-diodes and 20 pF capacitors) as a reference. Fig. 10 plots the rectified DC voltage of the integrated organic and assembled Si-based rectifiers versus the AC input voltage at a frequency of 13.56 MHz. At an amplitude of the AC voltage of 10.9 V, the rectified DC voltage of the organic rectifier is 14.9 V, while the Si-based rectifier produces 20.6 V. The slope of the measured V_{DCout} as a function of applied V_{ACin}, which we term the efficiency of the rectifier, is 1.64 for the organic rectifiers and 2 for the silicon rectifier. The onset voltage of the diodes can be estimated by extrapolation to an output voltage of 0 V and is 1.98 V for the organic rectifier and 0.42 V for the silicon rectifier. Each organic diode, therefore, has an onset voltage of only about 1 V.

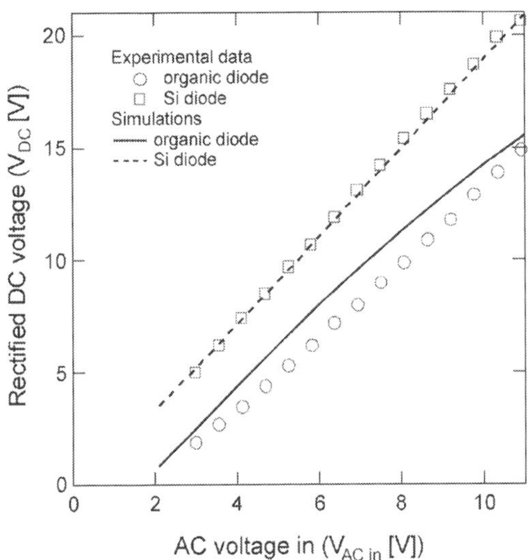

Fig. 10. Simulation results and experimentally obtained data points of the rectified DC output voltage versus the AC input voltage for an organic-based rectifier and a Si-based rectifier both in double half-wave rectifier configuration.

To get a better understanding of the efficiency limitations of the organic-based rectifier, in particular the slope efficiency of V_{DCout}/V_{ACin} = 1.64, we model the behavior of the circuit of Fig. 9 using the characteristics of Fig. 8. Both the forward and the reverse current density J can be expressed quite accurately as a function of the voltage V using the exponential curve $J(V) = J_0 + J_1 \exp(\alpha V)$. In the relevant forward current range (1.6 V to 3 V, see Fig. 8), J_0 = -0.311 A/cm², J_1 = 0.0189 A/cm² and α = 1.71 V⁻¹. In the relevant reverse current range (-15 V to -1 V, see inset Fig. 8), J_0 = 0 A/cm², J_1 = 0.001 A/cm² and α = -0.4 V⁻¹. These fits, extracted from the quasi-static characteristics, have been used to calculate the output voltage (V_{DC}) as a function of the AC input amplitude (V_{in}) by numerically matching the charging and discharging current of the rectifier's capacitors using Eq. (1), derived from the equations and conditions outlined in Steudel et al. (Steudel et al., 2005):

$$\frac{1}{2\pi}\int_{\pi-\sin^{-1}\left(\frac{V_{DC}}{2V_{in}}\right)}^{2\pi+\sin^{-1}\left(\frac{V_{DC}}{2V_{in}}\right)} J\left(V_{in}\sin(\theta)-\frac{V_{DC}}{2}\right)\partial\theta + \frac{V_{DC}}{2AR_L} = \frac{1}{2\pi}\int_{\sin^{-1}\left(\frac{V_{DC}}{2V_{in}}+\frac{1.6}{V_{in}}\right)}^{\pi-\sin^{-1}\left(\frac{V_{DC}}{2V_{in}}+\frac{1.6}{V_{in}}\right)} J\left(V_{in}\sin(\theta)-\frac{V_{DC}}{2}\right)\partial\theta \quad (1)$$

with A being the device area (0.001 cm²) and R_L being the load resistance (1 MΩ).
For the commercial silicon diode, the curves from the datasheet have been used.
Fig. 10 compares the calculated output voltage with the measured voltage. The slope efficiency is almost exactly reproduced. We can now trace back the origin of the slope efficiency to the presence of the reverse leakage current of the organic diodes: at higher input voltages, the leakage current increases exponentially, which discharges the capacitors, resulting in a lower DC output voltage. On the other hand, the modeled curve overestimates the DC output voltage by approximately a constant amount of 1 V, or 0.5 V for each diode. This has not been elucidated at present, and could be attributed to several factors, for example a slightly higher offset voltage or a slightly degraded mobility of the charge carriers by bias stress during the prolonged measurement time. We conclude from this analysis that the efficiency of the rectifier depends on the on/off ratio of the diode current.

5. Organic RFID tag using DC load modulation

In this part, the organic RFID transponder foils are integrated with the antenna coil, the HF-capacitor and the rectifier to form an organic RFID tag. In DC load modulation mode, the modulation transistor (W/L = 5040μm/5μm) is placed behind the rectifier, as can be seen in Fig. 1. All foils are placed into a socket and connected as depicted in Fig. 3. The RFID reader is a 7.5 cm radius antenna which emits the field at a base carrier frequency of 13.56 MHz. In Fig. 11, the internal rectified voltage of this double half-wave rectifier in the organic RFID tag is plotted as a function of the field generated by the reader, for the tag antenna placed in the near-field of the reader antenna, at a distance of about 4 cm from the coil generating the readers 13.56 MHz RF field. As can be seen in the graph, 10 V rectified voltage is obtained at a 13.56 MHz electromagnetic field of about 0.9 A/m, and 14 V at 1.26 A/m. The latter is the voltage currently required by our 64-bit organic transponder chips. The ISO 14443 standard states that RFID tags should be operational at a minimum required RF magnetic field strength of 1.5 A/m. The double half-wave rectifier circuit presented here therefore satisfies this ISO norm. After extrapolation of the measurement data, a DC voltage of 17.4 V can be obtained at a field of 1.5 A/m. If a single half-wave rectifier was used, the rectified voltage would be limited to 8 - 9 V, which is too low for current organic technology.

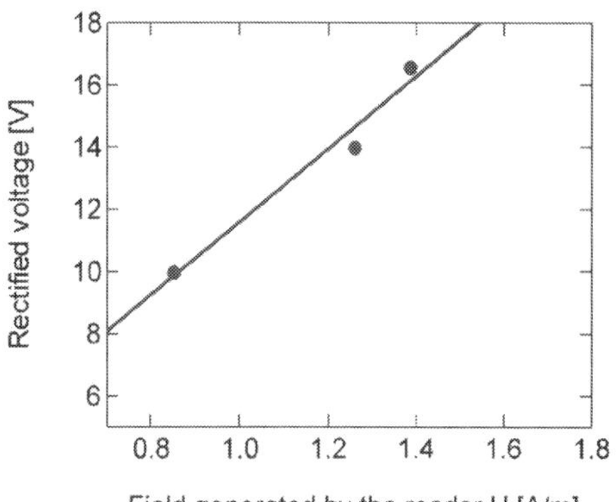

Fig. 11. Internal rectified voltage of a double half-wave rectifier generated in an organic RFID tag versus the 13.56 MHz magnetic field generated by the reader.

The obtained rectified 14 V drives the transponder chip, which sends the code to the modulation transistor. The signal sent from the fully integrated, plastic tag is received by the reader and subsequently visualized using a simple envelope detector (see inset Fig. 12) without amplification. The signal measured at the reader side is depicted in Fig. 12. This shows the fully functional, 64-bit RFID tag using an inductively-coupled 13.56 MHz RFID configuration with a data rate of 787 b/s. With a 0.7 V drop over the diode at the reader (envelope detector), a tag-generated signal of about 1.1 V is obtained, from which 30 mV is load modulation (modulation depth h = 1.4%).

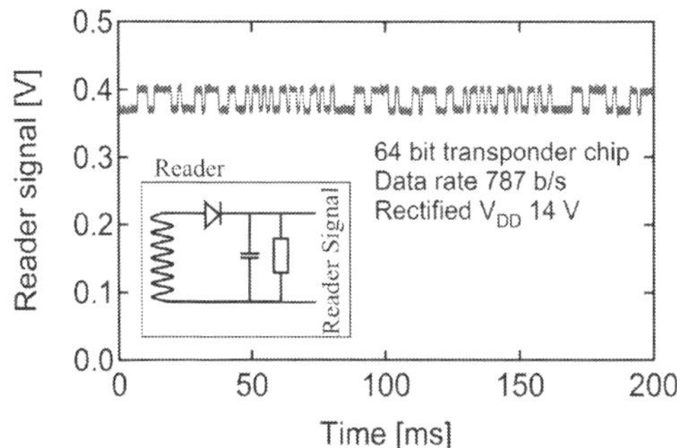

Fig. 12. Signal of the 64-bit RFID tag measured on the reader (unamplified reader signal). The envelope detector of the reader is depicted in the inset.

Two of the reader standards at 13.56 MHz base carrier frequency are the proximity (ISO 14443) and vicinity readers (ISO 15693). The main difference between them is the coil radius, being 7.5 cm for the proximity reader and 55 cm for the vicinity reader. This results in a maximum readout distance of 10 cm for the proximity and 1 m for the vicinity reader. As mentioned earlier, the standard (ISO 14443) states also that the tag should be operational at an RF magnetic field of 1.5 A/m, which is significantly lower than the maximum allowed RF magnetic field of 7.5 A/m. One can calculate the required magnetic field at the antenna centre in order to obtain the required field to operate the tag. In our case, the required field for an 8-bit organic RFID tag was 0.97 A/m. This is depicted in Fig. 13. The dots in this graph show the experimental data at distances of 3.75, 8.75 and 13.75 cm with respect to the field generating antenna. This graph shows that it is possible to energize the 8-bit organic RFID tag at maximum readout distances for proximity readers below the maximum allowed RF magnetic field. The signal detected by the reader during the same experiment is depicted in Fig. 14 at distances of 5 and 10 cm with respect to the sense coil.

6. Summary

In this chapter, we presented the technology, designs and implementation of an inductively-coupled passive 64-bit organic RFID tag which is fully functional at a 13.56 MHz magnetic field strength of 1.26 A/m. This RF magnetic field strength is below the minimum required RF magnetic field stated in the ISO standards. The 64-bit transponder chip employs 414 OTFTs. It is internally driven by a supply voltage of 14 V generated by an efficient organic, double half-wave rectifier. The obtained data rate of the organic RFID tag is 787 b/s. Also an 8-bit transponder chip was measured in DC load modulation configuration and could be readout at a distance of 10 cm, which is the expected readout distance for proximity readers.

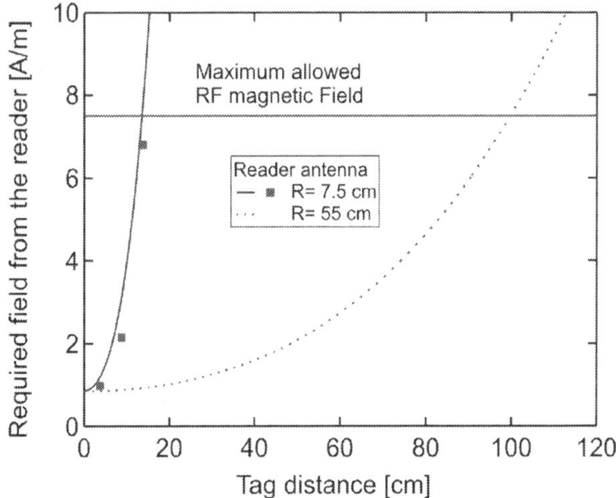

Fig. 13. Calculation and experimentally obtained data of the required RF magnetic field at the reader side as a function of the tag distance in order to generate the required RF magnetic field to operate the tag.

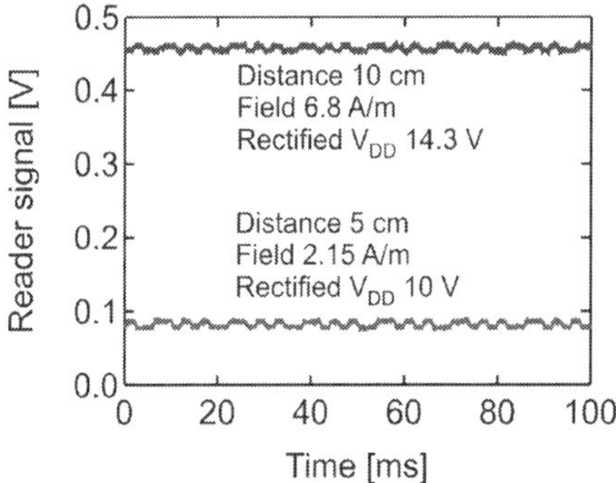

Fig. 14. Signal of the 8-bit RFID tag measured on the reader (unamplified reader signal) in DC load modulation mode at distances of 5 and 10 cm.

7. Future outlook

This work demonstrates that when taking proper care of the critical timing path in digital circuits to account for variability of parameters, state-of-the-art organic electronics technology allows fairly complex digital circuits. Furthermore, the performance of high-frequency analog front-end devices such as rectifiers and load modulators are sufficient for a full organic-based tag to match ISO standards in terms of allowed magnetic field and read-out distance.

A targeted application for organic HF tags is Electronic Product Coding (EPC). Some EPC specifications have already been met by organic tags in recent years, namely the transmission of a 64 bit or larger code (Cantatore et al., 2007; Ullman et al., 2007; Myny et al., 2009), the compatibility with regulations concerning human exposure to electromagnetic fields (Myny et al., 2009). Furthermore, EPC prescribes Manchester encoding of the data, several types of memory (i.e. write-once-read-many) and a basic anti-collision protocol. These have all been shown to date (Myny et al., 2009). Nevertheless, state-of-the-art organic transponder chips fall short of complying with other EPC specifications. Significantly higher data rates (26.48kb/s to 52.969kb/s) are needed as compared to the hundreds b/s up to 2000 b/s of the state-of-the-art organic RFID tags (Cantatore et al., 2007; Ullman et al., 2007; Myny et al., 2009). Furthermore, EPC specifications require a data clock on the tag which is derived from and synchronous with the RF carrier wave. Furthermore, the communication between reader and tag has to be bi-directional, in particular to implement anti-collision protocols where the reader talks first. All latter requirements are currently in research stage, but not demonstrated yet with organic electronics.

8. Acknowledgements

This work was performed in a collaboration between IMEC and TNO in the frame of the HOLST Centre. Part of it has been supported by the EC-funded IP POLYAPPLY (IST # 507143). The authors also thank Klaus Schmidegg of Hueck Folien GmbH for the antenna foils.

9. References

Blache, R.; Krumm, J. & Fix, W. (2009). Organic CMOS Circuits for RFID Applications, *IEEE International Solid-State Circuits Conference Digest Technical Papers*, pp. 208-9, USA, 2009, San Francisco

Cantatore, E.; Geuns, T. C.T.; Gelinck, G. H.; van Veenendaal, E.; Gruijthuijsen, A. F.A.; Schrijnemakers, L.; Drews, S. & de Leeuw, D. M. (2007). A 13.56-MHz RFID System Based on Organic Transponders, *IEEE Journal of Solid-State Circuits*, 42, 2007, 84-92

De Vusser, S.; Genoe, J. & Heremans, P. (2006). Influence of Transistor Parameters on the Noise Margin of Organic Digital Circuits, *IEEE Transactions on Electron Devices*, 53, 2006, 601-10

Gelinck, G. H.; Huitema, H. E.A.; van Veenendaal, E.; Cantatore, E.; Schrijnemakers, L.; van der Putten, J. B.P.H.; Geuns, T. C.T.; Beenhakkers, M.; Giesbers, J. B.; Huisman, B.-H.; Meijer, E. J.; Benito, E. M.; Touwslager, F. J.; Marsman, A. W.; van Rens, B. J.E. & de Leeuw, D. M. (2004). Flexible active-matrix displays and shift registers based on solution-processed organic transistors, *Nature Materials*, 3, 2004, 106-10

Huitema, H. E.A.; Gelinck, G. H.; van Veenendaal, E.; Cantatore, E.; Touwslager, F. J.; Schrijnemakers, L. R.; van der Putten, J. B.P.H.; Geuns, T. C.T. & Beenhakkers, M. J. (2003). A Flexible QVGA Display With Organic Transistors, *Proceedings of International Display Workshops*, pp. 1663-1664, Japan, 2003, Fukuoka

Myny, K.; Steudel, S.; Vicca, P.; Genoe J. & Heremans, P. (2008). An integrated double half-wave organic Schottky diode rectifier on foil operating at 13.56 MHz, *Applied Physics Letters*, 93, 2008, 093305

Myny, K.; Steudel, S.; Vicca, P.; Beenhakkers, M. J.; van Aerle, N. A.J.M.; Gelinck, G. H.; Genoe, J.; Dehaene, W. & Heremans, P. (2009). Plastic circuits and tags for HF radio-frequency communication, *Solid-State Electronics*, 53, 2009, 1220-1226

Pal, B. N.; Sun, J.; Jung, B. J.; Choi, E.; Andreou, A. G. & Katz, H. E. (2008). Pentacene-Zinc Oxide Verical Diode with Compatible Grains and 15-MHz Rectification, *Advanced Materials*, 20, 2008, 1023-1028

Steudel, S.; Myny, K.; Arkhipov, V.; Deibel, C.; De Vusser, S.; Genoe, J. & Heremans, P. (2005). 50 MHz rectifier based on an organic diode, *Nature Materials*, 4, 2005, 597-600

Steudel, S.; De Vusser, S.; Myny, K.; Lenes, M.; Genoe, J. & Heremans, P. (2006). Comparison of organic diode structures regarding high-frequency rectification behavior in radio-frequency identification tags, *Journal of Applied Physics*, 99, 2006, 114519

Ullmann, A.; Böhm, M.; Krumm, J. & Fix, W. (2007). Polymer Multi-Bit RFID Transponder, *International Conference on Organic Electronics*, abstract 53, The Netherlands, 2007, Eindhoven